Lecture Notes in Information Systems and Organisation

Volume 60

Lecture Notes in Information Systems and Organization—LNISO—is a series of scientific books that explore the current scenario of information systems, in particular IS and organization. The focus on the relationship between IT, IS and organization is the common thread of this collection, which aspires to provide scholars across the world with a point of reference and comparison in the study and research of information systems and organization. LNISO is the publication forum for the community of scholars investigating behavioral and design aspects of IS and organization. The series offers an integrated publication platform for high-quality conferences, symposia and workshops in this field. Materials are published upon a strictly controlled double blind peer review evaluation made by selected reviewers.

LNISO is abstracted/indexed in Scopus

Cinzia Dal Zotto · Afshin Omidi · Georges Aoun
Editors

Smart Technologies for Organizations

Managing a Sustainable and Inclusive Digital Transformation

Springer

Editors
Cinzia Dal Zotto
University of Neuchâtel
Neuchâtel, Switzerland

Afshin Omidi
University of Neuchâtel
Neuchâtel, Switzerland

Georges Aoun
Saint Joseph University of Beirut
Beirut, Lebanon

ISSN 2195-4968 ISSN 2195-4976 (electronic)
Lecture Notes in Information Systems and Organisation
ISBN 978-3-031-24774-3 ISBN 978-3-031-24775-0 (eBook)
https://doi.org/10.1007/978-3-031-24775-0

This Springer imprint is published by the registered company Springer Nature Switzerland AG
The registered company address is: Gewerbestrasse 11, 6330 Cham, Switzerland

Preface

In the present volume stemming from the Information and Communication Technologies in Organizations and Society (ICTO) 2020 conference on "*Smart Technologies for an Inclusive World*," 16 high-quality papers add to our understanding of a humanistic way to manage digital transformation toward inclusive organizations and societies. Organizing the papers into four sections, the present volume unveils different aspects of contemporary issues that organizations face when employing digital technologies with the aim of reaching an inclusive and sustainable society over the long run.

The part one, *Technology Adoption and Inclusivity*, comprises four papers, all of which address how digital technologies should be deployed to increase inclusivity within organizations and communities. In this vein, by conducting a systematic literature review and quantitative content analysis, Gierdien and Jokonya explore the factors affecting the adoption of financial technology at universities. In another paper Merhej Sayegh and Sidani reveal the factors that induce developing countries to employ information and communication technologies (ICTs) for sustainability ends. The authors found that social issues inside the workplaces can positively impact employees' intention to move toward a sustainable use of ICTs. Furthermore, Tognisse and Degila apply the technology acceptance model to study the factors influencing the adoption of mobile telephony in African rural areas, while Issami and Tandamba explore the potentialities of mobile payment for bringing inclusivity in the African finance industries.

The part two, *Technology Platforms and Inclusivity*, consists of three papers showing how different digital platforms could increase inclusivity in organizations. In this regard, Mallah Boustani and Merhej Sayegh address how commercial social platforms could have a positive effect on entrepreneurial outcomes in a developing country. Pimenta, Carvalho, and Silva explore how SMEs can use a low-code application development platform to develop an adaptable business performance management solution at low-cost. Moreover, Balozian and Curdy investigate antecedents leading users of popular social media platforms, such as Facebook, to switch to less popular but more reliable alternatives in the users' point of view.

The part three, *Digital Transformation, Value Creation and Sustainability*, includes five papers exploring how corporations can harness digital transformation to serve sustainable value creation. Concerning this issue, Diab and Ben Nasr investigate the contributions that digital transformation may bring to the improvement of Corporate Social Responsibility (CSR) outcomes. In their work, Hajar, Ibrahim, Darun, and Al-Sharafi address the role of emerging technologies in driving organizations toward a sustainable competitive advantage. According to the authors, by creating intangible resources and competencies, new technologies can bolster value innovations and achieve positive outcomes in terms of customer satisfaction and loyalty. Zein Eddine and Ben Nasr study the role of big data and big data analytics, concluding that organizations need to build a consolidated big data infrastructure if they want to move toward sustainable welfare creation. The fourth paper of this section conceptualizes how cities could be turning into sustainable urban communities in the future. Romanelli states that such a process could be enabled by smart technologies that encourage different urban stakeholders to engage in innovation initiatives. In the last paper Savastano, Bellini and D'Ascenzo explore how open production labs such as FabLabs can attract creators and users, boost entrepreneurship and sustainability. Authors identify strengths and weaknesses of current Fab-Lab models and suggest a framework for future research.

The part four, *Leadership and HRM for Inclusive Digital Transformation*, brings together five papers that conclude this volume. Sidani and Harb study the role of transformational leadership in driving innovation outcomes within the Lebanese banking sector. Cori, Sarti, and Torre address how remote working enhanced inclusiveness during the COVID-19 pandemic. In the third paper Veglianti, Trombin, Pinna and De Marco explore the contribution of Artificial Intelligence Technologies to increase the efficiency of recruitment processes and overcome human errors. Sherif and El-Masri investigate the ways organizations can reshape employees' perceptions of electronic performance monitoring. Finally, Pentassuglia, Bonomi, and Rossignoli unveil how the Italian pharmaceutical sector can benefit from digital transformation.

All in all, we are highly grateful to the authors for their contributions and their cooperation during the editing process of the present volume. We hope that this collection of papers will bring fresh insights on the ways in which we can employ digital technologies while keeping sustainability and inclusivity in mind as core principals guiding our business and social behaviors.

Neuchâtel, Switzerland — Cinzia Dal Zotto
Neuchâtel, Switzerland — Afshin Omidi
Beirut, Lebanon — Georges Aoun

Contents

Digital Transformation, Value Creation and Sustainability

Leadership and HRM for Inclusive Digital Transformation

Technology Adoption and Inclusivity

Factors Affecting the Adoption of Financial Technology at Universities

Atoofah Gierdien and Osden Jokonya

Abstract The current wave of technology advancement has sparked a rise in the cardless and cashless society. There are limited studies on factors that influence the adoption of financial technology at universities. This study explores the factors that influence the adoption of financial technology at universities. The primary objective of the study is to explore factors that influence the adoption of financial technology at universities. A systematic literature review was used to explore the factors that influence financial technology adoption at universities. The study adopted a quantitative content analysis to explore factors that influence financial technology adoption at universities. The study used the Technology-Organizational-Environmental (TOE) framework as a lens to explore factors that influence financial technology adoption at universities. The study results revealed that technology factors are the most important factors that influence FinTech adoption at universities. The study contributes to the body of knowledge on factors that influence FinTech adoption at universities.

Keywords Financial technology adoption · M-Pesa · FinTech · TOE framework · Smart payments

1 Introduction

The advent of the digital evolution of economies has prompted the banking sector to consider the transition from traditional physical cash services towards more cardless and cashless societies [1]. Cashless societies mitigate risks surrounding money laundering, terrorist funding, and tax evasion and reduce the financial exclusion of under-served populations worldwide. Approximately 40% of the world's adult population are not in possession of a formal bank account with registered institutions and

A. Gierdien (✉) · O. Jokonya
Department of Information Systems, University of the Western Cape, Bellville Campus, Cape Town 7535, South Africa
e-mail: 3510852@myuwc.ac.za

O. Jokonya
e-mail: ojokonya@uwc.ac.za

C. Dal Zotto et al. (eds.), *Smart Technologies for Organizations*, Lecture Notes in Information Systems and Organisation 60, https://doi.org/10.1007/978-3-031-24775-0_1

are affected by financial exclusion [1]. Financial Technology (FinTech) possesses the limitless potential to increase social and financial inclusion of various populations or demographic segments such as students [2]. FinTech facilitates the development of new transaction platforms, which, subsequently, opens new markets as it challenges traditional business models including financial services such as smart payments, which offer organisations and users with enhanced accessibility to financial services, greater accountability of transactions and improved efficiency [2].

Davies [3], states that universities are shifting away from the myopic view that they are merely learning institutions and are beginning to recognize the duality that they are both a learning institution and a business with a growing concern to both save on costs and compete globally. Universities can leverage smart payment technology to increase cost-savings from asset sharing, better management of their estate and improved energy-efficiency [4]. Although some universities are adopting smart payment solutions (cashless systems), the rate of adoption FinTech solutions is still low [5]. The slow adoption can be attributed to uncertainty, lack of business case clarity, difficulty in integrating existing functional systems, vendor lock-in, change resistance from both students and staff, security risk as well as reputational damage if the smart payment solution implementation fails [5].

Fintech, more specifically smart payments, is an emerging topic in the Information Systems domain with trending topics such as mobile payment (M-Pesa), crowdfunding and online peer-to-peer lending taking centre stage [6]. However, the majority of studies investigated conventional issues surrounding the adoption of smart payments and its underlying technology as well as infrastructure [6]. This paper therefore investigates the factors that influence the adoption of FinTech at universities. This paper conducted a systematic literature review to investigate factors that influence the adoption of financial technology (FinTech) at universities. The structure of the paper is as follows: Sect. 2 reviews the literature on factors influencing FinTech adoption in organisations. Section 3 discusses the research methodology of study, followed by Sect. 4, which presents the results of the study. Finally, Sect. 5 discusses the results and provides the conclusion of the study.

2 Literature Review

The advent of digital economies prompted the banking sector to consider the transition from traditional physical cash services towards more cardless and cashless societies [1]. The cardless and cashless societies constitute the application of innovative and novel technology (FinTech) that improves the efficiency of financial processes and services. FinTech comprises of emerging topics, which include mobile payment such as M-Pesa, crowdfunding and blockchain [6]. Manyika, et al. [7] define financial technology as "digital finance"- financial services distributed via mobile devices, the internet and smart cards. FinTech provides limitless potential to increase social and financial inclusion of various populations or demographic segments namely, students [2]. FinTech facilitates the development of new transaction platforms, which open new markets and challenges traditional business models inclusive of financial services such as smart payments, which offer organisations and users with enhanced accessibility to financial services, greater accountability of transactions and improved efficiency [2].

2.1 Technology Factors

The FinTech technology factors encompass both internal and external technology at the disposal of the organization [8]. Furthermore, the innate attributes i.e. complexity, usability and learnability of a particular technology play a pivotal role in its adoption [8]. Financial technology will have a lower probability of adoption if the technology is perceived as complex or difficult to use [9]. For example, if universities perceived financial technology as too complex with insufficient usability and learnability, then it is least probable that the university would adopt it [9]. An organisation is less likely to adopt an innovation it perceives as complex or complicated [10]. In addition, the costs (operating and capital expenditure) of financial technology may influence its adoption at universities [9]. The deployment of new technology is typically associated with high start-up costs and/or expensive Information Technology (IT) installations whereas organisations are primarily conservative about capital expenditure [10]. Therefore, the cost of technology can have a prohibiting effect on the adoption of financial technology at universities [9]. The relative advantage refers to the degree to which a technological innovation exceeds the tangible and intangible expectations of the initial idea [11]. Organisations are more likely to adopt an innovation with relative advantages associated with that particular innovation [10]. The compatibility defines the extent to which a technological innovation consistently aligns with the values of the incumbent organisation [11]. Thus, organisations are more likely to adopt an innovation that aligns with its values, practice, scope and IT infrastructure [10].

2.2 Organisational Factors

The FinTech organisational context provides a description of the characteristics (i.e. organisation size, structure and degree of centralisation) inherent of the organisation that influence technology adoption [8]. The adoption of financial technology is possible if it strategically aligns with the organisation objectives and has support from executive management [12]. Executive management is responsible for the approval and resource allocation for projects in the organisation. Therefore, the adoption of financial technology at universities requires executive management support and availability of resources, which include financial capital [12]. The support of executive management helps to secure resources, mitigating risk and overcoming employee resistance to change [10]. In addition, internal IT skills and infrastructure required to implement the innovation influences the adoption of FinTech at universities [12]. The lack of internal IT expertise to support IT infrastructure increases resistance in the adoption of FinTech at universities [10]. The firm's size also influences the adoption of technological innovations. The large organisations have more financial resources to both support the adoption and mitigate any risks associated with the adoption of the innovation [10].

2.3 *Environmental Factors*

The environmental context includes the extent to which the organisation's industry (i.e. industry structure, regulation and availability of IT vendors), customers and competitors that influence technology adoption [8]. The marketing landscape (including vendor capabilities) and perception of IT vendors influence FinTech adoption [9]. The availability of IT support and subsequent maintenance support after FinTech adoption mitigate perceived risk associated with the innovation is also an important factor [9]. The perceived competitive advantage of technology adoption and industry competitors influence the organisation to adopt new technology [9]. Therefore, organisations are more likely to adopt an innovation brings competitive advantage [10]. In addition, the interdependence and trust shared between a firm and its customers has a profound impact on its adoption of innovative technologies [8]. Finally, organisations adopt new technologies that help satisfy the needs of their customers [9].

2.4 *Related Studies*

Shaikh et al. [13] conducted a study on the adoption of mobile banking (m-banking). The study explored the major theories that researchers have used to predict technology innovation adoption. This study seeks to explore factors that influence FinTech adoption at universities using the TOE framework. Most related previous studies conducted FinTech studies using the TAM framework, whereas this study investigates the factors that influence FinTech adoption at universities using the TOE framework.

3 Research Methodology

The study used systematic literature review design to investigate factors influencing the adoption of Fintech in universities from articles published period 2013–2018. Systematic literature reviews provide researchers with a pre-defined literary tool to identify pertinent and credible literature by allowing the researcher to identify and discuss reliable evidence, inconsistent findings and gaps in extant literature [14]. A systematic literature review is rigorous due to its deliverance of a high-level summary of existing literature and evidence that focused on answering specific questions as concisely and accurately as possible [14].

3.1 Research Method

The study adopted the content analysis to validate the meaning of written texts by categorizing its content as well as to quantify and interpret its conclusions [15]. The content analysis is the quantitative analysis of qualitative data and entails viewing data from different viewpoints to identify key points that aid the understanding and interpretation of the data in its most raw form [16–18]. Furthermore, content analysis uses an iterative approach that can be either inductive or deductive in nature and thus, allowing the researcher to detect similar or dissimilar content in either written text that will approve or not the theoretical construct being examined [17]. The steps of a quantitative content analysis include developing a research question, formulating hypotheses, sampling, coding scheme development, data collection, statistical analysis, findings and conclusion [17]. The content analysis allows for the historical comparison of financial technology adoption, which offers valuable insights into how the major theme of financial technology adoption has evolved [19]. This method provides a simple and structured approach for the quantification of qualitative text that allows viewing the data through various lenses to identify patterns within the textual content that can be later categorized [20].

3.2 Data Sources and Sampling

The study used convenience sampling to collect, gather and measure data relevant search engines of selected scientific databases i.e. Science Direct, SCOPUS, AIS eLibrary and Taylor Francis Online. The entries into the search engines of these databases comprised of terms inclusive of "factors influencing financial technology adoption at universities", "factors affecting financial technology adoption" and keywords inclusive of "smart payments", "FinTech", "smart universities" and "smart campuses". The search criteria were peer-reviewed articles published from 2013 till 2018.

3.3 Data Analysis

Data analysis in the study constituted the categorizing and counting of the existence of pre-defined terms in the selected group of published articles. About 70 articles were identified and data manually coded based on subjective interpretation of key terms (words) in the qualitative data. Themes were used to categorize the coded qualitative data based on three TOE constructs (technology factors, organizational factors and environmental factors) that influence financial technology adoption. Reliability refers to the likelihood that a researcher(s) will produce similar results if the research process is repeated i.e. yielding consistent results [18]. Inter-coder reliability refers to the degree to which two or more coders agree upon the same method of coding the research data [21]. In terms of data analysis, the qualitative data was converted and coded into quantitative data. The SPSS (statistical analysis tool) assisted with the analysis of the quantitative data to produce statistical results on factors influencing the adoption of financial technology.

4 Research Results

This section presents the results from the selected 70 articles published during the period 2013–2018 on factors that influence Financial Technology adoption at universities. The section is categorized as follows: Sect. 4.1 presents demographic data of the study; Sect. 4.2 presents frequencies of TOE factors that influence FinTech adoption as cited in the 70 reviewed articles.

4.1 Demographic Data Articles Published by Year

Figure 1 presents the frequencies of published articles during the period 2013–2018 on factors that influence FinTech adoption at universities. The results show that 30% of related articles were published during 2013–2015 whereas 70% of related articles were posted later during 2016–2018. These results suggest that there was an increase in research output during 2013–2015, despite a downward trough in 2014. Furthermore, the lowest recorded research output occurred in 2014 at 3% whereas the highest research output was in 2017.

Articles by Region

Figure 2 presents frequencies of published articles during the period 2013–2018 by region on factors that influence FinTech adoption at universities. The results indicate that Asia had the highest recorded number of published articles at 44% followed by Europe at 29% and North America at 14%. Additionally, South America had the lowest number of articles published at 1% with Africa placing second last at 6%. The frequency depicts that Asia accounts for nearly half of all research based on factors that influence FinTech adoption at universities, published during the period 2013–2018.

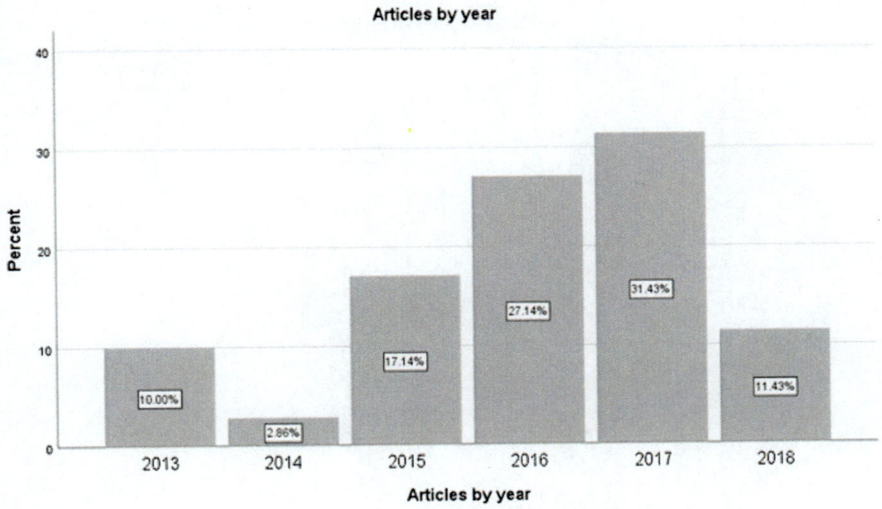

Fig. 1 Articles by year

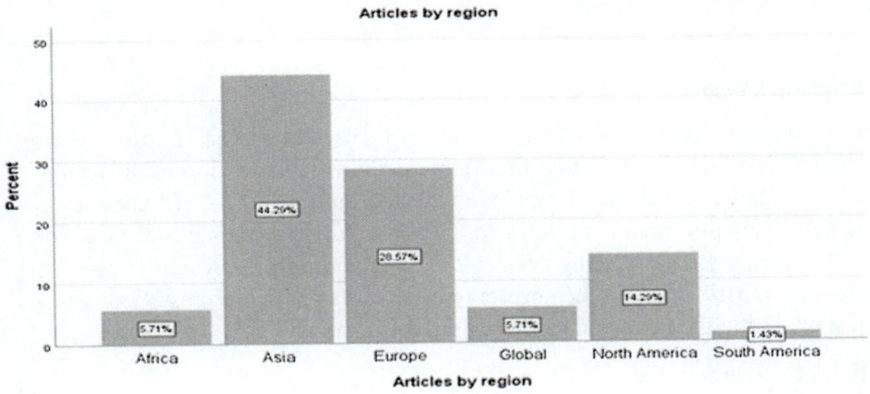

Fig. 2 Articles by region

Articles by Research Method

Figure 3 presents frequencies of the research methods used in published articles during the period 2013–2018 on factors that influence FinTech adoption at universities. The results show that most published articles were conducted using qualitative research studies at 40%, followed by quantitative research studies at 37% and finally, multi-methods research studies, with the lowest frequency, at 22%. The results suggest that qualitative research studies were the most used research method when conducting research based on factors that influence FinTech adoption at universities during the period 2013–2018.

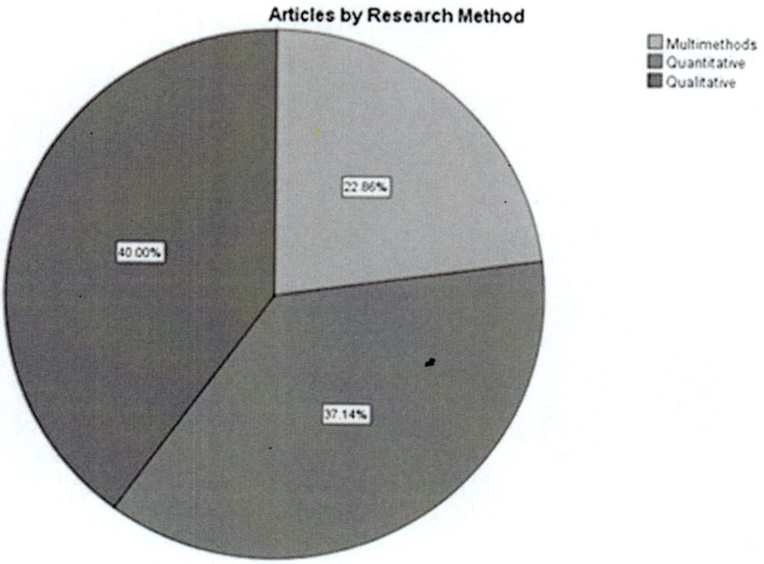

Fig. 3 Articles by research method

Articles by Research Type

Figure 4 presents frequencies of the research types used in published articles during the period 2013–2018 on factors that influence FinTech adoption at universities. The results indicate that most articles published during 2013–2018 conducted a systematic literature review at 36%, followed by surveys coming a close second at 33% and case studies placing third at 20%. Furthermore, the results suggest that conducting experiments was the lowest preferred research type as only 3% of articles published during the period 2013–2018 used this method.

Articles by Framework

Figure 5 presents frequencies of the research frameworks used in articles published during the period 2013–2018 on factors that influence FinTech adoption at universities. The frameworks include the Technology Acceptance Model (TAM), the Diffusion of Innovation Theory (DOI); Technology-Organisation-Environment framework (TOE) where NA represents articles that did not use a framework or rather suggested their own framework. The results depict that most articles, at 51%, did not use a framework or recommended a new framework (NA). Furthermore, the results suggest that the TAM framework was the second preferred research framework at 41%, followed by DIT at 4%. Additionally, the results suggest that the least used framework in articles, based on factors that influence FinTech adoption at universities, published during the period 2013–2018 used this method was the TOE framework at 3%.

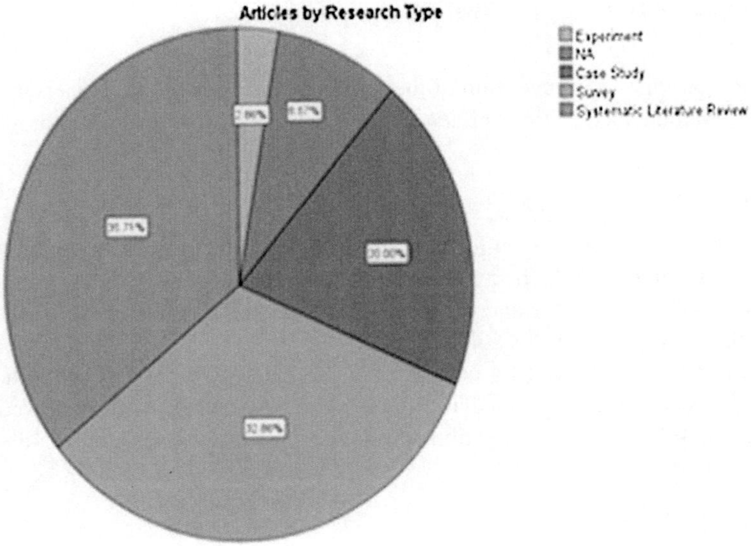

Fig. 4 Articles by research type

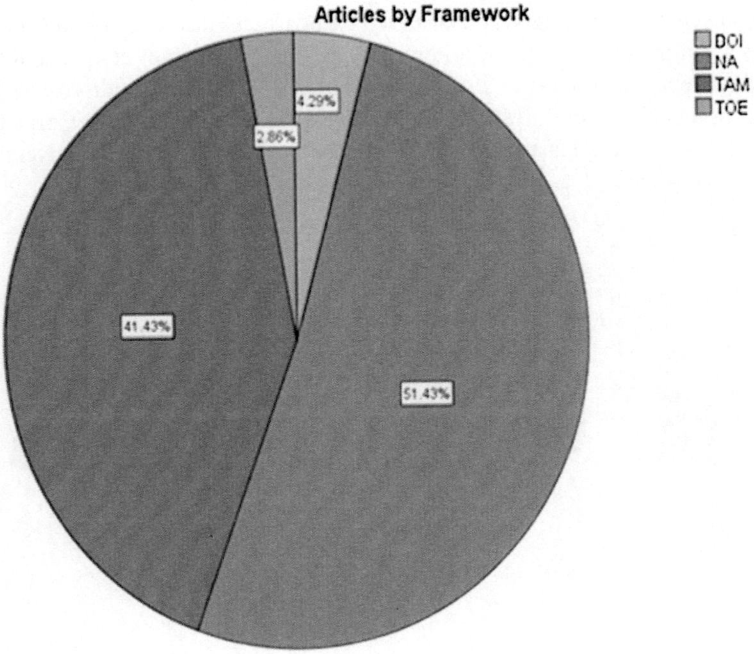

Fig. 5 Articles by framework

4.2 Factors Affecting the Adoption of FinTech

This section provides an illustration of the results of the technology, organisational and environmental factors that influence the adoption of financial technology at universities.

Technology Factors

Figure 6 presents frequencies on technology factors that influence the adoption of FinTech at universities, which include complexity, compatibility, cost, perceived usefulness, relative advantage and security. The results indicate that relative advantage was the most important technology factor that influences FinTech adoption at universities at 93%, followed by cost at 83% and security at 77%. Furthermore, 50% of articles discussed perceived usefulness as an influencing factor, followed by complexity at 46%. Finally, compatibility was the least discussed factor at 43% of the published articles.

Organisational Factors

Figure 7 presents frequencies on organisational factors that influence the adoption of FinTech at universities, which included factors such as organisational readiness, organisational resources, firm size, technical skills, managerial/executive support and strategic objectives. The results indicate that 59% of articles discussed organisational resources as an influencing factor, followed by strategic objectives at 43% and then managerial/executive support at 29%. Furthermore, 24% of articles discussed firm size as an influencing factor, followed by technical skills at 23%. Finally, organisational readiness was the least factor that influence FinTech adoption at universities at 13% of the articles.

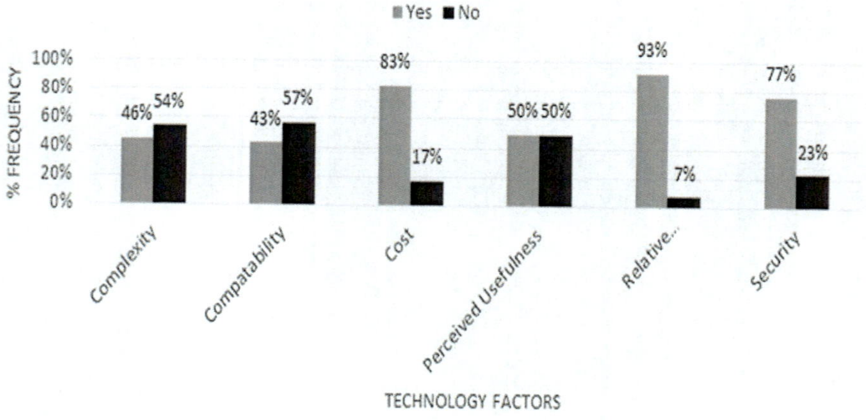

Fig. 6 Technology factors

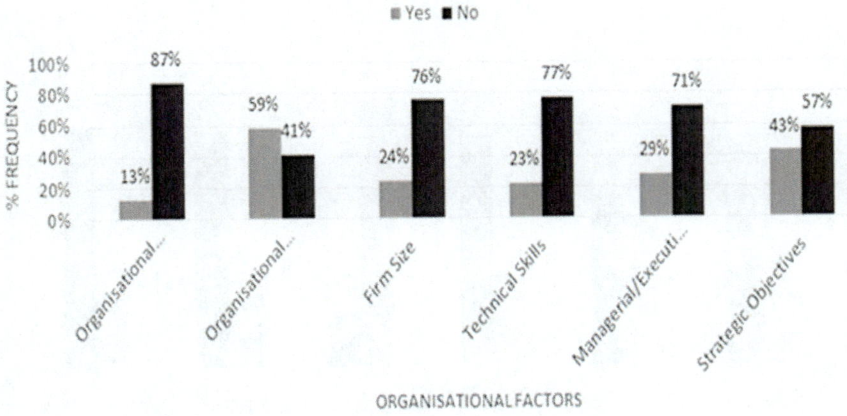

Fig. 7 Organisational factors

Environmental Factors

Figure 8 presents frequencies on environmental factors that influence the adoption of FinTech at universities, these included factors such as government, competition, IT policy/regulation, market structure, vendor capabilities and maintenance/support. The results indicate that 83% of articles discussed market structure as an influencing factor, followed by competition at 47% and then government pressure at 44%. Furthermore, 29% of articles discussed IT policy/regulation as an influencing factor, followed by vendor capabilities at 27%. Finally, maintenance/support was the least discussed factor that influences FinTech adoption at universities at 26% of the articles.

5 Discussion and Conclusion

The study conducted a systematic literature review on the factors that influence the adoption of Financial Technology at universities. The study used the TOE framework to investigate factors that influence FinTech adoption at universities. The study results indicate that technology factors inclusive of complexity, compatibility, cost, perceived usefulness, relative advantage and security were the prevailing factors that dominated the influence of Fintech adoption. More specifically, relative advantage, cost and security were the most cited technology factors to influence FinTech adoption. This suggests that universities are most likely to adopt FinTech if it directly benefits the organisation while simultaneously being cost-effective and offering secure payment solutions to end users.

The study results indicate that environmental factors were the second major factors whilst organisational factors were the third major factors affecting the acceptance

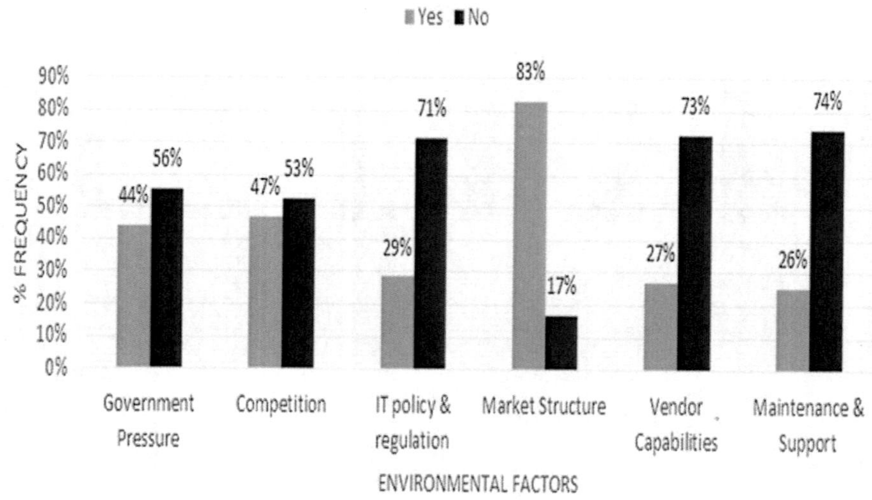

Fig. 8 Environmental factors

of FinTech adoption at universities. The market structure was cited as the principal environmental factor that affects FinTech adoption and this sheds light on the belief that organisations are only likely to accept FinTech adoption if it is considered suitable to the current market structure, especially in mature markets. The study showed that the availability of organisational resources is a major factor for FinTech adoption. The idea that the incumbent organisation should possess resources inclusive of financial, human and technology infrastructure to adopt financial technology solutions into their current business model.

In conclusion, the study tried to fill the gaps in understanding the factors understanding the factors that influence FinTech adoption at universities. The study also stimulates further research on FinTech adoption at universities. Despite some of the study-highlighted contributions to the body of knowledge, there are some limitations worth acknowledging. One of the limitations of this study is that it used convenience sampling to source data, which makes it hard to generalize the results to a large population on FinTech adoption at universities. However, the limitation of the study provides an opportunity for further research on the topic using other research design. Although, this study is not exhaustive it does attempt to bridge the gap in understanding the factors that influence FinTech adoption at universities.

References

1. Smart Payment Association. (2017). *Transitioning to a cashless world—The current state of play*. Munich: Smart Payment Association.
2. Ding, D., Chong, G., Chuen, D. L. K., & Cheng, T. L. (2018). From ant financial to Alibaba's rural Taoboa strategy—How Fintech is transforming social inclusion. *Handbook of Blockchain, Digital Finance and Inclusion, 1*, 19–35.
3. Davies, C. (2014). *Are cashless campuses the future?*. s.l.:Global Payments.
4. Universities, U. K. (2015). *Efficiency, effectiveness and value for money*. Univerties UK.
5. MonitorIT. (2018). *Going cashless in universities and colleges*. https://www.monitorit.co.uk/getmedia/7bf253bd-a155-4ecf-8361-a15e1d3d4205/MITWhitePaper-FinalVersion. Accessed April 2018.
6. Leong, C., Tan, B., Xiao, X., Tan, F. T. C., & Sun, Y. (2017). Nurturing a FinTech ecosystem: The case of a youth microloan startup in Chine. *International Journal of Information Management, 37*(2), 92–97.
7. Manyika, J., Lund, S., Singer, M., White, O., & Berry, C. (2016). *Digital finance for all: Powering inclusive growth in emerging economies*. McKinsey Global Institute.
8. Ngah, A. H., Zainuddin, Y., & Thursamy, R. (2017). Applying the TOE framework in the halal warehouse adoption study. *Journal of Islamic Accounting and Business Research, 8*(2), 161–181.
9. Alshamaila, Y., Papagiannidis, S., & Li, F. (2013). cloud computing adoption by SMEs in the North East of England: A Mulit-perspective Framework. *Journal of Enterprise Information Management, 26*(3), 250–275.
10. Wang, Y., Li, C., & Zhang, D. (2015). Factors affecting hotels' adoption of mobile reservation systems: A technology-organization-environment framework. *Elsevier, 53*, 163–173.
11. Rogers, E.M. (2003). *Diffusion of innovations*, 5th edn. New York: NY: Free Press.
12. Maduku, D., Mpinganjira, M., & Duh, H. (2016). Understanding mobile marketing adoption intention by South African SMEs: A multi-perspective framework. *International Journal of Information Management, 36*, 711–723.
13. Shaikh, A. A., & Karjaluoto, H. (2014). *Mobile banking adoption: A literature review* (p. 14). Elsevier.
14. Gupta, S., Rajiah, P., Middlebrooks, E. H., Baruah, D., Carter, B. W., Burton, K., Chatterjee, A. R., & Miller, M. M. (2018). Systematic review of the literature: Best practices. *Elsevier, 25*(11), 9.
15. Payne, G., & Payne, J. (2004). *Key concepts in social research* (1st ed.). Sage.
16. Myers, D. M. (2010). *Qualitative research in business & management*. Sage.
17. Maree, K. (2007). *First steps in research*, 1st edn. Pretoria: Van Schaik.
18. Jokonya, O. (2015). Investigating open source software benefits in public sector. In *Hawaii, 48th Hawaii International Conference on System Sciences*
19. Allen, M. (2017). Content analysis: Advantages and disadvantages. In M. Allen (Ed.), *The SAGE encyclopedia of communication research methods* (pp. 240–242). SAGE Publications Inc.
20. McNabb, D. E. (2002). *Research methods in public administration and nonprofit management: Quantitative and qualitative approaches*. Armonk(New York): M.E. Sharpe.
21. Lavrakas, P. J. (2008). *Encyclopedia of survey research methods*. Sage Publications.

Influencing Factors of ICT Use for Sustainability in Developing Countries—The Case of the Lebanese Private Sector

May Merhej Sayegh and Dina Sidani

Abstract Research has shown that information and communication technology (ICT) has a tremendous effect on organizations. This paper aims to investigate the ICT adoption within firms and propose a theoretical framework related to factors that influence employees' intentions to continue using it while working on their implications towards a sustainable economic, human and environmental world. Towards this end, a survey is designed and distributed to 218 Lebanese persons working in different sectors and industries. The findings of the structural equation model (SEM) indicate that the higher the hedonic value, facilitating conditions within the firms, social influence and team work in the working context have a strong positive impact on intentions to use ICT and continue using it in any context within the firm or outside it. The results offer several theoretical and practical implications, providing some insights for developing countries to commit to sustainable and inclusive world through ICT Adoption. The originality of this research is to study the correlation between ICT adoption in firms and a sustainable environment in the society. More importantly, the paper suggests a framework that focuses on different variables from the technology acceptance models and reveals their impact on employees and their intentions to use technology.

Keywords Information and communication technology ICT · Hedonic motivation · Sustainability · Social influence · Organizational culture · Developing country

M. M. Sayegh (✉) · D. Sidani
Faculty of Business and Management, Saint-Joseph University of Beirut, Beirut, Lebanon
e-mail: may.merhejsayegh@usj.edu.lb

D. Sidani
e-mail: dina.sidani@usj.edu.lb

C. Dal Zotto et al. (eds.), *Smart Technologies for Organizations*, Lecture Notes in Information Systems and Organisation 60, https://doi.org/10.1007/978-3-031-24775-0_2

1 Introduction

In the last decade, information and communication technologies (ICT) have gained an important place in the emerging economies. Nowadays, organizations are moving to be a sustainable business. ICT has been considered a boost for the economic capital of organizations and its dynamics.

Not only has it stood as the incremental agent of change toward effectiveness and efficiency in the organizational dynamics, but it has also boosted the economic capital of organizations [1, 2]. Changes in organizations and ways of working are essential key elements for business and strategies development to find a source of competitive advantage [3, 4]. As sustainability can ensure a competitive advantage and an opportunity for organizations in terms of economic, social and environmental goals [5, 6], integrating information technologies (IT) into business processes has been significant challenges for organizational managers.

This study examines the initiatives determined as Green ICT and reflected in the lifecycle of information and communication technologies (ICTs) in Lebanese Firms. The motivation for this research is to find answers, raised by the majority of the Lebanese Private sector, for worries related to ICT use. It was proven that, among all the recent technologies, the progress in ICT has no doubt and continues to have the greatest influence on the economy [7], making it possible to collect, process and transmit information at breath-taking speed. According to the McKinsey and Company report (2011), one euro invested in internet technology brings in two euros.

Therefore, our research contributes to a modest yet significant attempt to explore the importance of the role and impact of ICT on sustainability. Findings from this study will give us some insights related to factors that affect ICT use and adoption in the Lebanese Private Sector. It seems that the embracing of digitalization and the transformation of business organizations integrating digital technologies have become paramount for the survival of firms [8, 9].

Digitalization is considered an efficient tool to support sustainable environmental, social, and economic development [10]. Information and communication technologies (ICT) can enhance the strategy for a green environment and an ecosystem in which firms can operate [9]. It is about green sustainability which needs organizational change in firms connected to business processes, products and services [11]. Several researchers in the field of ICT adoption studied users' behaviors toward information systems [12, 13]. The ICT adoption in enterprises can have a positive impact on environmental preservation by increasing energy efficiency and equipment utilization [14] as well as it can influence social development by making information available to all society's stakeholders [15]. Our aim is to study the correlation between ICT adoption and sustainability.

Some variables from two acceptance technology models are adopted: The Technology acceptance model (TAM) [16] explaining the individual's beliefs and attitudes toward accepting technology in organization. To this model we include some of the variables of UTAUT [13] which has been used as a complementary model in our theoretical framework to measure technology use and adoption [17, 18] and the extension model UTAUT2 [19] to cover the behavioral intension of individuals and we have tested the performance expectancy, effort expectancy, social influence and facilitating conditions and behavioral intension of individuals (Hedonic motivation, price value and habit) [13, 19]. We tested many variables from the two models and their impact on ICT adoption and thus creating a competitive advantage in a sustainable economic and social environment. The study addresses the following questions:

RQ1. What is the relationship between the socio-demographic variables and their impact on IT adoption and intentional behavior towards a sustainable environment with its three pillars?

RQ2. What is the relationship between the organizational culture and their impact on IT adoption and intentional behavior towards a sustainable environment with its three pillars?

RQ3. What is the relationship between the facilitating conditions, social influence and Hedonic motivation and their impact on IT adoption and intentional behavior towards a sustainable environment with its three pillars?

As result of our study, in terms of the utilitarian motivation, the perceived enjoyment has a strong positive impact on using technology and can affect positively the intention to continue using ICT. Moreover, they match the results in the ICT adoption and intentions, which is considered to influence users' behaviors in firms [19, 20]. Additionally, facilitating conditions have been found to have a direct positive impact on users' intentions [13, 19]. Social influence also can have an impact on behaviors to use ICT [21]. Finally organizational culture has a strong influence on ICT Adoption, which is enhanced by teamwork, change adaptation and customer orientation. Also, the findings of the study show positive impact of awareness on adopting ICT on continue using ICT by users in the future. This result is somewhat consistent with Awareness of consequences that it can influence personal norm [22–24] related to economic, social and environmental sustainability.

2 Theoretical Framework

Numerous papers were advocated with the addition or elimination or combination of theories into an integrated model for exploring the adoption and usage of any technology [25]. Many researches have been developed in the field of Information Technology adoption to study users' awareness of consequences and behaviors toward information systems [12, 13]. The Technology acceptance model (TAM) [16] allows assessing the acceptance of a new technology. TAM is explaining the individual's beliefs and acceptance about usefulness and ease of use; these variables are the major determinants of adoption and use of Information technology in organization. Furthermore, TAM is rooted in the theory of reasoned action [26, 27] and has drawn on individual's attitude toward accepting technology. Usun [28] supports the importance of understanding the cultural context and traditions of individuals and organizations, and therefore for a successful IT integration, various practices must exist. It is argued that financial resources, policy and political factors, culture, attitudes and skills are important to support IT infrastructure such as hardware, broadband connectivity, backbone infrastructure and to realize effective technology adoption especially in developing countries.

In the same context, the Unified Theory of Acceptance and Use of Technology (UTAUT) [13] has been used as a complementary model in our theoretical framework to measure technology use and adoption [17, 18]. The four constructs (performance expectancy, effort expectancy social influence and facilitating conditions) from the UTAUT model were completed by three other constructs in the extension model UTAUT2 [19] to cover the behavioral intension of individuals (Hedonic motivation, price value and habit).

The present study has used a combination of the cited conceptual models to explain the adoption and implementation of ICT in an open innovation context. Keeping abreast of changes linked to technology implementation and its impact on individual awareness toward sustainability in the society, this study will try to predict the individual behavior and awareness of using ICT in their work and its impact on the environment from the three levels: economic, social and environmental. The conceptual model, based on a comprehensive review and synthesis of several theoretical models [13, 16–18, 26, 27], is a combination of many constructs in multi-level perspectives, such as social influences and Hedonic motivation (social level variable) and facilitating conditions, organizational culture (organizational level variable), and behavioral intention for predicting actual use behavior (individual-level variables) [13, 19], toward awareness and intention to use technology as a sustainable tool in the environment.

The construct Sustainability is a recent issue for the economy, the companies, and the individuals, being considered a systemic concept [29] related to the continuity of economic, social (including cultural), and environmental issues [30]. The construct sustainability is comprised of three dimensions: environmental, economic, and social [31]. In the context of technology and information system and according to [31]. The three dimensions are defined as follows: from the environmental terms, it can be an improving of poor environmental practices [32], such as the reduction of greenhouse gas emissions [31], so working online and using supply chain can reduce gas emissions; If viewing sustainability in economic terms, Information System can increase productivity [31]. Finally, Technology can enhance social sustainability in terms of monitoring, analyzing, and integrating citizens' consumptions and behaviors [33].

Nowadays, sustainable development is an important issue in organisations. Researchers have argued that the focus of the firm should be on environmental or social performance while enhancing the economic performance [34].

Corporations can also find new sources of income by increasing sustainability, cutting costs and maximizing resource efficiency. Thus, the technological and management innovation presents an excellent business opportunity for organizations [34] and a competitive advantage.

The question is: Does sustainability affect employees' behavioral intention to use technology in their work and to continue behaving using ICT?

A shortcoming we address by including some of the variables of UTAUT and UTAUT2. Our motivation for adding the sustainable construct to the current model is to improve its applicability and predictive power to address technologies entailing sustainability or being re-interpreted in a sustainable context. In addition, our aim is to bring together two of the most popular topics on the IS research—sustainability and technology acceptance by individuals [19, 32].

2.1 Factors Influencing ICT Adoption

Hedonic Motivation

A number of research studies on ICT use, at the firm level, were based on the model of diffusion of innovation [35].

The associated fun or pleasure experienced in using a technology is explained as individuals' hedonic motivation. Hedonic motivation has been conceptualized as perceived enjoyment during technology and information system (IS) research and usage and has been found to have a direct effect on ICT use [36–38]. Also, reported hedonic motivation as a key predictor of behavioral intention to use technology. Hedonic motivation has a direct influence on behavioral intentions [19]. It is considered an important factor predictor to determine and affect user's acceptance to use technology [20, 36–39].

Sustainability is beneficial to the welfare of the community while Hedonic motivation is characterized by a beneficial individual end-state goal [19, 40]. Thus, sustainability captures a motivational facet distinctly different from hedonic motivation. Thus, we propose the following hypothesis:

H1. The Hedonic Motivation will positively influence the employees' ICT Adoption within the organization.

Social Influence

Researchers have noted that individuals may adopt a particular technology not because of their own personal convictions but because of the influence of others [21]. The unified theory of acceptance and use of technology (UTAUT) has proposed that social influence is a key element in determining user acceptance of an information technology [13] and a driver of user behavior.

Social influence reflects the impact of other people's (peers, instructors and friends) beliefs on individuals' intention or use behavior [13]. Al-Shafi et al. [41] reported that employees are socially influenced by their peers' beliefs, which subsequently influences their behavioral intention to use. Similarly, Fidani and Idrizi [42] study asserts that students' intention towards Learning Management System use is influenced by their friends 'or teachers' beliefs about LMS. Taiwo and Downe [43] mentioned that social influence might affect the administrative staff of the ICT through the influence of colleagues and senior management, as they motivate administrative staff to utilize ICT by explaining the benefit and the importance to use ICT before and during the use the particular ICT.

Social influence is affecting the plan of the users to use the ICT if other users perceive the importance of ICT as influential to their decision to use and adopt ICT [44]. Social influence may have an impact on the users when they see other people using ICT tools as they benefit from it. We therefore suggest testing the following hypothesis:

H2. The Social influence positively influences employees' ICT Adoption within the organization.

Facilitating Conditions (Resources and Knowledge)

Facilitating conditions relate to the availability of enough resources and support for individuals to use the technology [13]. While some studies demonstrated that facilitating conditions can impact behavioral intention to use technology [45–47], other documents reported that there is no significant influence of facilitating conditions on the intention to use technology [25, 48]. We posit that facilitating conditions accelerates the intention to use of technology and information system from a UTAUT model point of view [45].

Lack of availability of assistance, no timely support, incomplete information, and limited resources could hinder users in their acceptance of web-based technology [48]. Deng et al. [49] also reported a significant relationship between facilitating conditions and actual use of web-based question-answer services. Facilitating conditions have a direct influence on behavior (actual use of technology) [13] because invariant support and training was available to each individual [19]. In the mobile technology acceptance context, a new link between facilitating conditions and intention towards mobile technology use was proposed [19]. In compliance with these concepts, we hypothesize the following link:

H3. Facilitating Conditions will positively influence employees' ICT Adoption within the organization.

Education and Experience

Researchers in ICT adoption [50, 51] found that educated and experienced employees were more likely to use ICT. Moreover, previous research found a clear relationship between the level of education and technology use and adoption [51–54]. Thus, we propose the following hypothesis:

H4a. The Individual Background of employees in terms of Education will positively influence their ICT Adoption within the organization.

H4b. The Individual Background of employees in terms of Experience will positively influence their ICT Adoption within the organization.

Organizational Culture

The biggest challenge that organizations can face is to create a culture that supports and embraces innovation [55]. Adopting a culture of innovation by organization means willing to embrace new ideas or techniques required by this innovation [55].

Further, according to Kellick and Wilson [56] and Zhu and Engels [55], appropriate identification of the culture facilitate the creation, storage, sharing, adoption and application of information technology that enables a corporate strategic advantage for the organizations by transforming its fundamental business processes into digital ones [53]. Change management is required in such transformation integrated within the organizational culture [53]. In their study, Cooper and Quinn [57] and Smith et al. [58] found that there is an important relation between IT and the culture.

Moreover, an important key for ICT adoption and behavioral intention to continue of using it is to understand the cultural context and traditions of individuals and organizations [28]. In the context of our study, we find relevant to adopt the work of Talcott Parsons, a sociologist at Harvard. Parsons [59] developed a framework and theory of action in social systems. He argued that all organizations must carry out four crucial functions if they are to survive for any substantial length of time. These four crucial functions are labelled: managing change, achieving goals, coordinating teamwork, and building a strong culture. In our study, we adopted the "Organizational Culture Assessment Questionnaire (OCAQ)", which is based on these four functions of Organizational Culture [60].

It is important in such a research to identify the impact of these functions of Organizational Culture so that we can understand and analyze the means of communication, the means of coordination and the way to manage change in the context of ICT use in order to understand and lead to decision making concerning ICT Adoption and behavioral Intention to continue of using it. Thus, the following hypothesis could be suggested:

H5. The Organizational Culture will positively influence employees' ICT Adoption within the organization.

ICT Adoption and Behavioral Intention to Continue Using ICT

The Technology Acceptance Model [16] theorizes that individual's beliefs about usefulness and ease of use are the major determinants of the use of ICT in any organization. TAM is rooted from another well-known theory in human psychology, the theory of reasoned action (TRA) [24, 61]. TRA proposed that individual's beliefs will influence their attitudes which in turn influence their intention and then generate the behavior. Several researchers in the field of ICT adoption studied users' behaviors toward information systems [12, 13].

As the aim of this study is to investigate the factors that might influence the ICT adoption among employees of the organizations in a developing country, we have also considered the theory of reasoned action that was designed by [62]. According to those authors, the attitude is actually the degree to which an individual expresses a positive attitude towards ICT use. Attitude usually plays an important role in people's behavior towards their behavioral process to integrate ICT and to continue using it. There have been many empirical evidence of a relation between intention and behavior [63–66]. Moreover, the study of Hallikainen et al. [67] gives us an interesting contribution regarding the behavioral intention of employees to continue using ICT. Like any other proposed technology, the successful adoption of ICT depends largely on long-term viability and users' intention to continue using it [68]. The concept of

Behavioral Intention to continue using ICT (Continuance Intention, CI) is used to identify the changes in users' perception toward the proposed technology after using it. Behavioral intention implies the probability that an individual will lock in within the genuine behavior.

Concurring to Miladinovic and Hong [69], whether an individual performs a specific behavior or not, is chosen by the intentional of the individual to perform the conduct, known as behavioral intention. The Behavioral intention of continue using ICT is emphasizing on users' long-term viability [70]. In this regard, Mosavi and Ghaedi [71] communicated that high-benefit quality as observed by customers habitually prompts incredible behavioral intention, while low-benefit quality will in general reason trouble some social expectation. In addition, individual contrasts of age, sex, and involvement, coordinate the impacts of these builds on behavioral intention [19]. The present study aims at exploring the intentional behavior of employee to continue using ICT in the future and to analyze if they adopted ICT due to current circumstances only or because they become convinced over time, that they have to adopt it in the future. Hence, we suggest the following hypothesis:

H6: ICT Adoption have a positive influence on ICT intentional behavior to continue using ICT.

Awareness of Consequences

Awareness of consequences is defined as people's beliefs towards an environmental condition which will lead to an acceptance of responsibility [72]. Awareness of consequences influences personal norm [73]. In this study, we propose to study the people awareness of consequences if "practicing" green use of environmental tools and materials (like not using cars, working online…).

Sustainable innovation has become lately a part of business models in several companies and it is related to technologies helping in designing green products and allowing to create new business processes [74]. The open innovation referred to minimize costs by companies, adopting technologies to reach new markets [75].

Through open innovation and change management, leverage on digitalization is an opportunity. Collective action within the firm needs to be based on shared understandings of the benefits of ICT and digitalization on sustainability [53]. Hart and Dowell [76] position sustainable development among the three key strategic capabilities following pollution-prevention and product stewarddship. Firms adopting a sustainable development strategy do less environmental damage and consider the wider social as well as the economic sustainability aspects of their operations [77]. ICT can foster the adaptation of sustainability strategies. Thus, in line with the above considerations, the following hypothesis is proposed:

H7: ICT adoption has a positive influence on the awareness of its impact on sustainability.

2.2 Impact of ICT on Sustainability

The Three Pillars of Sustainability

The concept of sustainability has attracted attention and has been developed when environmental problems caused by various human activities require serious solutions. Although many definitions abound, the evolution of the concept of sustainability involves common and different imperatives regarding their author and the country the studies and researches relied on. Most definitions integrate many buildings under development: expansion, growth, progress, development and satisfaction. Therefore, all conceptualizations show a directly proportional relationship between economic growth and population satisfaction.

In 1987, the Bruntland Commission published its report, Our Common Future, in an effort to link the issues of economic development and environmental stability. In doing so, the World Commission on Environment and Development (WCED) provided the oft-cited definition of sustainable development as development that meets the needs of the present without compromising the ability of future generations to meet their own needs. For Harwood [78], Sustainable development is a "…system that can evolve indefinitely toward greater human utility, greater efficiency of resource use and a balance with the environment which is favorable to humans and most other species." In 1999, Sustainable Development constitutes, in South Africa context, a program for changing the process of economic development so that it ensures a basic quality of life for all people and at the same time protects the ecosystems and community systems that make life possible and worthwhile." In United Kingdom "Sustainable development is a process of change, where resources are being gathered, an investment direction is chosen, the development technologies directed and various institutions have convergent actions, increasing the potential for human needs and desires." In Romania's context, Sustainable development as "…. maintaining system stability by developing a balance of responsibilities: economic, social, environmental and technological support technique without compromising the needs of future generations."

Today, the most often used definition of sustainable development is that proposed by the Brundtland Commission [79]. The key principle of sustainable development underlying all others is the integration of environmental, social, and economic concerns into all aspects of decision-making. Therefore, sustainable development requires the elimination of fragmentation; that is, environmental, social, and economic concerns must be integrated throughout decision-making processes in order to move towards development that is truly sustainable.

ICT use for Sustainable Development

The literature on ICT for Sustainable Development is at nascent stages, academic literature is widespread and attempted to examine a wide variety of topics Ziemba [80] proposed an ICT framework that addresses the Sustainable Development Goals. He investigated the contribution of ICT adoption in enterprises and how does it lead to a sustainable information society. Gouvea et al. [81] studied the link between ICT and human development on environmental sustainability and found that ICT and human development has strong significant main effects on environmental sustainability. Moreover, Mulamula and Amadi-Echendu [82] had recognized the importance of ICT for Sustainable Development in the least developing countries and examines the relationship between ICT transfers and Sustainable Development through case studies of developing countries. Latif et al. [83] had empirically affirmed the positive influence of ICT on environmentally Sustainable Development in South Asian countries for a period of 2005–2015.

It is important to address this question as the members of the United Nations are bound to address the issue of Sustainable Development as a part of the international Sustainable Development Goals. Most international organizations such as ITU and UN rely on ICT to accelerate the achievement of Sustainable Development.

Economic Growth Dimension

Economic sustainability refers to produce a maximum flow of income in terms of rational use, resource efficiency, and thus by producing goods and services continuously to maintain manageable levels of government and avoid extreme sectorial imbalances affecting agriculture and production industrial. Instruments that lead to economic purpose is achieved education, scientific and intellectual potential that is available [84].

The contribution of ICT on economic growth is through the advancements in innovative products, processes and business models [85]. Literature has provided concrete pieces of evidence of the positive relation of various factors of ICT on economic growth in various contexts such as OECD countries [86]. In this information era, using ICT becomes an inevitable choice and the effort is to make the usage fruitful in best of the means. In accordance with previous studies, it is hypothesized as a positive association of ICT usage on the economic growth of a nation.

H8: The level of ICT adoption in a country is associated with the nation's economic growth.

Human Development

Human sustainability has regard to social interactions, relationships, behavioral patterns and values of humanity [87]. Realizing the need for further social development (without harming the environment) is imperative to protect and improve the state of the environment. It represents the only possibility to create and maintain the welfare of both the present generation and those to come. This balance was the factor that can and should ensure the development of the whole company. Minica and France [88] synthesize the human aspect of sustainable development worldwide involves education, training and public support for the environment, the Protecting and promoting human health, the fight against poverty through promoting human development and integrated policy investment in human capital.

Similarly, ICT also transforms, completes and enriches the education system for the better. UNESCO supports the ministry of states to formulate policies and plans for effective imparting of education to the remote corners of the world. ICTs are transforming the world, it creates new job opportunities making the labor markets innovative, comprehensive and universal [86].

H9: The level of ICT adoption in a country is associated with the nation's human development.

Environmental Dimension

Environmental sustainability is oriented towards satisfying specific practical requirements, and long, proposing harmony. It is the capacity to grow and to bring the environment and its peculiarities, while ensuring the protection and renewal of natural resources and environmental heritage. It stimulates the development of knowledge based on consumption, but subordinates planning opportunities; expected a harmonious development. Thus, economic growth should not affect the environment in order to talk about sustainable development.

The concept of Green Information Systems views information systems as a potential solution to reduce pollution in terms of emissions or waste [89]. The environmental consciousness of the IT industry in its day-to-day business operations can ensure pollution prevention and product stewardship. This reduces the burden on the environment and increases the environmental performance of the nation [86]. In light of the above statements, it is hypothesized:

H10: The level of ICT Adoption in a country is associated with the nation's environmental performance.

Socio-Demographic Variables

Regarding Socio-demographic variables, Szymkowiak and Garczarek-Bąk [90] highlight the gender differences towards ICT use through e-commerce. Thus, in line with the results of this research, the following hypothesis is proposed:

H11: Gender will influence employees' ICT Adoption within the organization.

As younger administrative staff would be more likely to copy the behavior of others [91], we will propose the following hypothesis:

H12: Age will influence employees' ICT Adoption within the organization.

It is therefore interesting to test these relationships within the context of ICT adoption. Building on the above literature the following model is proposed in Fig. 1.

3 Research Methodology

3.1 Sample and Data Collection

In order to validate the proposed theoretical model, a survey using a questionnaire consisting of 34 items was developed. All items we measured using a five-point Likert-scale varying from 1 = Totally disagree to 5 = Totally agree. The questions in the survey were adapted from previous researches as shown in Appendix A. the respondents who answered the questions were in the categories of age 20 and above

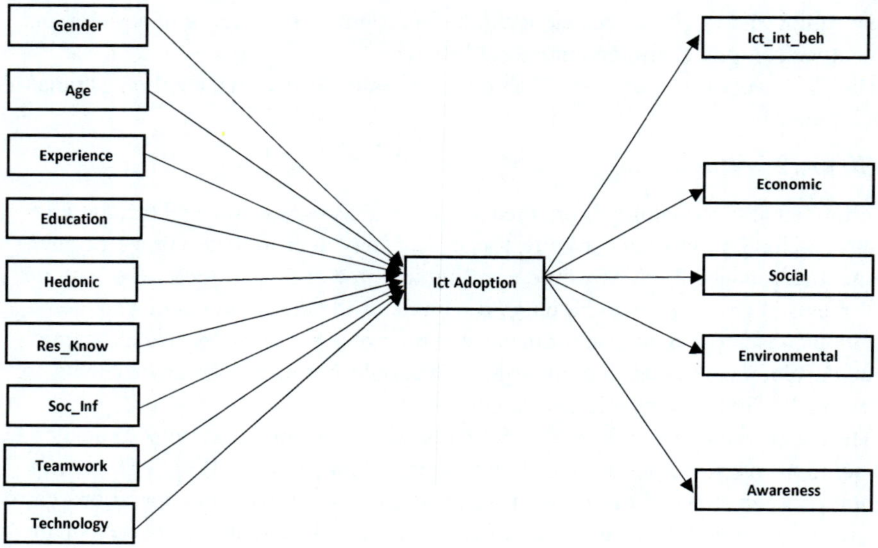

Fig. 1 Proposed model

and were using PC and Internet and have some knowledge of technologies. The survey was placed online and a link was sent by email to a database of alumni of a prominent Lebanese university whose graduates are working in different sectors and industries. A total of 450 questionnaires were sent, of which 218 were analyzed after excluding respondents who did not have an experience using internet or computers and whose age is under 20, considering that the employability percentage of under age 20 is not meaningful as they are still in the university.

The sample consisted of 61.2% females and 38.8% males. Using technologies can enhance and transform the working methods, it creates new job opportunities making the labor markets innovative, comprehensive and universal (World Bank 2013; [86]). By selecting age categories above 20, the choice increases the likelihood that the respondents will be characterized by a good level of education which increases the chances to engage in a certain kind of work using technologies. The sample also consisted of 32.2% from the category 41–50 years, 20.6% from the category 31–40 years, 18.2% from the category 51 and above, 15.9% from the category 20–25 years and 13.1% from the category 26–30 years. As for the education the sample is 64.5% master degree's holders and 34.6% Bachelor degree's holders.

Furthermore, by selecting a certain subgroup of the overall population, we can achieve higher consistency in sample characteristics as demonstrated by Taras et al. [92]. The chosen sample therefore is considered homogenous.

3.2 Methodological Details and Model Building

The study selected the structural equation modeling (SEM) technique to simultaneously test the relationships between the model variables [93]. IBM SPSS 20 and IBM Amos 23 statistical packages were used. Following the accepted procedure for confirmatory factor analysis, as a first step, a CFA was conducted on all variables to ensure the reliability and validity of individual constructs. Next, SEM was conducted on the whole model to validate the proposed hypotheses.

The research model was examined using the two-step procedure: a measurement model and a structural model. First, exploratory and confirmatory factor analyses (CFA) were conducted to test the reliability and validity of the individual constructs. Second, SEM was performed to validate the proposed hypotheses. The data was first screened for any missing values. Collected data was checked for normality prior to analysis. Multivariate normality could be assumed as the values of skewness and kurtosis obtained were between −2 and +2 (Byrne, 2010). Additionally, to avoid any bias due to possible non-normality of the variables, this study used the maximum likelihood (ML) estimation method with nonparametric bootstrapping to estimate the model parameters. Nevitt and Hancock [94] shows that bootstrap estimates for the model are generally less biased when compared with those obtained from standard ML estimation under conditions of non-normality and for sample sizes of $N \geq 200$.

4 Analysis and Results

4.1 Principal Components Analysis (PCA)

First, we used the Principal Component Analysis (PCA) to extract the items that do not fit the conditions for each concept. Principal Component Analysis, or PCA, is a dimensionality-reduction method that is often used to reduce the dimensionality of large data sets, by transforming a large set of variables into a smaller one that still contains most of the information in the large set. Because smaller data sets are easier to explore and visualize and make analyzing data much easier and faster for machine learning algorithms without extraneous variables to process. The PCA process provides a list of factor loadings where a factor loading represents the amount of variance captured by a given component. The analysis continues until all variance in the dataset has been accounted to determine how many meaningful components should be retained to interpret [95, 96]. Reducing the number of variables of a data set naturally comes at the expense of accuracy, but the trick in dimensionality reduction is to trade a little accuracy for simplicity. So to sum up, PCA is meant to reduce the number of variables of a data set, while preserving as much information as possible, while extracting the items that do not fit for each dimension. The following approach is then implemented:

Table 1 Constructs reliability

Construct	Cronbach's α
Hedonic motivation Q1-Q2-Q3	0.940
Facilitating conditions Q4-Q5	0.798
Social influence Q7-Q8-Q9	0.875
ICT adoption Q10-Q11-Q12	0.782
Organizational culture Q13-Q14-Q15-Q16	0.866
ICT intentional behavior Q20-Q21-Q22	0.946
Awareness Q23-Q24-Q25	0.873
Economical sustainability Q26-Q27-Q28	0.881
Social sustainability Q29-Q30-Q31	0.875
Environmental sustainability Q32-Q33-Q34	0.850

- Extraction of items where h2 (the sum of squared factor loadings) is less than 0.6.
- Use the Varimax rotation method which is a statistical technique used at one level of factor analysis as an attempt to clarify the relationship among factors.
- Cumulative percentage that explains the axis must be greater than 70%.

So after we applied the PCA to the direct and indirect variables we will continue the study with the questions illustrated in Table 1.

4.2 *Reliability and Validity Analysis: Measurement Model*

The reliability of the model constructs was assessed based on two criteria, Cronbach's Alpha. Results are displayed in Table 1. The Cronbach's alpha values ranged from 0.75 to 0.940 for all constructs. Hence, internal consistency is ensured [97].

4.3 *Assessment of the Structural Model*

In previous sections we utilized the SPSS for dimensionality-reduction and to measure the internal consistency. In this section we will use AMOS where we can apply the structural equation model (SEM) to verify the models fit and the validity of the hypothesis.

We used the absolute model fit, the incremental fit and the parsimonious fit to test the hypothesis. The measures are the following:

- Absolute model fit:
  - Chi square value and p-value $p > 0.05$
  - Root Mean Square Error of Approximation (RMSEA) < 0.05
  - Goodness of Fit Index (GFI) > 0.9

- Incremental fit
  - Adjusted Goodness of Fit Index (AGFI) > 0.9
  - Comparative Fit Index (CFI) > 0.9
  - Normed Fit Index (NFI) > 0.9
  - Tucker Lewis Index (TLI) > 0.9
- Parsimonious Fit
  - Chi square/DF
  - Minimum discrepancy (CMIN/DF) < 3.0
  - The test of hypotheses is verified in this study by the normality (exhibit C).

In this section we will use AMOS where we can apply the structural equation model (SEM) to verify the models fit and the validity of the hypothesis (Chin, 1998). For all hypotheses we have Chi square value and p-value $p > 0.05$, Root Mean Square Error of Approximation (RMSEA) < 0.05, Goodness of Fit Index (GFI) > 0.9.

The results are shown in Table 2 and Fig. 2.

5 Discussion

This section discusses the theoretical and practical implications of the results. The present study expands the ICT adoption literature by empirically exploring the influence of direct and indirect variables and its impact on sustainability. The findings offer several theoretical and practical implications. Within this particular context of remote work, and as a relevant topic, the results could give hints to ICT practitioners in organizations, in order to enhance ICT adoption and intention to continue using it in any circumstances.

5.1 Theoretical Implications

This study contributes to the literature by investigating the role of hedonic motivation, facilitating conditions, social influence and teamwork in adopting ICT in firms and influencing intentions to continue using it in order to enhance a sustainable world. From a practical point of view, the study highlights the need for firms engaged in sustainability to enhance the derived economic benefits from ICT adoption by setting by reinforcing employees hedonic motivations and facilitating conditions. Several studies were conducted to explore ICT adoption in organization but few explored the impact of a combination of variables from 2 different technology adoption models: The Technology acceptance model (TAM [16], UTAUT [13] and UTAUT2 [19]) using many variables and their impact on ICT adoption and thus creating a competitive advantage in a sustainable economic and social environment.

Table 2 Results of the structural model

Relationships	Hypothesis	Estimate	SE	P value	Hypothesis verification
Hedonic motivation → ICT adoption	H1	0.487	0.110	***	Supported
Social influence → ICT adoption	H2	0.365	0.057	***	Supported
Facilitating conditions → ICT adoption	H3	0.547	0.095	***	Supported
Educational background (indirect variable) → ICT adoption	H4a	0.195	0.129	**0.131**	Not supported
Experience (indirect variable) → ICT adoption	H4b	0.159	0.104	0.124	Not supported
Knowledge (indirect Variable) → ICT adoption	H4c	0.591	0.160	***	Supported
Organizational culture → ICT adoption	H5	0.817	0.074	***	Supported
ICT adoption → ICT intentional behavior	H6	0.428	0.088	***	Supported
ICT adoption → awareness on impact on sustainability	H7	0.454	0.079	***	Supported
ICT Use and adoption → impact on economic sustainability	H8	0.323	0.059	***	Supported
ICT Use and adoption → impact on Social sustainability	H9	0.468	0.082	***	Supported
ICT Use and adoption → impact on environmental sustainability	H10	0.416	0.063	***	Supported
Gender (indirect variable) → ICT adoption	H11	−0.053	0.133	0.690	Not supported
Age (indirect Variable) → ICT adoption	H12	−0.046	0.049	0.339	Not supported

This table reports the results for the Structural Equation Model (SEM) estimation. The first and second columns show the model hypotheses. The third column indicates the standardized estimated coefficients β's. The fourth and fifth columns display the standard errors (SE) and the corresponding p-values. *** indicates that the coefficient is highly statistically significant ($p < 0.001$), ** shows that the coefficient is significant at the 1% significance level ($p < 0.01$). The last column indicates if the hypotheses are supported or not

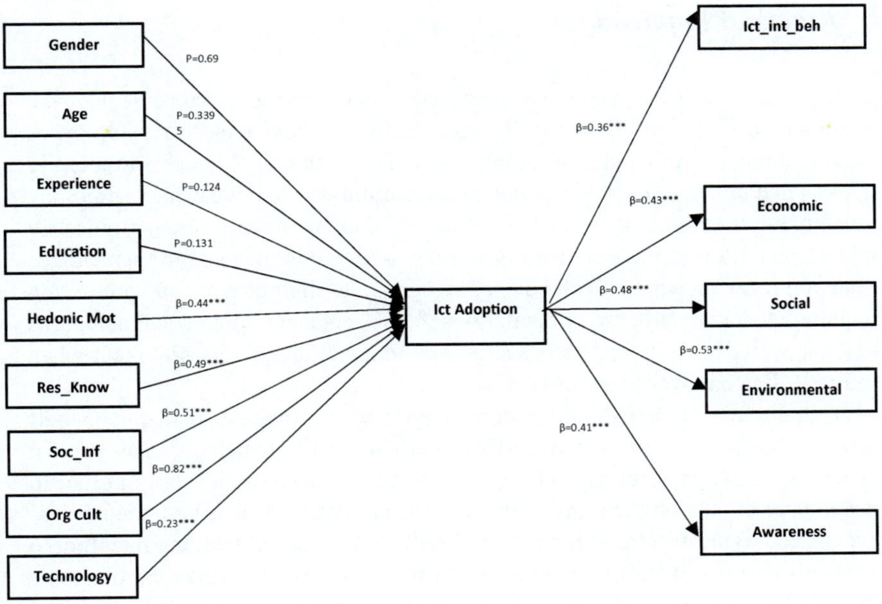

P means that hypothesis is not validated.
Beta*** means the hypothesis is validated and p<0.001

Fig. 2 Results

This study contributes to the literature by investigating the role of hedonic motivation, social influence, organizational culture on adopting ICT and their impact on a sustainable environment.

First, in terms of the utilitarian motivation, the perceived enjoyment was found to have a strong positive impact on using technology that affect positively the intention behavior to continue using ICT. Moreover, they match the results in the ICT adoption and intentions, which is considered to influence users' behaviors in firms [19, 20]. Additionally, facilitating conditions were found to have a direct positive impact on users' intentions which is consistent with [12, 18]. Social influence also can have an impact on behaviors to use ICT [16]. Finally, organizational culture has a strong influence on ICT Adoption, and we can then talk about "ICT Culture", which is enhanced by teamwork, change adaptation and customer orientation.

Second, the results did not provide support for the positive relationships between education and ICT adoption or between experience and ICT adoption and intention to continue using it. Unlike the influence of teamwork and communication, adopting a new process is very important to the organization [59].

Third, regarding the awareness of consequences on adopting ICT, findings show the positive impact of this variable on continue using ICT in the future. This result is somewhat consistent with Awareness of consequences that it can influence personal norm [22, 24] related to economic, social and environmental sustainability.

5.2 Practical Implications

From a practical point of view, this study contributes to the enhancement of the management of ICT adoption within the organization and how to motivate employees to use it in an innovation and a sustainable way. They can help managers to identify the factors that influence employees and their intentions to continue using technology for sustainable environment. As such, they provide them with the necessary tools they should emphasis on to increase the employees use for this mean from their offices and later on from anywhere. Such businesses can focus on improving and promoting the economic, social and environmental benefits. Users can realize when using this channel as well as deploying various strategies to strengthen their enjoyment when adopting ICT in any work they are realizing.

More elaborated studies should be conducted to help companies focusing on their facilitating conditions and their team work in their culture which can have a great influence on employees perceptions to adopt technologies in their work in order to have a greater impact on their environmental sustainability. And as hedonic motivation is proven to have a strong impact on adopting ICT and influencing intention to continue using it for sustainable purposes, firms are advised to utilize various tools to try to increase users' enjoyment when using technology inside and outside the work environment.

6 Conclusion, Limitations and Future Research Venues

This study sheds light on the importance of considering the many variables related to personal users' norms and to firms' factors when investigating factors that affect intentions to adopt and use ICT for a better sustainable world. Findings reveal that hedonic motivation, facilitating conditions, social influence, organizational culture as team work and awareness of consequences on a sustainable environment have a direct effect on intentions to adopt and continue using ICT while age, gender, experience, education doesn't have any impact. Hence, this study helps to understand the influence that firms can use to motivate employees on intention to use ICT in their work inside and outside their offices and confirms the important role that ICT plays in sustainable economic, social and environmental world.

Related to a larger and broader approach, ICT adoption in developing countries could be a way to commit to a sustainable and inclusive world, in alignment with the eight Goals of the Millennium Development Goals.

This outcome has important implications for future scholarly research alike for practitioners. Considering the context of remote work nowadays, the organizational culture towards adopting ICT becomes a must. Performance is a key concept regarding the socio-economic context the world crosses actually. The concept of sustainability is then a guide thread while considering efficiency and profitability on the economic level. Further analyses could present a multi-cases approach dealing with other realities. In addition, new scientific contributions could conduct a comparative study between the public and the private sector in Lebanon.

Exhibit A: Items and Questionnaire

Socio-Demographic Variables

H11—Gender

☐ Female

☐ Male

H12—Age

☐ 20–25 years

☐ 26–30 years

☐ 31–40 years

☐ 41–50 years

☐ 51 and above

H4a—Experience in using PC

☐ None

☐ Low

☐ Moderate

☐ High

H4b—Educational background

☐ High school or below

☐ Bachelor's degree

☐ Master's degree or above

H4c—Access to Technology

Please indicate your level of access to different types of technologies (outside University) by checking only one answer.

Types of TechnologyNon AccessLimited or No Problem.

Types of Technology	*Non Access*	*Limited or Inconvenient Access*	*No Problem with Access*
Broadband (ADSL/Optic Fibre/Cable)	□	□	□
Wireless Internet Access	□	□	□

Model Variables

H1—Hedonic motivation to use ICT

1. I feel fun using online platform to do my work from a distance
2. I enjoy using online platform to do my work from a distance
3. Using online platform to do my work from a distance is very entertaining to me

H3—Facilitating Conditions to use ICT

4. I have resources to use online platform to do my work from a distance
5. I have knowledge to use online platform to do my work from a distance
6. A specific person (or group) is available to assist when difficulties arise while using online platform to do my work from a distance

H2—Social Influence

7. My peers who influence my behavior think that I should use online platform to do my work from a distance
8. My friends who are important to me think that I should use online platform to do my work from a distance
9. My directors whose opinions that I value prefer that I should use online platform to do my work from a distance

ICT Adoption

10. Employees are flexible and ready to adopt ICT when they are convinced change is necessary
11. ICT will succeed when employees are convinced that it will make their work faster and more efficiently
12. ICT adoption rely mainly on availability and accessibility to ICT Infrastructure

H5—Organizational Culture

13. Employees and teams participate in defining specific goals
14. Employee believe in teamwork, the "what's in it for us" approach rather than "what's in it for me."
15. Employee know what is expected of them and understand their impact on other people, teams, and functions
16. Each one of us give the highest priority and support to meeting the needs of clients and customers and solving their problems.

17. Employee who do the best job of serving customers are more likely than other employees to be recognized or rewarded
18. Business decisions are most often made on the basis of facts, not just perceptions or assumptions.
19. Employees have access to timely and accurate information about what's really happening in the organization and why.

H6—Behavioral Intention

20. In the future, I intend to continue using online platform to do my work from a distance
21. In the future, I plan to use online platform to do my work from a distance
22. I will recommend to others to use online platform to do their work from a distance

H7—Awareness

23. Online Working could contribute to an environmental sustainability
24. Online work using technology can promote living a better and healthy life
25. Online work using technology is an environment-friendly who can contribute to a healthy environment

Perception of ICT use on Sustainability

H8—Economic

26. Online work using technology work produces economical solutions in economic sectors
27. Online work using technology can help to increase productivity of organizations 24/7
28. Online work using Technology can contribute to economical sustainability

H9—Social

29. Online work using Technology can be the main driver that can have social influence
30. Online work using technology can provide automation to various business sectors to contribute to society
31. Online work using technology contributes to society by providing information and learning

H10—Green

32. Online working can contribute to on environmental sustainability
33. Online work using technology can promote living a better and healthy life.
34. Online work using technology is an environment-friendly who can contribute to a healthy environment.

Exhibit B: Constructs and Measuring Scales Adopted

Construct	Items	Sources
Hedonic motivation	I feel fun using online platform to do my work from a distance I enjoy using online platform to do my work from a distance Using online platform to do my work from a distance is very entertaining to me	Adapted from Venkatesh et al. [18], Ul Ain et al. (2015)
Facilitation conditions (resources and knowledge)	I have resources to use online platform to do my work from a distance I have knowledge to use online platform to do my work from a distance FC3: A specific person (or group) is available to assist when difficulties arise while using online platform to do my work from a distance	Adapted from Venkatesh et al. [12, 18], Ul Ain et al. (2015)
Social influence	My peers who influence my behavior think that I should use online platform to do my work from a distance My friends who are important to me think that I should use online platform to do my work from a distance My directors whose opinions that I value prefer that I should use online platform to do my work from a distance	Venkatesh et al. [12, 18]

(continued)

(continued)

Construct	Items	Sources
Behavioral intention	In the future, I intend to continue using online platform to do my work from a distance In the future, I plan to use online platform to do my work from a distance I will recommend to others to use online platform to do their work from a distance	Adapted from Venkatesh et al. [12]
Economical sustainability	Online work using technology work produces economical solutions in economic sectors Online work using technology can help to increase productivity of organizations 24/7 Online work using Technology can contribute to economical sustainability	Adapted from Malik et al. [86]
Green sustainability	Online working can contribute to on environmental sustainability Online work using technology can promote living a better and healthy life Online work using technology is an environment-friendly who can contribute to a healthy environment	Adapted from Malik et al. [86]

(continued)

(continued)

Construct	Items	Sources
Social sustainability	Online work using Technology can be the main driver that can have social influence Online work using technology can provide automation to various business sectors to contribute to society Online work using technology contributes to society by providing information and learning	Adapted from Malik et al. [86]
Organizational culture achieving goals	Employees and teams participate in defining specific goals	Adapted from Saskhin [60]
Coordinated work	Employee believe in teamwork, the "what's in it for us" approach rather than "what's in it for me." Employee know what is expected of them and understand their impact on other people, teams, and functions	
Customer orientation	Each one of us give the highest priority and support to meeting the needs of clients and customers and solving their problems Employee who do the best job of serving customers are more likely than other employees to be recognized or rewarded	

(continued)

(continued)

Construct	Items	Sources
Cultural strength	Business decisions are most often made on the basis of facts, not just perceptions or assumptions Employees have access to timely and accurate information about what's really happening in the organization and why	
Awareness of consequences	Online working could contribute to an environmental sustainability Online work using technology can promote living a better and healthy life Online work using technology is an environment-friendly who can contribute to a healthy environment	Hoyos and Hess [22]. Incorporating environmental attitudes in discrete choice models: An exploration of the utility of the awareness of consequences scale, Science of the Total Environment, 505, pp. 1100–1111
ICT adoption	Employees are flexible and ready to adopt ICT when they are convinced change is necessary ICT will succeed when employees are convinced that it will make their work faster and more efficiently ICT adoption rely mainly on availability and accessibility to ICT Infrastructure	Constraints for ICT Adoption—The EFITA Questionnaire Case Study. The Journal of Agricultural Education and Extension, 2008

Exhibit C

Model	Absolute model fit					Parsimonious fit	Incremental fit			
	Chi-2	Df	P-value	RMSEA	GFI	CMIN/DF	AGFI	CFI	NFI	TLI
H1	6.060	5	0.300	0.032	0.991	1.212	0.961	0.999	0.993	0.996
H2	1.488	3	0.685	0.000	0.997	0.496	0.986	0.999	0.996	1.012
H3	4.428	6	0.619	0.000	0.993	0.738	0.976	1.000	0.994	1.006
H4a	1.483	1	0.223	0.048	0.997	1.483	0.965	0.988	0.994	0.998
H4b	0.526	1	0.468	0.000	0.999	0.526	0.988	1.000	0.998	1.012
H4c	2.812	4	0.589	0.000	0.995	0.703	0.980	1.000	0.992	1.009
H5	11.689	9	0.231	0.037	0.985	1.299	0.952	0.997	0.987	0.993
H6	9.104	8	0.333	0.025	0.986	1.138	0.963	0.999	0.990	0.998
H7	9.886	7	0.195	0.044	0.985	1.412	0.995	0.995	0.985	0.990
H8	6.929	8	0.544	0.000	0.989	0.866	0.972	1.000	0.989	1.003
H9	6.043	6	0.418	0.006	0.991	1.007	0.968	1.000	0.991	1.000
H10	7.067	8	0.529	0.000	0.989	0.883	0.972	1.000	0.988	1.003
H11	0.043	1	0.043	0.000	1.000	0.836	0.999	1.000	1.000	1.023
H12	2.268	2	0.322	0.025	0.995	1.134	0.974	0.999	0.990	0.997
All	1212.128	511	0.000	0.080	0.767	2.372	0.712	0.878	0.809	0.858

References

1. Kucukvar, M., Egilmez, G., & Tatari, O. (2014). Sustainability assessment of U.S. final consumption and investments: Triple-bottom-line input–output analysis. *Journal of Cleaner Production, 81*, 234–243.
2. Akram, M. S., Shakir Goraya, M. A., Malik, A., & Aljarallah A. M. (2018). Organizational performance and sustainability: Exploring the roles of IT capabilities and knowledge management capabilities, sustainability, MDPI.
3. Parida, V., & Wincent, J. (2019). Why and how to compete through sustainability: A review and outline of trends influencing firm and network-level transformation. *The International Entrepreneurship and Management Journal, 15*, 1–19. https://doi.org/10.1007/s11365-019-00558-9
4. Lahti, T., Wincent , J., & Parida, V. (2018). A definition and theoretical review of the circular economy, value creation, and sustainable business models: Where are we now and where should research move in the future?, *Sustainability, 10*(8). https://doi.org/10.3390/su10082799.
5. Reim, W., Parida, P., & Örtqvist, D. (2015). Product–service systems (PSS) business models and tactics—A systematic literature review. *Journal of Cleaner Production, 97*, 61–75. https://doi.org/10.1016/j.jclepro.2014.07.003.
6. Gimenez, C., Sierra, V., & Rodon, J. (2012). Sustainable operations: Their impact on the triple bottom line. *International Journal of Production Economics, 140*, 149–159.

7. Gumport, P. J., & Chun, M. (2005). Technology and higher education: Opportunities and challenges for the new era. In P.G. Altbach, R.O. Berdahl, & P.J. Gumport (Eds.), American Higher Education in the Twenty-First Century: Social, Political, and Economic Challenges, 2nd edn (pp. 315–339). Baltimore, MD: Johns Hopkins University Press.
8. Peter, M. K., Kraft, C., & Lindeque, J. (2020). Strategic action fields of digital transformation. An exploration of the strategic action fields of Swiss SMEs and large enterprises. *Journal of Strategy Management, 13*, 160–180.
9. Irimiás, A., & Mitev, A. (2020). Change management, digital maturity, and green development: Are successful firms leveraging on sustainability? *MDPI, Sustainability 12*, 4019. https://doi.org/10.3390/su12104019. Seele, P., & Lock, I. (2017). The game-changing potential of digitalization for sustainability: Possibilities, perils, and pathways. *Sustainability in Science, 12*, 183–185.
10. Tripsas, M., & Gavetti, G. (2000). Capabilities, cognition and inertia: Evidence from digital imaging. *Strategic Management Journal, 2000*(21), 1147–1161.
11. Wallace, L., & Sheetz, S. (2014). The adoption of software measures: A technology acceptance model (TAM) perspective. *Information & Management, 51*(2), 249–259. ISSN 0378-7206. https://doi.org/10.1016/j.im.2013.12.003.
12. Venkatesh, V., Davis, G. B., Davis, F. D., & Morris, M. G. (2003). User acceptance of information technology: Toward a unified view. *MIS Quarterly, 27*(3), 425–478.
13. Hilty, L. M., & Aebischer, B. (2015). ICT for sustainability: An emerging research field. *Advances in Intelligent Systems and Computing, 310*, 1–34.
14. Schauer, T. (2003). *The sustainable information society—Vision and risks.* The Club of Rome-European Support Centre.
15. Davis, F. (1989). Perceived usefulness, perceived ease of use, and user acceptance of information technology. *MIS Quarterly, 13*(3), 319–340. https://doi.org/10.2307/249008
16. Maldonado, U. P. T., Khan, G. F., Moon, J., & Rho, J. J. (2011). E-learning motivation and educational portal acceptance in developing countries. *Online Information Review, 35*(1), 66–85.
17. Uddin, M. A., & Fan, L. (2019). The influence of transformational leadership on creative process engagement: A multilevel analysis. *Management Decision, 57*(3), 78–83. https://doi.org/10.1108/MD-07-2017-0707
18. Venkatesh, V., Thong, J. Y. L., & Xu, X. (2012). Consumer acceptance and use of information technology: Extending the unified theory of acceptance and use of technology. *MIS Quarterly, 36*(1), 157–178.
19. Yang, K., & Lee, H. J. (2010). Gender differences in using mobile data services: Utilitarian and hedonic value approaches. *Journal of Research in Interactive Marketing, 4*(2), 142–156. Emerald Group Publishing Limited. https://doi.org/10.1108/17505931011051678. Ifinedo, P. (2016). Applying uses and gratifications theory and social influence processes to understand students' pervasive adoption of social networking sites: Perspectives from the Americas. *International Journal of Information Management, 36*(2), 192–206.
20. Harland, S. J., Kynaston, H., Grigor, K., Wallace, D. M., Beacock, C., Kockelbergh, R., Clawson, S., Barlow, T., Parmar, M. K. B., & Griffiths, G. O. (2007). A randomized trial of radical radiotherapy for the management of pT1G3 NXM0 transitional cell carcinoma of the bladder. *National Cancer Research Institute Bladder Clinical Studies Group.* https://doi.org/10.1016/j.juro.2007.05.024
21. Klöckner, C. A. (2013). comprehensive model of the psychology of environmental behaviour—A meta-analysis, Global Environmental Change, 23(5). *ISSN, 1028–1038*, 0959–3780. https://doi.org/10.1016/j.gloenvcha.2013.05.014
22. Hoyos, M., & Hess,. (2015). Incorporating environmental attitudes in discrete choice models: An exploration of the utility of the awareness of consequences scale. *Science of the Total Environment, 505*, 1100–1111.
23. Barrane, F. Z., Karuranga, G. E., & Poulin, D. (2018). Technology adoption and diffusion: A new application of the UTAUT model. *International Journal of Innovation and Technology Management, 15*, 1950004–1950023.

24. Ajzen, I., & Fishbein, M. (1980). *Understanding attitudes and predicting social behavior*. Prentice-Hall.
25. Fishbein, M., & Ajzen, I. (2005). Theory-based behavior change interventions: Comments on Hobbis and Sutton. https://doi.org/10.1177/1359105305048552.
26. Usun, S. (2004). Instructional technologies in the world and Turkey (A Comparative Review). *AACE Review (formerly AACE Journal), 12*(3), 249–279. Norfolk, VA: Association for the Advancement of Computing in Education (AACE). Retrieved 18 Aug 2021.
27. Buys, L., Mengersen, K., Johnson, S., Buuren, V., & Chauvin, A. (2014). Creating a sustainability scorecard as a predictive tool for measuring the complex social, economic and environmental impacts of industries, a case study: Assessing the viability and sustainability of the dairy industry. *Journal of Environmental Management, 133*, 0301–4797. https://doi.org/10.1016/j.jenvman.2013.12.013
28. Riddell, W. C., & Song, X. (2017). The role of education in technology use and adoption. *ILR Review, 70*(5), 1219–1253. https://doi.org/10.1177/0019793916687719
29. Melville, N. P. (2010). Information systems and environmental sustainability. *MIS Quarterly, 34*(1), 1–21. Management Information Systems Research Center, University of Minnesota.
30. Watson, H. J., & Wixom, B. H. (2010). The BI-based organization. *International Journal of Business Intelligence Research, 1*(1), 13–28.
31. Elliot, A. J., Murayama, K., & Pekrun, R. (2011). A 3 × 2 achievement goal model. *Journal of Educational Psychology, 103*(3), 632–648. https://doi.org/10.1037/a0023952
32. Kumar, V., & Kumar, U. (2017). Introduction: Technology, innovation and sustainable development. *Transnational Corporations Review, 9*(4), 243–247. https://doi.org/10.1080/19186444.2017.1408553
33. Harwood, A. M., Hahn, C. L. (1990). Controversial Issues in the Classroom, ERIC Clearinghouse for Social Studies/Social Science Education Bloomington IN.
34. Vandermerwe, S., & Oliff, M. (1990). Customers drive corporations. *Long Range Planning, 23*(6), 10–16. ISSN 0024-6301. https://doi.org/10.1016/0024-6301(90)90096-M
35. Thong, J. Y. L., Hong, E. J., & Tam, K. Y. (2006). The effects of post-adoption beliefs on the expectation-confirmation model for information technology continuance. *International Journal of Human-Computer Studies, 64*(9), 799–810.
36. Heijden, H. (2004). User acceptance of hedonic information system. *MIS Quarterly, 28*, 695–704. https://doi.org/10.2307/25148660
37. Brown, S. A., Venkatesh, V. (2005). A model of adoption of technology in the household: A baseline model test and extension incorporating household life cycle. *MIS Quarterly, 29*(3).
38. Childers, T. L., Carrb, C. L., Peck, J., & Carson, S. (2001). Hedonic and utilitarian motivations for online retail shopping behavior. *Journal of Retailing, 77*(4), 511–535.
39. Savary, J., Goldsmith, K., Dhar, R. (2015). Giving against the odds: When tempting alternatives increase willingness to donate. *Journal of Marketing Research, 52*(1), 27–38. https://doi.org/10.1509/jmr.13.0244.
40. Fidani, A., & Idrizi F. (2012). Investigating students' acceptance of a learning management system in university education: A structural equation modeling approach. *ICT Innovations 2012 Web Proceedings*. ISSN 1857-7288.
41. Taiwo, A. A., & Downe, A. G. (2013). The theory of user acceptance and use of technology (UTAUT): A meta-analytic review of empirical findings. *Journal of Theoretical and Applied Information Technology., 49*(1), 48–58.
42. Daniel, B. (2015). Big data and analytics in higher education: Opportunities and challenges. *British Journal of Educational Technology, 46*(5), 904–920. https://doi.org/10.1111/bjet.12230
43. Alam, M. S., & Uddin, M. A. (2019). Adoption and implementation of enterprise resource planning (ERP): An empirical study. *Journal of Management Research, 6*, 84–116.
44. Dwivedi, Y. K., Rana, N. P., Jeyaraj, A., Clement, M., & Williams, M. D. (2019). Re-examining the unified theory of acceptance and use of technology (UTAUT): Towards a revised theoretical model. *Information Systems Frontiers, 21*, 719–734.
45. Raza, S. A., Shah, N., & Ali, M. (2019). Acceptance of mobile banking in Islamic banks: Evidence from modified UTAUT model. *Journal of Islamic Market, 10*, 357–376.

46. Bhatiasevi, V. (2016). An extended UTAUT model to explain the adoption of mobile banking. *Information Development, 32*, 799–814.
47. Nanayakkara, C. A model of user acceptance of learning management systems: A study within Tertiary Institutions in New Zealand.
48. Deng, Y., Ezyske, C. (2011). Sulfate radical-advanced oxidation process (SR-AOP) for simultaneous removal of refractory organic contaminants and ammonia in landfill leachate. *Water Research, 45*(18), 6189–6194. ISSN 0043-1354. https://doi.org/10.1016/j.watres.2011.09.015.
49. John, S. P. (2015). The integration of information technology in higher education: A study of faculty's attitude towards IT adoption in the teaching process. *Contaduría y Administración, 60*(S1), 230–252.
50. Krueger, N. (1993). The impact of prior entrepreneurial exposure on perceptions of new venture feasibility and desirability. *Entrepreneurship Theory and Practice, 18*(1), 5–21. https://doi.org/10.1177/104225879301800101.
51. Riddell, W. C, & Song, X. (2017). *The role of education in technology use and adoption: Evidence from the Canadian workplace and employee survey*. SAGE Publications Inc. https://doi.org/10.1177/0019793916687719.
52. Lleras-Muney, A., & Lichtenberg, F. (2002). The effect of education on medical technology adoption: are the more educated more likely to use new drugs? In *NBER Working Paper No. 9185*.
53. Wozniak, G. (1987). Human capital, information, and the early adoption of new technology. *The Journal of Human Resources, 22*(1), 101–112. JSTOR. www.jstor.org/stable/145869.
54. Zhu, D., & Engels, F. (2014). Organizational culture and instructional innovations in higher education. *Educational Management Administration and Leadership, 42*(1), 136–158. Sage.
55. Kellick, B., & Wilson, J. W. (2000). *Information technology for schools: Creating practical knowledge to improve student performance*. Jossey-Bass.
56. Ukko, J., Nasiri, M., Saunira, M., Rantala, T. (2019). Sustainability strategy as a moderator in the relationship between digital business strategy and financial performance. *Journal of Cleaner Production, 236*.
57. Cooper, R. B., & Quinn, R. E. (1993). Implications of the competing values framework for management information systems. *Human Resource Management, 32*(1), 175–201.
58. Smith, M., Then, A., Wor, C., Ralph, G., Pollock, K., & Hoenig, J. (2012). Recommendations for catch-curve analysis. *North American Journal of Fisheries Management, 32*(5), 956–967. https://doi.org/10.1080/02755947.2012.711270
59. Parsons, D. (1972). Specific human capital: An application to quit rates and layoff rates. *Journal of Political Economy, 80*(6). https://doi.org/10.1086/259961.
60. Sashking, Rosenbach, Sashkin, M., & Rosenbach, W. E. (1996, 2002, 2013). Organizational assessement questionnaire.
61. Ajzen, I., & Fishbein, M. (2005). The influence of attitudes on behavior. In D. Albarracín, B. T. Johnson, & M. P. Zanna (Eds.), *The handbook of attitudes* (pp. 173–221). Lawrence Erlbaum Associates Publishers.
62. Fishbein & Ajzen. (1975). *Belief, attitude, intention, and behavior: An introduction to theory and research*. Reading, MA, USA: Addison-Wesley Pub. Co. ISBN 0201020890.
63. Shu, W., & Chuang, Y. (2011). The perceived benefits of six-degree-separation social networks. *Internet Research, 21*(1), 26–45. https://doi.org/10.1108/10662241111104866
64. Tsai, Y., Chien, C., & Tsai., T,. (2011). Curcumin and its nano-formulation: The kinetics of tissue distribution and blood–brain barrier penetration. *International Journal of Pharmaceutics, 416*(1), 331–338.
65. SivaKumar, A., & Gunasekaran, A. (2017). An empirical study on the factors affecting online shopping behavior of millennial consumers. *Journal of Internet Commerce, 16*(3), 219–230. https://doi.org/10.1080/15332861.2017.1317150
66. Wei, C. (2020). Agroecology, information and communications technology, and smallholders' food security in Sub-Saharan Africa. *Journal of Asian and African Studies., 55*(8), 1194–1208.
67. Hallikainen, H., Paesbrugghe, B., Laukkanen, T., Rangarajan, D., & Gabrielsson, M. (2017). How individual technology propensities and organizational culture influence B2B customer's

behavioral intention to use digital services at work? In *Proceedings of the 50th Hawaii International Conference on System Sciences 2017* (pp. 4577–4585). https://doi.org/10.24251/HICSS.2017.552.
68. Terzis, V., & Economides, A. A. (2011). The acceptance and use of computer based assessment. *Computers & Education, 56*(4), 0360–1315. https://doi.org/10.1016/j.compedu.2010.11.017
69. Miladinovic & Hong. (2016). A study on factors affecting the behavioral intention to use mobile shopping fashion apps in Sweden. http://www.diva-portal.org.
70. Jasperson, J., Carter, P., & Zmud, R. W. (2005). A comprehensive conceptualization of post-adoptive behaviors associated with information technology enabled work systems. *MIS Quarterly, 29*(3), 525–557.
71. Mosavi, S., & Ghaedi, M. (2013). Behavioral intention in the luxury fast food restaurant. *African Journal of Business Management, 7*(18), 1845–1853.
72. Stern, P. C., Dietz, T., & Guagnano, G. A. (1995). The new ecological paradigm in social-psychological context. *Environment and Behavior, 27*, 723–743.
73. Harland, C. M., Caldwell, N. D., Powell, P., & Zheng, J. (2007). Barriers to supply chain information integration: SMEs adrift of eLands. *Journal of operations management, 25*(6), 1234–1254.
74. Niehaves, B., & Krause, A. (2010). Shared service strategies in local government—A multiple case study exploration. *Transforming Government: People, Process and Policy, 4*(3), 266–279. https://doi.org/10.1108/17506161011065235
75. Nidumolu, R., Prahalad, C. K., & Rangaswami, M. R. (2009). Why sustainability is now the key driver of innovation. *Harvard business review, 87*(9), 56–64.
76. Hart, S., & Dowell, G. (2011). A natural-resource-based view of the firm: Fifteen years after. *Journal of Management, 37*, 1464–1479.
77. Hart, S. L. (1995). A natural-resource-based view of the firm. *Academy of Management Review, 20*, 986–1014.
78. Harwood, R. R. (1990). A history of sustainable agriculture. In C. A. Edwards, R. Lal, R. Madden, R. Miller, & G. House (Eds.), *Sustainable agricultural systems* (pp. 3–19). Soil and Water Conservation Society.
79. Mayer, J., Mooney, B., Gundlapalli, A., Harbarth, S., Stoddard, G. J., Rubin, M. A., & Samore, M. H. (2011). Dissemination and sustainability of a hospital-wide hand hygiene program emphasizing positive reinforcement. *Infection Control & Hospital Epidemiology, 32*(1), 59–66.
80. Ziemba, E. (2019). The contribution of ICT adoption to the sustainable information society. *Journal of Computer Information Systems, 59*(2), 116–126.
81. Gouvea, R., Kapelianis, D., & Kassicieh, S. (2018). Assessing the nexus of sustainability and information & communications technology. *Technological Forecasting and Social Change, 130*, 0040–1625. https://doi.org/10.1016/j.techfore.2017.07.023
82. Mulamula, G., & Amadi-Echendu, J. (2017). An examination of the potential links between ICT technology transfer and sustainable development. *International Journal of Technology Management & Sustainable Development, 16*(2), 119–139.
83. Latif, H. H., Gopalakrishnan, B., Nimbarte, A., & Currie, K. (2017). Sustainability index development for manufacturing industry. *Sustainable Energy Technologies and Assessments, 24*, 82–95.
84. Ştefănescu-Mihăilă, R. O. (2015). Social investment, economic growth and labor market performance: Case study—Romania. *Sustainability, 7*(3), 2961–2979.
85. Czernich, N., Falck, O., Kretschmer, T., & Woessmann, L. (2011). Broadband infrastructure and economic growth. *The Economic Journal, 121*(552), 505–532.
86. Malik, K., Rahman, S. M., Khondaker, A. N., Abubakar, I. R., Aina, Y. A., & Hasan, M. A. (2019). Renewable energy utilization to promote sustainability in GCC countries: Policies, drivers, and barriers. *Environmental Science and Pollution Research, 26*(20), 20798–20814.
87. Dempsey, N., Bramley, G., Power, S., & Brown, C. (2011). The social dimension of sustainable development: Defining urban social sustainability. *Sustainable development, 19*(5), 289–300.
88. Miemczyk, J., Johnsen, T. E., & Macquet, M. (2012). Sustainable purchasing and supply management: a structured literature review of definitions and measures at the dyad, chain and network levels. *Supply Chain Management: An International Journal.*

89. Dedrick, J. (2010). Green IS: Concepts and issues for information systems research. *Communications of the Association for Information Systems, 27*(1).
90. Szymkowiak, A., & Garczarek-Bąk, U. (2018). *Gender Differences in E-commerce, 4*, 250–261.
91. Nassar, A. A. M., Othman, K., & Nizah, M. A. B. M. (2019). The impact of the social influence on ICT adoption: Behavioral intention as mediator and age as moderator. *International Journal of Academic. Research in Business and Social Sciences, 9*(11), 963–978.
92. Taras, V., Rowney, J., & Steel, P. (2009). Half a century of measuring culture: Review of approaches, challenges, and limitations based on the analysis of 121 instruments for quantifying culture. *Journal of International Management, 15*, 357–373.
93. Hair, J., Black, W., Babin, B., & Anderson, R. (2010). *Multivariate data analysis* (7th ed.). Pearson.
94. Nevitt & Hancock. (2001).
95. O'Rourke, N., Psych, R., & Hatcher, L. (2013). A step-by-step approach to using SAS for factor analysis and structural equation modeling. SAS Institute. https://support.sas.com/publishing/pubcat/chaps/55129.pdf.
96. Nunnally, J. (1978). *Psychometric methods*. McGraw-Hill.
97. Tavakol, M., & Dennick, R. (2011). Making sense of Cronbach's alpha. *International journal of medical education, 2*, 53–55. https://doi.org/10.5116/ijme.4dfb.8dfd

Factors to the Adoption of GSM Telephony in Rural Areas in West Africa

Sèmévo Ida Tognisse and Jules Degila

Abstract Mobile phone has revolutionized communication and comes with many services. However, its adoption is slow in some regions, particularly in rural, poor, and remote areas in West Africa. The purpose of this study is to examine the mobile telephony adoption in these regions. We use a research model based on the TAM (Technology Acceptance Model) and the moderating effects of age and user experience. Six hundred twenty persons from four villages and two representative countries in West Africa participated in the survey. The quantitative data analysis shows that the factors involved in the research model explain the technology adoption at 44%. In addition, the multigroup analysis showed significant gaps in technology adoption according to gender, country of study, and whether the person subscribes to the technology or not.

Keywords Technology acceptance model · Adoption · GSM telephony · Rural areas

1 Introduction

Information and Communication Technology (ICT) has invaded all aspects of our lives. Mobile telephony is one of the best evidence of such a revolution. The number of unique mobile subscribers has increased, reaching 67% of the world population at the end of 2019, and is expected to reach 70% in 2025 [1]. This rate shows the high adoption of telephony worldwide, but it also shows the digital gap, with 33% of the world population not connected. Previous studies have shown that almost 70% of mobile phone owners are in the developing world [2]. In developing countries, particularly in the West African region, by the end of 2018, the overall penetration rate of unique mobile subscribers was only 48% [3]. Despite this remarkable increasing

S. I. Tognisse (✉) · J. Degila
Institut de Mathématiques et de Sciences Physiques, Porto-Novo, Benin
e-mail: ida.tognisse@imsp-uac.org

J. Degila
e-mail: jules.degila@imsp-uac.org

C. Dal Zotto et al. (eds.), *Smart Technologies for Organizations*, Lecture Notes in Information Systems and Organisation 60, https://doi.org/10.1007/978-3-031-24775-0_3

adoption rate compared to previous years, it shows that half of the population of the sub-region does not yet use the telephone and is not connected to the modern world [4]. Studies conducted by the International Telecommunication Union (ITU) revealed that the majority of the unconnected peoples are living in uncovered regions, particularly in rural, poor, and remote areas [5].

Indeed, bringing communication infrastructure cost-effectively in rural, poor, and remote areas is a big challenge. As a result, public network operators limit their investment to urban and more populated areas than in rural areas [6]. Thus, there is a substantial digital gap between rural and urban populations. In some cases, although the infrastructure exists and the telephony services are available, their adoption by the community is not systematic, and the return on investment is not guaranteed for sustainable and commercially viable telephony services. An experiment in a village of 5.000 people in the Democratic Republic of the Congo (survey conducted in 2017 in Louabo, Katanga) showed, six months after the technology was made available, less than three hundred single users connected to the network. It is, therefore, vital to understand and test the factors that can predict the intention of users on adopting telephony in rural Africa, particularly in West Africa, the terrain of this study. A literature review shows a considerable number of studies addressing the adoption of services and technology such as 3G, 4G, 5G, smartphone, mobile banking, Internet… [7–14]. But there is very little literature on the adoption of mobile telephony in general, and in the context of rural and remote areas in West Africa. That is the purpose of this study. We first review the relevant literature to identify the main determinants for rural telephony adoption. Secondly, we explain the research model and methodology in Sects. 3 and 4. In the remaining part, we present the results, aiming at helping the network operators, planning at expanding their infrastructure in the remaining uncovered areas, to understand the key factors shaping the intent and behavior of villagers in West African regions. Our work results aimed at supporting operators in developing a more appropriate marketing strategy that can ensure a return on their investment.

2 Literature Review

The mobile telephony sector has evolved exponentially over time, from 1 to 2G, 3G, 4G, and now 5G. At first, 2G has been widely deployed globally over the past decades and still leads the deployment in rural areas [15]. As the first telephony digital generation, 2G is considered the primary standard for mobile phones and coexists with the newer generations. In Africa, the most widely deployed 2G standard is still GSM. Our study considers 2G services adoption in rural areas.

In terms of adoption, previous studies and researchers have developed and proposed various models to explain the factors that influence the adoption of a technology or innovation by individuals. These studies used established models such as the Theory of Reasoned Action (TRA) [16] and Theory of Planned Behavior (TPB) [17] or specialized technology acceptance models such as the Technology

Acceptance Model (TAM) [18], Technology Acceptance Model 2 (TAM 2) [19], or Unified Theory of Acceptance and Use of Technology (UTAUT) [20]. Among all these theories, TAM is considered the most powerful and popular method in information systems' fields and is ranked among the most effective tools for empirical research [21]. It has become the dominant model of ICT acceptability and adoption [22] and has been praised for its predictive power, which makes it easy to apply to a variety of cases [23]. TAM has been extensively used to evaluate the adoption of services such as mobile banking [24], telemedicine service [25], mobile Internet [26], mobile learning [27], Dance learning [28], m-shopping [29], and more.

In the mobile telephony area, the TAM model has been used to explain the drivers of 3G adoption in Vietnam, China, and Taiwan [22, 24, 30]. Much research has studied the factors affecting mobile telephony in developed countries [31, 32]. Still, very little has focused on trends in developing countries [33–35] and a limited number have gone out into the rural, poor, and remote areas of these countries. Wei and Zhang [36] studied the adoption of mobile telephony in rural China in 2008. In 2011, the RuTAM approach was developed by Islam [37] to address the scarcity of a model for technology use in rural areas. This conceptual model is a combination of TAM with several variables, namely enabling conditions, social influence and demographic factors, and individual factors. De Silva, Ratnadiwakara, and Zainudeen have also studied the adoption of mobile telephony in Bangladesh, Pakistan, India, Sri Lanka, the Philippines, and Thailand [38]. In 2013, Khalil Moghaddam and Khatoon-Abadi conducted a study on ICT adoption among rural users in Iran. They found a significant relationship between ICT adoption and individual user characteristics (such as age and gender) and innovation attributes [39]. In 2014, Gupta and Jain identified factors influencing mobile phone adoption in rural India [40], and in 2015, they conducted a multigroup analysis to identify differences in the perceptions of different groups [41]. However, in the context of West Africa, there is a lack of literature to explain the adoption of telephony in rural areas.

3 Research Model

The Technology Acceptance Model is a theory developed to determine the reasons and factors for which a person accepts or refuses to use technology. This theory postulates that ICT acceptance or adoption is driven by three factors: Perceived Usefulness (PU), Perceived Ease of Use (PEOU), and Attitudes (ATT) [18]. It is based on the following assumptions:

- Hypothesis H1: Perceived Ease of Use (PEOU) has a positive effect on Perceived Usefulness (PU).
- Hypothesis H2: Perceived Ease of Use (PEOU) has a positive effect on Attitudes towards using (ATT).
- Hypothesis H3: Perceived Usefulness (PU) has a positive effect on Attitudes towards using (ATT).

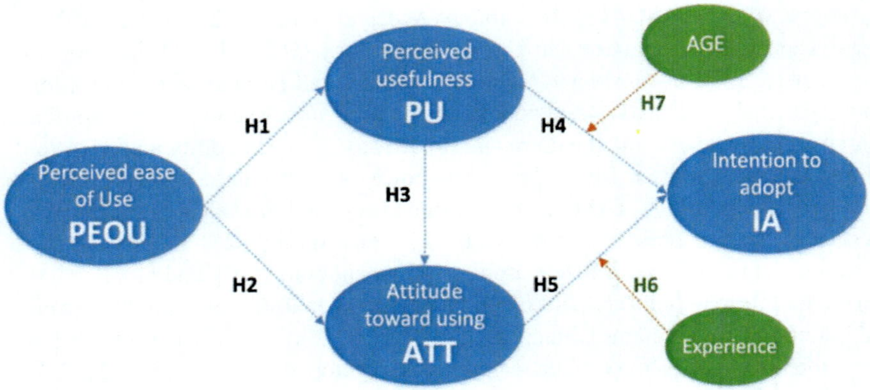

Fig. 1 Research model

- Hypothesis H4: Perceived Usefulness (PU) has a positive effect on Intention to Adopt (IA).
- Hypothesis H5: Attitude toward using (ATT) has a positive effect on Intention to Adopt (IA).

Also, several studies demonstrated other assumptions, indicating that there may be a direct impact of external variables such as age and ICT experience on attitude and use behavior [42, 43]. Other studies also showed that the experience of use is significantly different when it comes to the relationship between (ATT) and (IA) [44]. By analogy, in the context of the GSM telephony adoption, and in addition to the assumptions of the TAM, we make the following assumptions:

- Hypothesis H6: The experience is significantly different with regards to the relationship between (ATT) and (IA).
- Hypothesis H7: Age is significantly different with regards to the relationship between (PU) and (IA).

This results in the following research model (Fig. 1).

4 Methodology

To test the assumptions mentioned above, we use a quantitative approach. Our research was conducted among people living in unconnected areas of two countries: Guinea Conakry and Nigeria. These countries are two representative candidates for our study in West Africa. Guinea has a low mobile network penetration rate, which was 46.5% [45] in 2018. Nigeria has the largest mobile telephony market in the West African region [4]. We select two unconnected villages in each of these two countries. The first criterion is that they all had a population of more than a thousand inhabitants to justify a significant addressable market. The second criterion is that there is a plan to roll out a mobile network in these villages in the next six months, to allow a post-study.

The survey instrument is based on constructs validated by Davis [18] and adapted to the context of this study. All questionnaire items were measured using a seven-point Likert-type scale ranging from “1 strongly disagree” to “7 strongly agree”. For reliability and credibility, the data collection was carried out by ourselves in the field through contact with the inhabitants. No priority choices were made on respondents other than that their age should be greater than or equal to 14 years and that the respondents should be permanent residents of the considered area. The total sample size at the end of the survey was 620 respondents. The demographic characteristics of our respondents are shown in Table 1.

Table 1 Demographic characteristics of respondents

Profile	Description	Frequency	Percentage
Gender	Male	348	56.13
	Female	272	43.87
Age	14–18 years old	75	12.09
	18–25 years old	181	29.19
	26–35 years old	153	24.68
	36–45 years old	100	16.13
	46–60 years old	71	11.45
	Over 60 years old	40	6.45
Level	No degrees	317	51.13
	Primary school	147	23.71
	BECE	89	14.35
	WASSCE SSCE	50	8.06
	ABCE, HND	11	1.77
	Bachelor degree	06	0.97
	Master degree and more	00	00
Subscription	Having a phone	422	68.06
	Not having a phone	198	31.94
Experience	Never	135	21.77
	Under 6 months	31	0.5
	6–12 months	31	0.5
	12–18 months	24	3.87
	18–24 months	12	1.94
	Over 2 years	387	62.42

Out of the 620 respondents, 51.13% have no education at all. This highlights the level of lack of literacy/education in rural areas

5 Results

Structural equation modeling was adopted in this study to test both the measurement model (confirmatory factor analyses) and the structural model. The test procedure of the model is conducted following the methodology used in PLS-Approach-based studies. We rely, in particular, on the recommendations in [46–48]. The validation and estimate procedure consists of three parts:

- In the first place, the aim is to examine the general characteristics of the variables of the model (descriptive statistics, level, and significance of the correlations) to ensure that the postulated links between the variables exist.
- Secondly, it is necessary to ensure the validity of the measurement model (quality of the measurement of latent variables), by using validation procedures adapted to the type of variables of the model.
- Thirdly, we test the structural model by gauging the assumptions as in [49].

For the implementation of PLS regression analyses, we decided to use Smart PLS software (SmartPLS: Ringle, CM, Wende, S. and Becker, J.-M. 2015. "SmartPLS 3."), because of its easiness to use, its interface, and the possibility of obtaining graphical representations of the estimated models. The following results are the outcome of the model analysis.

5.1 *Analysis of the Measurement Model*

Table 2 presents the item loadings and the descriptive statistics of all the manifest variables of the model.

To evaluate the measurement model and internal reliability, we use the convergent and discriminant validity for each construct. According to the literature, the acceptable values of Cronbach Alpha, Composite Reliability (CR), and Average Variance Extracted (AVE) measurements must be greater than 0.7, 0.7, and 0.5 [50–52], respectively. Table 3 shows all Cronbach Alpha, Composite Reliability, and Average Variance Extracted values.

A first analysis shows that the Cronbach Alpha values of AVE and the composite reliability obtained prove to be entirely satisfactory. The table shows that the values of CR are between 0.909 and 0.968. They are then higher than 0.700. Also, the Cronbach Alpha values of the constructs are between 0.867 and 0.956, which are higher than 0.700. There is then a strong internal consistency and reliability of our constructs. As for AVE, its values vary from 0.688 to 0.884 > 0.5. With these results, we can conclude that convergent validity is assured.

Discriminatory validity: We test discriminant validity by examining all correlation matrices with the square root of the AVE in the diagonal. All values exceeded inter-correlationships (Table 4).

Table 2 Descriptive statistics of the measurement model

Construct	Items	Loadings	Min	Max	Average	Standard deviation
Age	Age	1.000	1.00	6.00	3.05	1.40
ATT	ATT_1	0.916	2.00	7.00	6.35	0.85
	ATT_2	0.957	2.00	7.00	6.28	0.95
	ATT_3	0.957	1.00	7.00	6.21	1.01
	ATT_4	0.930	1.00	7.00	6.24	0.99
Experience	Experience	1.000	1.00	6.00	4.47	2.12
IA	IA_1	0.843	1.00	7.00	6.45	0.86
	IA_2	0.846	1.00	7.00	6.45	0.86
	IA_3	0.856	1.00	7.00	5.79	1.80
	IA_4	0.856	1.00	7.00	5.80	1.80
PEOU	PEOU_1	0.832	1.00	7.00	5.29	1.56
	PEOU_2	0.794	1.00	7.00	5.14	1.57
	PEOU_3	0.807	1.00	7.00	5.18	1.56
	PEOU_4	0.861	1.00	7.00	5.27	1.54
	PEOU_5	0.852	1.00	7.00	5.53	1.39
PU	PU_1	0.833	1.00	7.00	6.09	1.08
	PU_2	0.875	1.00	7.00	6.01	0.97
	PU_3	0.875	1.00	7.00	6.06	1.01
	PU_4	0.797	1.00	7.00	6.19	0.90

This table shows that the measurement model is valid, with a loading greater than 0.7

Table 3 Reliability and validity of the construct

	Cronbach alpha	rho_A	Composite reliability	Average variance extracted (AVE)
ATT	0.956	0.957	0.968	0.884
IA	0.873	0.873	0.913	0.723
PEOU	0.887	0.891	0.917	0.688
PU	0.867	0.869	0.909	0.715

Table 4 Fornell-Larcker criterion

	ATT	IA	PEOU	PU
ATT	0.940			
IA	0.543	0.851		
PEOU	0.454	0.462	0.830	
PU	0.510	0.467	0.498	0.846

Table 5 Heterotrait-Monotrait ratio (HTMT)

	ATT	IA	PEOU	PU
ATT				
IA	0.595			
PEOU	0.489	0.522		
PU	0.557	0.534	0.563	

The values corresponding to the HTMT correlation ratios between the constructs are shown in Table 5.

These are acceptable because they are below the threshold of 0.90 [51]. Values in Tables 4 and 5 thus confirm the reliability and validity of the constructs.

5.2 *Analysis of the Structural Model*

The Bootstrapping method is used to test the importance of the relationship between the constructs in the search model through the interpretation of t-statistics, as well as the correlation between these constructs by looking at the values of the path coefficient. For it to be meaningful, the value must be higher than 1.96 [53].

Table 6 shows that the normalized path coefficients for all assumptions are significant, and the relationships defined by H4 PU -> IA (t = 4.229) and H5: ATT -> IA (t = 5.095) have significant effects on the adoption of GSM telephony. Assumptions H6 and H7 are substantial at a level of 0.01 and 0.05, respectively. We then confirm that experience and age are, respectively, significantly different for the relationships between (ATT) and (IA) on the one hand and (PU) and (IA) on the other hand.

Table 7 shows the R^2 values and adjusted R^2 of the latent constructs. It shows that the variable "Intention to Adopt" is explained at 44% by the variables "Attitude

Table 6 Testing the assumptions of the measurement model

	Original sample (O)	Sample mean (M)	Standard deviation (STDEV)	T statistics (|O/STDEV|)	p-values	Decision
H1	0.498	0.497	0.033	15.089	0.000****	Supported
H2	0.267	0.270	0.040	6.745	0.000****	Supported
H3	0.377	0.376	0.047	8.058	0.000****	Supported
H4	0.191	0.191	0.045	4.229	0.000****	Supported
H5	0.249	0.249	0.049	5.095	0.000****	Supported
H6	−0.096	−0.096	0.036	2.691	0.007***	Supported
H7	0.083	0.079	0.039	2.100	0.036**	Supported

*p < 0.1; ** p < 0.05; *** p < 0.01; ****p < 0.001

Table 7 Quality criteria "R^2" and "adjusted R^2"

	R^2	Adjusted R^2
ATT	0.313	0.311
AI	0.440	0.434
PU	0.248	0.247

toward using" and "Perceived Usefulness" in the model. For R^2 values above 0.25, we can conclude that the TAM model is good and relevant enough to explain the adoption of GSM telephony in rural areas [53].

From all of the above, we deduce that the search model is valid and explains at 44% the factors in the adoption of GSM telephony.

5.3 *Multigroup Analysis*

A multigroup analysis is used to evaluate whether predefined parameters have significant differences for the estimates of the group-specific model [53].

We use this method here to assess the differences that might exist between parameters such as country, subscription, and gender. A p-value smaller than 0.05 or larger than 0.95 indicates a significant difference from the probability of 0.05 [53] (Table 8).

The table shows that, for the two country groups, results are significantly different as regards the relationships defined by hypotheses H1, H2, H3, H4, and H7. For the relationship defined by hypothesis H4 (PU -> IA), we notice that the value of the path coefficient of GROUP_ Nigeria is higher than the value of the path coefficient of GROUP_ Guinea, as well as for hypothesis H1, H3, and H7.

Table 9 shows for both groups significantly different results with regards to the relationships H1, H2, and H4. For all these hypotheses, the analysis of the path

Table 8 Multigroup analysis according to the "Country" group

	Path coefficients original (GROUP_ Guinea)	Path coefficients original (GROUP_ Nigeria)	Path coefficients diff (|GROUP_ Guinea – GROUP_ Nigeria|)	p-value (GROUP_ Guinea vs. GROUP_ Nigeria)
H1	0.374	0.769	0.395	1.000
H2	0.333	−0.174	0.507	0.000
H3	0.289	0.914	0.625	1.000
H4	0.168	0.467	0.299	0.954
H5	0.248	0.077	0.171	0.117
H6	−0.127	0.010	0.137	0.935
H7	0.030	0.205	0.175	0.965

Table 9 Multigroup analysis following "Subscription"

	Path coefficients original (GROUP_ Subscriber)	Path coefficients original (GROUP_ NonSubscriber))	Path coefficients-diff (|GROUP_ Subscriber – GROUP_ NonSubscriber |)	p-value (GROUP_ Subscriber vs. GROUP_ NonSubscriber)
H1	0.314	0.448	0.134	0.966
H2	0.107	0.345	0.238	0.997
H3	0.286	0.292	0.006	0.521
H4	0.034	0.301	0.267	0.996
H5	0.298	0.239	0.059	0.279
H6	0.023	−0.078	0.102	0.126
H7	0.082	0.133	0.051	0.736

Table 10 Multigroup analysis according to gender

	Path coefficients original (GROUP_ Man)	Path coefficients original (GROUP_ Woman)	Diff-path coefficients (| GROUP_Man – GROUP_Woman |)	p-value (GROUP_ Man vs. GROUP_ woman)
H1	0.420	0.496	0.075	0.855
H2	0.213	0.273	0.060	0.761
H3	0.385	0.327	0.058	0.269
H4	0.077	0.287	0.209	0.995
H5	0.338	0.197	0.141	0.082
H6	−0.073	−0.148	0.076	0.147
H7	0.019	0.095	0.076	0.854

coefficient values shows that the value of the (GROUP_ NonSubscriber) is higher than the value of the (GROUP_ Subscriber).

Table 10 shows significantly different results regarding the relation: PU -> IA (p-value = 0.995 greater than 0.95) for the two groups GROUP_Man and GROUP_ Woman.. A comparison of the values of the Path Coefficients shows that the value of the Path Coefficients of GROUP_Woman is higher than the value of the path coefficient of the GROUP_Man.

6 Discussion and Implications

As stated in the introduction, this study responds to the need to understand better the main factors that determine the intention to adopt GSM telephony in rural, poor, and remote areas of West Africa. The empirical results confirm the validity of the measurement and the structural models with an explanatory variance of 44% for adoption intention. Although this explanatory variant is significant, it has led to what we can consider as a low intention to adopt. Another model must then be coined to explain the adoption better. However, since all the assumptions of the model are significant, we can retain that these results are valid and are consistent with Davis' findings [18], which confirms the validity of the TAM in the context of the adoption of GSM telephony in rural areas. Also, in line with the results obtained by Gupta and Jain in 2015 [35] in the context of ICT adoption in rural India, our results confirm that perceived usefulness is a crucial factor influencing mobile phone adoption in West Africa, and age has a moderating effect.

Moreover, more than the PU effect, our results show that ATT, a parameter that was not taken into account by Gupta and Jain, is more significant than PU in adoption in West Africa. Thanks to the multigroup analysis, it can be seen that the adoption of GSM telephony is different from one country to another. Almost all the hypotheses are significantly different, except for the relationship between ATT and AI and the moderating effect of the experience, which remains the same in both countries. The inference is that, in general, individuals adopt telephony if they see its attractive potential, and their experience of use is a factor moderating this factor. Furthermore, it can be said that in Nigeria (one of the most developed countries in West Africa), Perceived Usefulness defines much more the intention to adopt compared to a less wealthy country like Guinea Conakry. The level of economic development of the country and thus the financial autonomy of the inhabitants could indeed be a factor, which explains the fact that the intention to adopt is not always the same, despite the awareness of mobile phone usefulness.

Multigroup analysis by gender shows that the AI of women was more strongly influenced by PU. Indeed, in all low- and middle-income countries, 15% of adults do not yet have a mobile phone. As a result, these people often belong to the most marginalized population groups and are found disproportionately among rural, illiterate, and older people [54]. Thus, in rural areas, women, who are generally uneducated, are more motivated to adopt GSM telephony when they are aware of its usefulness, which is not necessarily the case for men who naturally opt much more for the adoption of new technology. In line with the findings of Gupta and Jain [41], men and women in rural West Africa seem to perceive communication differently. Men accept the use of telephony easily because it is almost included in their lives and activities. At the same time, women look for specific useful features because the usefulness of mobile phones for women lies in their ability to connect and keep them in touch with family members. They see it as a means of personal communication. Subscription to the technology has also proved to influence some TAM relationships in rural West Africa differently.

The results of the multigroup analysis following subscription show that the relationships between PU on AI, PEOU, and PU and PEOU and ATT have a more substantial effect at the subscriber level than at the nonsubscriber level. So, in an unconnected region, among those who do not yet use a GSM phone, these factors have a significant influence on their intention to adopt it soon. In order to increase the adoption rate, it will, therefore, be necessary to act on the perceived usefulness and ease of use of nonsubscribers. To attract nonsubscribers in rural Africa, simple devices that are easy to learn to use and adapt to their needs are recommended. This study thus makes an essential contribution by highlighting factors influencing the adoption of GSM telephony in West Africa. The proposed conceptual model contributes to existing knowledge by including dimensions such as the moderating effect of age and usage experience that influence mobile phone adoption, and by trying to observe the differences that might exist between two different countries. In practice, this study provides elements on which a network operator can base its marketing strategy in rural areas. The factors that need to be acted upon are different for different groups, suggesting that the decision-making process is context-specific. Overall, closing the digital gap requires addressing the perceived usefulness of nonsubscribers and how they find the technology attractive, targeting specific age groups as well as those with telephony experience.

7 Conclusion and Limitations

In this work, we deal with the adoption of GSM telephony in rural areas. Using a TAM-based model, we determined factors that influence the adoption of mobile phones in the West African context. Our study was conducted in two representative countries. Our analysis considers different groups of people and has led to understanding the differences in the adoption by country, gender, and previous subscription. In general, bridging the usage gap in rural West Africa will require addressing the perceived usefulness of nonsubscribers, targeting specific age groups, as well as those with previous experience in using telephony. The cost of the technology, people's income, social influence, expected performance, and cultural factors are other factors that influence the adoption. This research, however, has some limitations. The first one is that it is only based on two countries. To fully understand the empirical generalization of the extended TAM model, additional research on other countries in the sub-region would be of interest. Besides, factors such as social influence, cultural factors, expected performance, and evaluating or predicting future use of the technology will need to be incorporated. The support for TAM in this initial application suggests that it provides a useful starting point for further research on telephony adoption. Thus, in our future studies, we hope to find a model that better explains the adoption of GSM telephony. We plan to extend our research to several other African countries and also make a comparative study between the actual use of the technology by the populations before and after connectivity.

Acknowledgements We are grateful to the African Center of Excellence in Mathematical Science and Applications, IMSP, Dangbo, Republic of Benin, which supported this work. We would like to thank especially Prof. Samuel Fosso WAMBA for his contribution and all the field surveyors: Essohanam DJEKI, Aristide BAMAZI, James ADEOLA, Aziz SAIBOU, Gilles MBOGNE, and Charly GNOGUEM.

Annex

Constructs	Definition	Items
Perceived usefulness **PU**	This is the degree to which you think that GSM telephony is useful for you	PU_1: My exchanges with family and business partners would be difficult to achieve without GSM telephony
		PU_2: The use of GSM telephony gives me better control over my exchanges with family and business partners
		PU_3: The use of GSM telephony improves the exchanges with family and business partners
		PU_4: All in all I find the GSM telephony useful for my exchanges with family and business partners
Perceived Ease Of Use **PEOU**	It is your perception of the degree of effort related to the use GSM telephony	PEOU_1: My interaction with the GSM telephony is easy for me to understand
		PEOU_2: I often get confused when I use GSM telephony
		PEOU_3: It is easy for me to remember how to do things using GSM telephony
		PEOU_4: Interaction with GSM telephony is simple and flexible enough
		PEOU_5: All in all I find GSM telephony easy to use
ATTitude toward using **ATT**	It is about measuring your perception of using the GSM	ATT_1: I think using GSM is a good idea
		ATT_2: GSM makes exchanges more interesting
		ATT_3: It is fun using
		ATT_4: I like to use GSM
Intention to Adopt **IA**	It is about measuring your intention to subscribe and use the GSM	IA_1: I will use GSM regularly in the future
		IA_2: I will urge others to use GSM
		IA_3: I intend to use GSM regularly in the future
		IA_4 I intend to use GSM the next month
		IA_5: I predict that I will use GSM the next month

1. Do you have a phone?

Yes	No

2. I am

A man	A woman

3. My age range is

14–18 years	18–25 years	26–35 years	36–45 years	46–60 years	More than 60 years

4. My level of university education

None	Primary school	BECE	WASSCE SSCE	ABCE, HND	Bachelor degree	Degree masters and more

5. My experience in using GSM

None	Less than 6 months	Between 6 and 12 months	Between 12 and 18 months	Between 18 and 24 months	More than 2 years

References

1. GSM Association: The Mobile Economy 2020. (2020).
2. FAO: La téléphonie mobile en milieu rural. (2011).
3. GSM Association: The Mobile Economy Sub-Saharan Africa 2020. (2020).
4. GSM Association: The Mobile Economy Sub-Saharan Africa 2018. (2019).
5. ITU: Connecting The Unconnected Working together to achieve Connect 2020 Agenda Targets. (2017).
6. Simba, F., Mwinyiwiwa, B. M., Mjema, E. M., Trojer, L., & Mvungi, N. H. (2011). Broadband access technologies for rural connectivity in developing countries. *International Journal of Research and Reviews in Computer Science (IJRRCS), 2.*
7. Tapanainen, T. J., Dao, K. T., Nguyen, H. T. T. (2018). Investigating adoption factors of 3G services: An empirical study in Vietnam. *The Electronic Journal Of Information Systems In Developing Countries, 84.*
8. Liao, C.-S., Chun-Wang, C. K., Shou, I., & Ming-Feng, K. (2007). Factors influencing the usage of 3G mobile services in Taiwan. *Online Information Review, 31.*
9. Tapanainen, T., Dao, T. K., Nguyen, T. H., Nguyen, T. T. H. (2019). 4G adoption: A survey of Vietnam market. *Journal of Information Technology Applications & Management, 26*, 1–19
10. Awni, R. (2015). Adoption of 4G Mobile Services from the Female Student's Perspective: Case of Princess Nora University. *Malaysian Online Journal of Educational Technology, 3.*
11. Jeong, M. S., Hong, D. S., Ji, Y. G. (2020). A study on the factors affecting the usage intentions of 5G mobile communication service. *The Journal of Society for e-Business Studies, 25*, 135–176.
12. Ali, A., Yogesh, K., Nripendra, P.: Factors influencing adoption of mobile banking by Jordanian bank customers: Extending UTAUT2 with trust. *International Journal of Information Management, 37*, 9–110.
13. Kim, S. H. (2008). Moderating effects of job relevance and experience on mobile wireless technology acceptance: Adoption of a smartphone by individuals. *Information & management, 45*, 387–393.
14. Alalwan, A. A., Baabdullah, A. M., Rana, N. P., & Tamilmani, Kuttimani, Dwivedi, Y. K. (2018). Examining adoption of mobile internet in Saudi Arabia: Extending TAM with perceived enjoyment, innovativeness and trust. *Technology in Society. 55*, 100–110.

15. Adediran, Y., Opadiji, J. F., Faruk, N., & Bello, O. W. (2016). On issues and challenges of rural telecommunications access in Nigeria. *African Journal of Education, Science and Technology, 3.*
16. Ajzen, F. (1977). Attitude behavior relations theoretical analysis and review of empirical research. *Psychological Bulletin, 84.*
17. Ajzen. (1991). The theory of planned behavior. *Organizational Behavior and Human Decision Processes, 50*, 179–211.
18. Davis, F. D. (1993). User acceptance of information technology: System characteristics, user perceptions and behavioral impacts. *International Journal of Man Machine Studies, 3*, 475–487.
19. Venkatesh, V., & Davis, F. D. (1996). A model of the antecedents of perceived ease of use: Development and test. *Decision Sciences, 7*, 451–481.
20. Venkatesh, V., Morris, M. G., Davis, G. B., & Davis, F. D.: User acceptance of information technology: Toward a unified view. *MIS Quarterly, 27*, 425–478.
21. Chang, S.-H., Chou, C.-H., & Yang, J.-M. (2010). The literature review of technology acceptance model: A study of the bibliometric distributions. *PACIS.*
22. Hsiao, C. H., Yang, C. (2011). The intellectual development of the technology acceptance model a co-citation analysis. *International Journal of Information Management, 31*, 128–136.
23. Agarwal, R., Prasad, J. (1999). Are individual differences germane to the acceptance of new technologies? *Decision Sciences, 30*, 361–391.
24. Al-Jabri, I. M., & Sohail, M. S. (2012). Mobile banking adoption: Application of diffusion of innovation theory. *Journal of electronic Commerce Research, 131*, 379–391.
25. Syeda, A. K., Muhammad, S., & Priyanka, K. (2019). Investigating acceptance of telemedicine services through an extended technology acceptance model (TAM). *Technology in Society.*
26. Ali, A. A., Abdullah, B., Nripendra, R., & Kuttimani, T. (2017). Examining adoption of mobile internet in Saudi Arabia: Extending TAM with perceived enjoyment, innovativeness and trust. *International Journal of Information Management, 37*, 99–110.
27. Mostafa, A., Ibrahim, A., & Said, A. S. (2020). An empirical examination of continuous intention to use m-learning: An integrated model. *Education and Information Technologies.*
28. Hong, J.-C., Chen, M.-L., Y, J.-H. (2020). Acceptance of YouTube applied to dance learning. DOI: https://doi.org/10.18178/ijiet.2020.10.1.1331.
29. Wong, C. H., Lee, H. S., Lim, Y. H., Chai, B. H., & Tan, G. W. H. (2012). Predicting the 'consumers' intention to adopt mobile-shopping: An emerging market perspective. *International Journal of Network and Mobile Technologies, 3.*
30. Yee-Loong Chong, A., Ooi, K.-B., Lin, B., Bao, H. (2012). Chong An empirical analysis of the determinants of 3G adoption in China. *Computers in Human Behavior.*
31. Frank, L. D. (2004). An analysis of the effect of the economic situation on modeling and forecasting the diffusion of wireless communications in Finland. *Technological Forecasting and Social Change,71*, 391–403.
32. Lee, M., & Cho, Y. (2007). The diffusion of mobile telecommunications services in Korea. *Applied Economics Letters, 14*, 477–481
33. Gupta, R., & Jain, K. (2012). Diffusion of mobile telephony in India: An empirical study. Technology Forecasting & Social Change. In: *Technological Forecasting and Social Change* (vol. 79, pp. 709–715). Elsevier.
34. Liu, Y., & Li, H. (2010). Mobile Internet diffusion in China: An empirical study. *Industrial Management & Data Systems.*
35. Singh, S. K. (2008). The diffusion of mobile phones in India. *Telecommunications Policy, 32*, 642–651.
36. Wei, L., & Zhang, M. (2008). The adoption and use of mobile phone in rural China: A case study of Hubei.
37. Sirajul, I. M., & Åke, G. (2011). Factors influencing the adoption of mobile phones among the farmers in Bangladesh: Theories and practices. *International Journal on Advances in ICT for Emerging Regions, 4*, 4–14.
38. De Silva, H., Ratnadiwakara, D., & Zainudeen, A. (2011). Social influence in mobile phone adoption: Evidence from the bottom of the pyramid in emerging Asia.

39. Khalil Moghaddam, B., & Khatoon-Abadi, A. (2013). Factors affecting ICT adoption among rural users: A case study of ICT Center in Iran. *Telecommunications Policy, 37*, 1083–1094.
40. Gupta, R., & Jain, K. (2014). Adoption of mobile telephony in rural India: An empirical study. *Decision Sciences, 45*.
41. Gupta, R., Jain, K. (2015). Adoption behavior of rural India for mobile telephony: A multigroup study. *Telecommunications Policy, 39*, 691–704. Elsevier.
42. Burton-Jones, A., & Hubona, G. S. (2006). The mediation of external variables in the technology acceptance model. *Information and Management*.
43. Atarodi, S., Anna-Maria, B., & Toniolo, A. (2018). Le modèle d'acceptation des technologies depuis 1986: 30 ans de développement. *Psychologie du Travail et des Organisations*.
44. Wamba, S. F. (2018). Social media use in the workspace: Applying an extension of the technology acceptance model across multiple countries. *WorldCIST*.
45. GSMA. (2018). Réformer la fiscalité de la téléphonie mobile en Guinée pour libérer les avantages socio-économiques de la connectivité mobile.
46. Haenlein, M., & Kaplan, A. M. (2004). A beginner's guide to partial least squares analysis—Understanding statistics. https://doi.org/10.1207/s15328031us0304_4.
47. Sosik, J., Kahai, S., & Piovoso, M.: Silver bullet or voodoo statistics? A primer for using the partial least squares data analytic technique in group and organization research. *Group & Organization Management, 34*, 5–36.
48. Bruhn, M., Georgi, D., & Hadwich, K. (2008). Customer equity management as a formative second-order construct. *Journal of Business Research, 61*, 1292–1301.
49. Lacroux, A. (2010). L'analyse des modèles de relations structurelles par la méthode pls: une approche émergente dans la recherche quantitative en GRH. *XXème congrès de l'AGRH, Toulouse*.
50. Chin, W. W. (1998). The partial least squares approach for structural equation modeling. In: Marcoulides, G. A. (Ed.). (1998). *Modern methods for business research* (vol. 8, pp. 295–336).
51. Hair, J. F., Hult, G. T. M., Ringle, C. M., & Sarstedt, M. (2017). *A: Primer on partial least squares structural equation modeling (PLS-SEM)* (2nd ed.). Sage.
52. Fornell, C., & Cha, J. (1994). Partial least squares. In R. P. Bagozzi (Ed.), *Advanced methods of marketing research* (pp. 52–78). Blackwell.
53. Fobang, A. N., Wamba, S. F., & Kamdjoug, J. R. K. (2019). Exploring factors affecting the adoption of HRIS in SMEs in a developing country: evidence from Cameroon. *ICT for a Better Life and a Better World*, 281–295.
54. GSM Association. (2019). The Mobile Gender Gap Report.

A Disruptive Fintech for Inclusive Finance in Africa: Role and Contribution of Mobile Money

Mohamed Amine Issami and Boukary Tandamba

Abstract Technological innovations applied to finance have given rise to a new industry: Fintech. One of the main issues this industry is addressing, especially in developing countries like Africa, is financial inclusion. However, despite the rapid development of fintech firms in Africa, particularly the mobile payment (mobile money) segment, over the past two decades, two-thirds of adults are still unbanked on the continent. This paper reviews the fintech landscape in Africa and the state of financial inclusion on the continent, focuses on the impact of fintech companies on three fundamental dimensions of financial inclusion (access, usage, and quality of financial products and services offered), and concludes with a discussion of the critical role that these fintech firms could play in the continent's future.

Keywords Fintech · Disruption · Mobile money · Financial inclusion · Africa

1 Introduction

Financial technology (Fintech) has experienced a meteoric rise over the past two decades, driven by the constant evolution of new technologies and customer expectations, favorable regulatory environments, and an unmet demand for financial services. This new industry holds promise for greater financial inclusion especially in developing countries [1] like Africa. The continent is considered a "promised land" for fintech firms [2], especially in the mobile payment (mobile money) segment. This segment is growing rapidly on the continent and is demonstrating disruption in the provision of services to individuals and businesses, taking advantage of the increasing penetration of mobile phones. Although Africa accounted for only 2% of the total of nearly 13,000 fintech companies identified worldwide in 2018 [3], the continent has the highest population of adults with a mobile money account, at 21% compared

M. A. Issami (✉) · B. Tandamba
Groupe ISCAE, Casablanca, Morocco
e-mail: aissami@groupeiscae.ma

B. Tandamba
e-mail: btandamba_doct21@groupeiscae.ma

C. Dal Zotto et al. (eds.), *Smart Technologies for Organizations*, Lecture Notes in Information Systems and Organisation 60, https://doi.org/10.1007/978-3-031-24775-0_4

to a global average of 4% [4]. Mobile money could therefore be a solution toward financial inclusion for the 370 billion adults still unbanked on the continent.

This paper explores fintech and financial inclusion trends with a focus on Africa, notably the Sub-Saharan region; and then proposes to conduct tests to determine the impact of mobile money fintech on the financial inclusion of Africans by focusing on three main dimensions, namely access, usage, and quality of financial products and services offered to the population. The main questions addressed are as follows: *what are the dynamics of the fintech ecosystem in Africa? What are the trends in financial inclusion in Africa compared to the global level? Does fintech have an impact on the access, usage, and quality dimensions of financial inclusion in Africa? What approaches can be taken to harness the potential of fintech firms so that they better contribute to the inclusive development of the continent?*

These questions are the main concern of this paper, which is divided into four main parts. The first part is dedicated to the literature review. It includes defining key concepts and outlining their current dynamics, as well as presenting some of the currents of thought regarding the disruption of fintech for financial inclusion. The second part of this paper describes the research methodology used. The third part is dedicated to the empirical study which measures the impact of mobile money on the different dimensions of financial inclusion in Africa. In addition, a study on the contribution of some innovative, even disruptive, business models is carried out. Finally, the last part of the paper is about discussions and conclusions on how disruptive fintech companies can better contribute to financial inclusion and inclusive development in Africa.

2 Disruptive Fintech for Financial Inclusion: Concepts and Theories

The term Fintech is usually associated with other buzzwords such as the digital revolution phenomenon, digital transformation, uberization, and disruption. These concepts are among the most used terms and expressions during this decade and continue to arouse the interest and curiosity of the general public. As an illustration, Internet searches for the word Fintech have followed an exponential trajectory since 2014 on Google Trends.

However, the current scientific literature on Fintech is less abundant, especially concerning the African region. In addition, the available literature often lacks completeness and depth. In fact, the role and contribution of Fintech are often treated marginally. Our paper proposes to address the research question on the impact of fintech on financial inclusion to complement this scientific literature.

Before addressing the different currents of thought on disruptive Fintech and financial inclusion, it is worth exploring the different concepts.

2.1 *Definitions and Current Dynamics of Fintech, Disruption, and Financial Inclusion*

Fintech. It is a concept that comes from the contraction between *finance* and *technology*. It refers to the use of new technologies to improve the existing offer of financial products and services and to offer new tools or approaches to address consumers' needs in terms of financial products and services. One of the commonly accepted succinct definitions of the concept is "Fintech is a new financial industry that applies technology to improve financial activities" [5].

The generalization of the concept is linked to the fundraising records in the fintech industry in 2015 [6].

Before reaching its current stage, Fintech has experienced four main eras in its evolution [7] from 1866 to 1967 to the global financial crisis of 2008, which has been a turning point in the development of Fintech. Indeed, the reforms at the end of the global financial crisis, the mistrust of populations toward banks and banks toward each other, the reinforcement of prudential measures toward the latter, and their willingness to reduce costs through the use of technology were catalysts for the emergence and development of fintech firms and initiatives such as Wealthfront (2008), Bitcoin (2009), Kickstarter (2009), and Wise (2011). The ecosystem is taking full advantage of the penetration of mobile telephony, smartphones and broadband, the advent of big data, artificial intelligence, the cloud, open banking, and the new regulations (PSD1&2, GDPR in Europe, etc.). The advent of 5G and the adoption and expansion of current and future technologies herald an imminent new era of Fintech.

Fintech's various sectors include mobile payments and transfers, peer-to-peer lending, neo-banking, blockchain, asset management, etc. In different sectors, Fintech is fragmenting the value chain of traditional players. They specialize in these niches with very innovative approaches and agile methods and generally use advanced technologies. Given the preponderance of technology in the fintech industry, Skinner [8] argued that the starting point of the deployed entity can determine whether it is a Fintech or a Techfin. The difference is that Fintech takes financial processes and applies them to the technology, while Techfin consists of adopting the technology and applying it to its financial processes.

Geographically, the Fintech ecosystem is booming in each region. Fintech firms play a crucial role in the regional or national financial landscapes where they operate. For example, in Kenya, the mobile payment Fintech M-Pesa, launched in 2007, has become a national and regional champion in its field. M-Pesa is used by more than 70% of Kenya's adult population, increasing the number of bank accounts held by this population by 58% and contributing significantly to its GDP (6.5% in 2016).

In the overall African continent, fintech firms are present in various sectors and address various issues. Multiple factors, including the under-banking of populations and the rapid penetration of mobile phones on the continent, have contributed to the development of the sector in the continent.

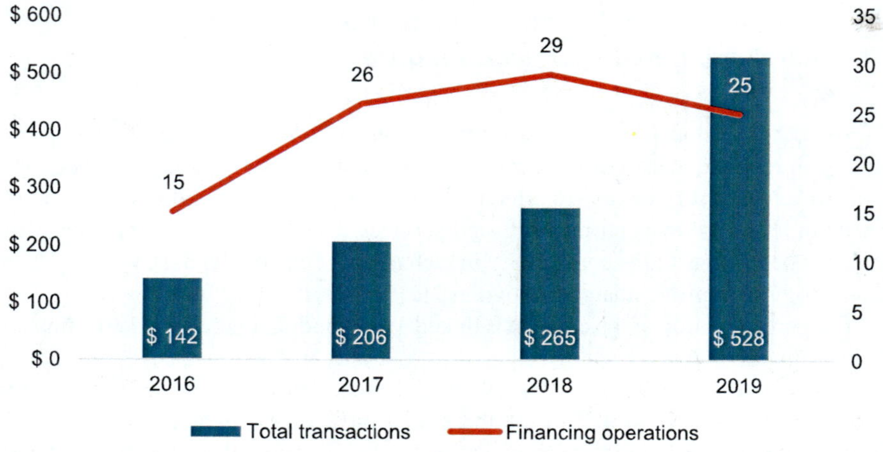

Fig. 1 Venture capital financing of African fintech startups, in millions of dollars. *Source* FT Partners, 2020 [2, 3, 10]

In 2018, Fintech accounted for up to 39% of fundraising within the African startup ecosystem and continued to attract investors' interest [9]. In terms of venture capital financing for Fintech, an upward trend has been observed over the last four years. In 2019, the 471 Fintech identified on the continent have mobilized according to different considerations: $528 million [1–3, 10], $678.73 million [11], or $836 million [12] (Fig. 1).

In terms of sectors, fintech firms that offer mobile payment services are the leading fundraiser [10]. This report shows these fintech companies have raised a cumulative amount of more than $656 million from 2011 to 2019. Moreover, mobile payment Fintech accounted for nearly 9% of Africa's GDP in 2018 according to the Global System for Mobile Communications Association (GSMA). These fintech startups are followed by those offering banking, loans, and credit services and by Insurtech firms (offering insurance and reinsurance services). Several innovations from these different categories of Fintech have bypassed traditional processes and adopted solutions centered on mobile telephony to take advantage of the windfall that the latter represents. In Sub-Saharan Africa, the mobile penetration rate is over 80% with an Internet access rate of 36% [13].

African fintech firms operating in mobile payment services are pioneers at the global level in terms of registered mobile accounts (45.09%), active accounts (49.65%), transaction volume (64.15%), and transaction value (66.12%) (GSMA, 2020). The holding of mobile accounts reached 469 million mobile accounts. Of the region's total, 53% of accounts are registered in East Africa, 35% in West Africa, 10% in Central Africa, and 2% in South Africa. In terms of transaction volume, an increase of 19.7% was recorded in 2019 compared to 2018, for a total value of $456.3 million. About 64% of this amount was realized in East Africa, 28% in West Africa, 7% in Central Africa, and 1% in South Africa. In addition, the majority of active mobile payment services are located in Sub-Saharan Africa (Fig. 2).

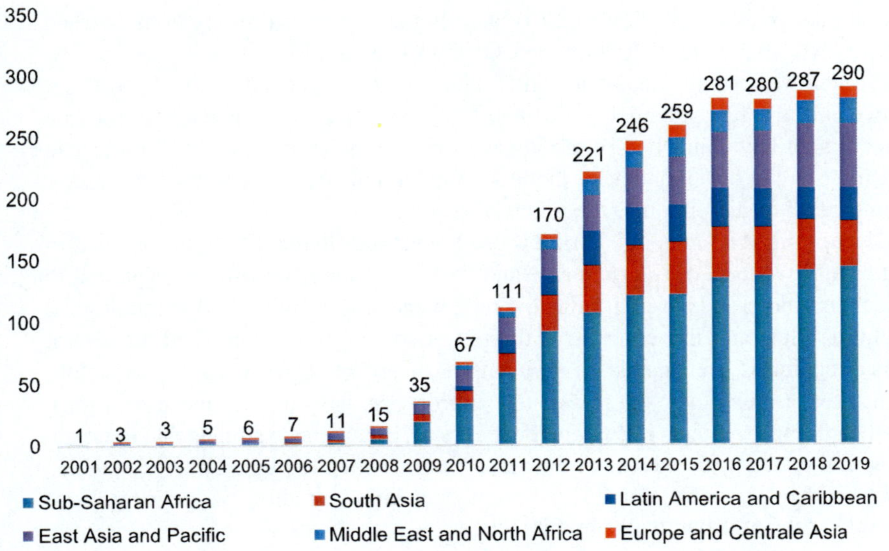

Fig. 2 Geographical evolution of active mobile payment services. *Source* GSMA Database, 2020 [1–3, 14]

Of the 290 services identified worldwide, nearly half (144) are in Sub-Saharan Africa, including 59 in West Africa, 57 in East Africa, 17 in Central Africa, and 14 in South Africa. The proliferation of sales outlets has enabled the various players to serve even populations in remote geographical areas, thus filling the void left by traditional financial players. In fact, for every 100,000 inhabitants, there are more than 228 mobile payment agent outlets compared to only 11 bank branches and 33 ATMs. Mobile payment agents, therefore, have a reach seven times greater than that of ATMs and twenty times greater than banks.

One of the peculiarities in deploying mobile payment solutions in Africa, as in most developing economies, is the massive use of USSD (Unstructured Supplementary Service Data) technology compared to other technologies such as mobile applications and SMS, due to the low democratization of the smartphone, the lack of Internet connection infrastructure, the high cost of mobile broadband, etc. However, this technology remains a preserve of telecommunication operators who have a historical monopoly in most countries. Concerted initiatives should be taken by key authorities in African countries or sub-regions, including financial regulators, competition authorities, and telecom regulators, to ensure free and fair competition and improve the customer experience for payment services on the continent. Countries such as Cameroon, Mali, and Senegal have acted to liberalize USSD technology. Furthermore, in order to facilitate exchanges, several interoperability initiatives have been launched or have been underway since 2014 in several countries, including Tanzania, Madagascar, Uganda, Ghana, Kenya, etc. since late 2018. In 2019 and for the first time, digital mobile cash flows (57%) were higher than cash transactions, a trend that was accentuated in 2020 with the COVID-19 pandemic.

In order to encourage innovation and fill the regulatory gap that is a real constraint to revealing the full potential of African fintech companies, a dozen countries on the continent have launched or are engaged in a "regulatory sandbox[1]" project to contribute to the development of the local fintech industry that could be the vector of innovation or disruption of the financial sector.

Disruption. It is a neologism that is used in various fields. The term comes from the Latin "*disruptus*" or "*disrumpere*" and evokes a rupture, a revolution, a paradigm shift that is both sudden and radical. The advent of the digital and technological revolution illustrates this concept. Entire players or systems, established for several years or centuries, are, against all expectations, overtaken by new players who enter the market or create new ones with new approaches, new and advanced solutions, significant improvements in customer experience, etc. Some examples that illustrate disruption concern the advent of the smartphone, which disrupted players such as Nokia and Kodak; the hotel industry disrupted by Airbnb, traditional media by social networks, cabs by Uber, banks by Fintech, etc.

The pioneers of this concept are Bower and Christensen [1, 2, 15] and Dru [1, 2, 16]. According to Christensen [17], disruptive innovations always start at the low-end or new markets. Traditional players focus on the most profitable and demanding customer segments and neglect the low-end segments. Thus, on a disruptive trajectory, new entrants enter the market by offering products and services to low-end customers and gradually move up the range until they compete with the dominance of the historical players. For Bowe and Christensen [15], while disruption may be less profitable in the short term, it is beneficial in the long term following a process of "creative destruction" as enunciated by Schumpeter [18]. However, Dru [16] finds Clayton M. Christensen's vision of disruption very reductive. For his part, he proposes three steps for the disruptive methodology, namely first identifying cultural conventions around a brand in a given market, then finding a disruptive strategy to challenge existing conventions, and finally defining a new vision. Authors such as Lepore [19] consider that specific sectors, notably health and education, should not be open to disruption. Stiegler [20] contests the positive impacts of disruption. He describes disruption as "a new form of barbarism" and considers that "the acceleration of innovation short-circuits everything that contributes to the development of civilization". He, therefore, hopes that an alternative model will emerge and proposes a negative entropy[2] as cited by Schrödinger [21] in order to reorganize the current situation.

Innovations in the Fintech ecosystem are the archetype of disruption in financial services. Does this Fintech disruption play a role in financial inclusion?

---

[1] A regulatory sandbox is a testing device set up and supervised by a regulatory authority to allow Fintech startups and other innovators to operate under real-world conditions.

[2] Negative entropy, as opposed to the natural tendency to disorganization, is a factor in the organization of systems, whether biological, physical, human, or social. This was evoked in his book.

Financial Inclusion. This concept is defined as the access by individuals and businesses to useful and affordable financial products and services that meet their needs and are provided in a responsible and sustainable way [22]. Four determinants must be brought together to achieve successful financial inclusion:

- Access to financial products and services (affordability, distance, etc.);
- Use of a wide range of products and services adapted to the needs of individuals and businesses;
- Quality of products and services offered (transparency, convenience, etc.);
- Impact of financial products and services on the population as a whole, particularly on the low-income segment of the population.

Financial inclusion holds the promise of poverty reduction, especially for African economies that are confronted with different economic challenges, and has three-dimensional implications that create a virtuous circle. Firstly, through financial inclusion, individuals can enjoy socio-economic well-being through their ability to carry out secure and less costly transactions, to better manage their daily expenses, invest in their education and health, and save for the future [23]. Secondly, for businesses, financial inclusion has a competitive implication. It makes it easier for them to move out of the informal sector, finance their growth, make new investments, and create jobs [24]. Finally, through financial inclusion, economies ensure inclusive growth. They can easily mobilize more resources, ensure financial stability and integrity, create jobs, and reduce socio-economic disparities [25].

Several trends have been observed over the past few decades in some key indicators of financial inclusion globally, and in Sub-Saharan Africa in particular. Globally, there has been an increase in bank penetration rate (16 percentage points in three years) and in the use of accounts held that are either used for savings or for applying for credit from different financial institutions. On average there are more accounts held by Sub-Saharan African populations (21%) compared to the global average (4%) in 2017 (Fig. 3).

Overall, there has been substantial growth in mobile payment accounts (mobile money) holding, which rose from 12% in 2014 to 21% in 2017. The banking and the use of the accounts held are growing but less than mobile money (Fig. 4).

While many authors or entities highlight the virtues of financial inclusion, others point out its limits. They consider financial inclusion as a pretext for the financialization of economies in the strict sense of the term, with the risk of deepening the impoverishment of the most vulnerable segments of society. Mader [27] refutes the argument that financial inclusion is a sufficient condition for poverty reduction. These contrasting views are also found in the disruption of fintech in favor of financial inclusion.

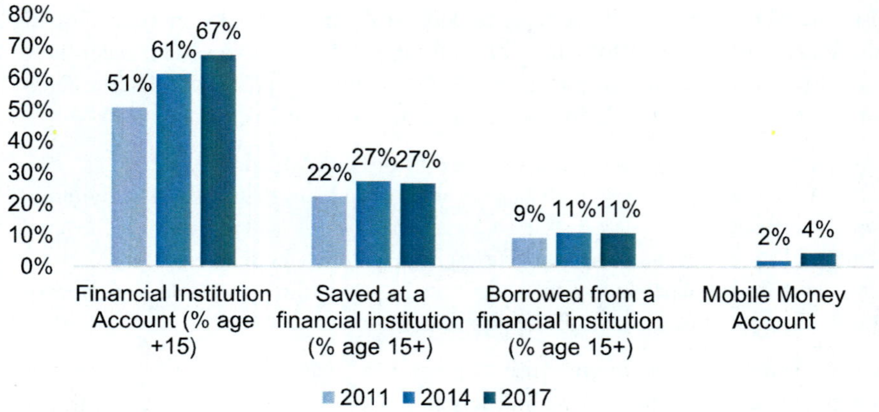

Fig. 3 Three-year global trends in selected indicators of financial inclusion. *Source* WB Global Findex, 2017 [1–4, 26]

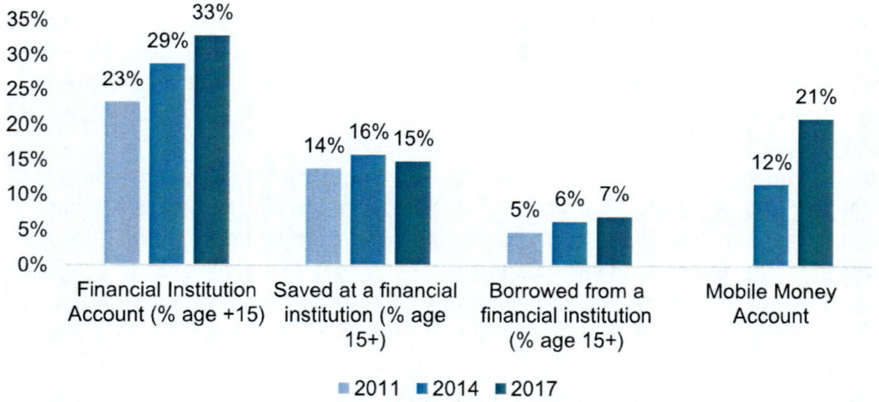

Fig. 4 Three-year trends in selected indicators of financial inclusion in Sub-Saharan Africa. *Source* WB Global Findex, 2017 [2–4, 26]

2.2 *Some Currents of Thought on Fintech's Disruption for Financial Inclusion*

Fintech, as financial innovation, can be of a disruptive or incremental degree,[3] on one or more categories of innovations such as business model, products and services, organization, processes, and the system as a whole, with an inter- or intra-organizational scope [28].

[3] Incremental innovation evokes an improvement of what already exists, while disruptive innovation concerns a change in approaches or paradigms through a proposal of novelties or inventions.

For the past few decades, the idea that the disruption of Fintech may or may not play a critical role in financial inclusion has been discussed by various authors, including academics, research groups, and even business leaders. The currents of thought on the subject can be classified into two groups.

The first school of thought believes that fintech can play a key role in financial inclusion because it goes off the beaten track to provide financial products and services tailored to the needs of people and businesses de facto excluded by the traditional financial system. This is revealed by Lee and Low [29] who show, for example, that through their independence and disintermediation, fintech firms restore confidence in the financial system (mistrust being one of the causes of exclusion). In addition, the two authors believe that inclusiveness is the next frontier of growth in the banking and financial sector. By neglecting those excluded from the conventional financial system, banks have left room for fintech firms to make financial inclusion a viable and sustainable business model. They have since innovated more efficient alternative systems to provide large-scale banking and financial services to the unbanked. Starting with payments, they have turned to financial services such as savings, loans, microinsurance, and investments. Multilateral institutions such as the World Bank Group, the International Monetary Fund, and other organizations and initiatives such as the United Nations' Sustainable Development Goals (SDGs), the Consultative Group to Assist the Poor (CGAP), and the Global Partnership for Financial Inclusion (GPFI) have identified fintech as a key means to achieve financial inclusion for poverty reduction and the socio-economic development of different nations, especially low- and middle-income countries.

The second school of thought, however, suggests that not all fintech companies are disruptive [30]. Indeed, the authors of this school consider evolution to be a perfectly logical element. As there have been discoveries in the past that have advanced science and economics, so the continuity of human history is ensured by the changes and developments that are taking place today. As a result, this school states that while some fintech firms may have brought revolutionary developments, not all of them should be considered disruptive in all markets.

Furthermore, some proponents of this current of thought consider that innovation or disruption is not a sufficient condition to achieve financial inclusion for all segments of society. Instead, fintech companies are a source of new risks and digital exclusion. Authors such as Buckley et al. [31] are alert to the new risks abounding in the fintech industry, including the issue of digital exclusion and the digital divide, which are new constraints for populations and businesses. They sound the alarm on the fraud and cybersecurity challenges faced by so-called disruptive fintech firms. The latter are victims of cyber-attacks, which compromise their action, their usefulness, and their viability. In addition to the risks coming from outside, other risks are inherent to the business model of fintech firms. They generally have a riskier customer base than financial institutions in the conventional financial system, which undermines their viability. Thus, multiple customer failures or the occurrence of a black swan event[4] may result in negative systemic impacts on the economy as a whole. Other studies suggest distinguishing between the availability and access dimensions of financial services and their use. Indeed, despite the availability of financial services through fintech firms and other financial institutions, individuals and businesses may voluntarily exclude themselves from using them for personal, religious, or cultural reasons [32]. To go further, the use of fintech financial services is also not synonymous with their quality and their positive impact on the lives of the population. Taking all these thoughts into account, it is therefore worth considering measuring Fintech firms' real contributions to financial inclusion.

3 Research Methodology on the Inclusive Dimension of Fintech

Considering the problem that this paper addresses, which is the impact of fintech on financial inclusion, the hypothetico-deductive approach is the most appropriate. Indeed, this methodology responds to the objective of testing the hypotheses of positive, negative, or neutral correlation that may exist between fintech and financial inclusion in Africa in order to project the possible contribution and role of this technological innovation for the continent.

The chosen hypothetico-deductive approach follows five main steps (Fig. 5).

The hypotheses to be tested are formulated based on the dimensions and indicators of mobile money and financial inclusion. Data for fintech indicators are collected from the Global System for Mobile Communications Association (GSMA) databases and the International Monetary Fund's Financial Access Surveys. Data for selected financial inclusion indicators on its three main dimensions (access, usage, quality) are collected from the World Bank's Global Findex database and the International Monetary Fund's Financial Access Surveys (Table 1).

4 Empirical Study on the Role and Contribution of Mobile Money to Financial Inclusion in Africa

After defining the theoretical and conceptual framework of the two key variables, namely the mobile payment Fintech and financial inclusion, it is appropriate to tackle the subject from an empirical point of view. This consists of testing and interpreting the correlation between fintech indicators and financial inclusion indicators.

[4] The Black Swan theory refers to a "rare event" with a low probability of occurrence but with considerable impact if it does occur.

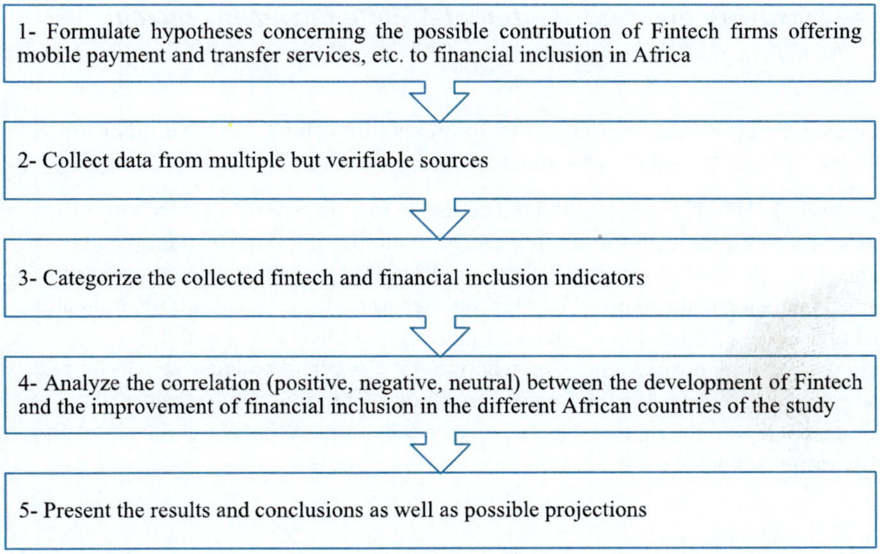

Fig. 5 The analytical approach followed to test the impact of fintech on financial inclusion in Africa. *Source* Authors' analysis

Table 1 Hypotheses and indicators of mobile money and financial inclusion

Hypothesis	Dimension	Mobile money	Financial inclusion
H1: The availability and access to mobile payment fintech in africa promote financial inclusion	Availability and access	Number of active agents through which customers access mobile money per 100,000 inhabitants Data source: IMF financial access surveys [1, 2, 33]	Total number of service points (Number of agencies, number of MFIs, mobile money points) per 100,000 inhabitants Data source: IMF financial access surveys [33]
H2: The Adoption and Use of Mobile Payment Fintech Contributes to Increase Financial Inclusion in Africa	Use	Number of active money mobile accounts held by people over 15 years old Data source: GSMA [2, 3, 14]	Number of accounts held by people over 15 years old Data source: WB global findex triennial database [3, 4, 26]
H3: The quality of the services offered by Fintech mobile payment services enables greater financial inclusion on the African continent	Quality	Mobile money regulatory index including authorization (30%), consumer protection (15%), transaction limit (15%), agent network (15%), KYC (15%), and infrastructure and investment environment (10%) Data source: GSMA [14]	Indicator of financial behavior (e.g., saving for any potential emergency) Data source: WB global Findex triennial database (2017) [26]

4.1 Measuring the Contribution of Mobile Payment Fintech to Financial Inclusion

The measurement of the impact of mobile money fintech on financial inclusion is based on the three hypotheses formulated.

Hypothesis 1 (H1) The availability and access to mobile payment Fintech in Africa promote financial inclusion in the region. For this first hypothesis, the objective is to determine whether deploying Fintech in different parts of Africa contributes to financial inclusion in the region. Does the mobile money Fintech availability indicator impact the access dimension of financial inclusion (Tables 2, 3, 4 and 5)?

There is a strong positive correlation between the availability and access of mobile money active service points and access to financial products and services. This is the first step toward the financial inclusion of populations. What about the usage and quality dimensions (Fig. 6)?

Table 2 Descriptive statistics for mobile money agents and total point of services

Number of active mobile money agent outlets per 100,000 adults		Total point of services	
Mean	122,8,580,427	Mean	149,654,057
Standard error	32,56,438,356	Standard error	32,2,044,442
Median	0	Median	63,502,815
Standard deviation	216,0,076,836	Standard deviation	213,620,116
Sample variance	46,659,31,938	Sample variance	45,633,5539
Kurtosis	4,345,250,299	Kurtosis	4,65,786,404
Skewness	2,143,716,098	Skewness	2,18,681,954
Range	847,5819	Range	870,218,942
Minimum	0	Minimum	1,9,484,969
Maximum	847,5819	Maximum	872,167,439
Count	44	Count	44

Table 3 Regression statistics for mobile money agents and total point of services

Regression statistics	
Multiple R	0,992,288,625
R square	0,984,636,715
Adjusted R square	0,984,270,922
Standard error	26,79,129,911
Observations	44

Table 4 Descriptive statistics of mobile money accounts and total accounts

Total accounts (% age 15+)		Mobile money accounts (% age 15+)	
Mean	0,41,446	Mean	0,202,681
Standard error	0,029,754	Standard error	0,027,038
Median	0,404,989	Median	0,185,553
Standard deviation	0,188,178	Standard deviation	0,171,003
Sample variance	0,035,411	Sample variance	0,029,242
Kurtosis	0,335,032	Kurtosis	0,790,545
Skewness	0,655,659	Skewness	0,889,369
Range	0,812,725	Range	0,729,317
Minimum	0,0857	Minimum	0
Maximum	0,898,425	Maximum	0,729,317
Count	40	Count	40

Table 5 Regression statistics of mobile money accounts and total accounts

Regression statistics	
Multiple R	0,568,840,313
R square	0,323,579,302
Adjusted R square	0,305,778,757
Standard error	0,156,790,057
Observations	40

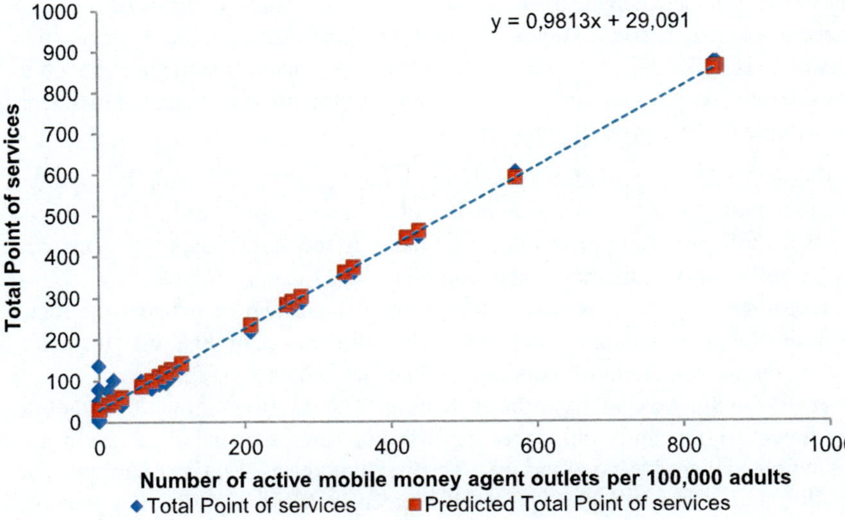

Fig. 6 The regression curve: active mobile money services and service points. *Source* Authors' analysis

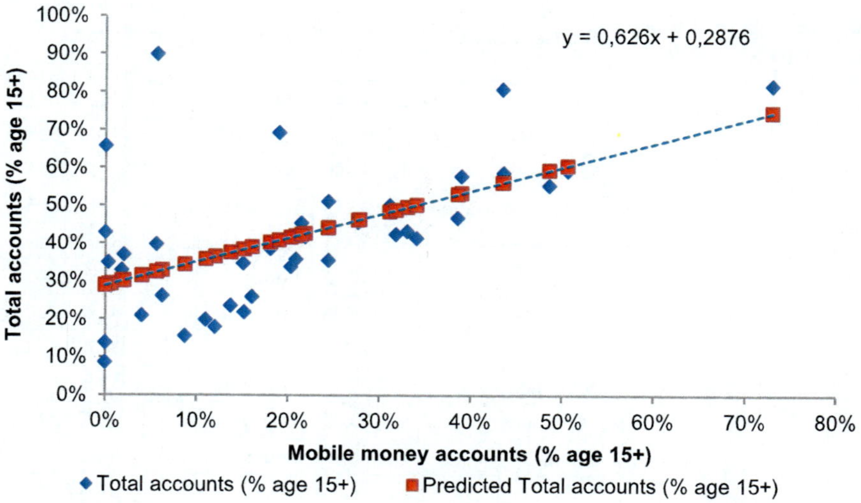

Fig. 7 The regression curve between money mobile active accounts and total accounts. *Source* Authors' analysis

Hypothesis 2 (H2) The adoption and use of mobile payment Fintech contribute to increasing Africa's financial inclusion. This hypothesis aims to go beyond access to financial services and address the actual use of financial products and services by the population.

The correlation between the use of a mobile money account and the use dimension of financial inclusion is less significant than the access dimension. However, this dimension is very decisive for financial inclusion. This means that the use of mobile money does not yet have enough impact on the use dimension of financial inclusion for the sample under consideration (Fig. 7).

Hypothesis 3 (H3) The quality of services offered by Fintech mobile payment services enables greater financial inclusion of Africa's populations. This hypothesis tests the influence of improved service quality in the mobile money fintech area on the overall quality of financial inclusion (Tables 6, 7 and 8).

As represented by the regression curve, there is no correlation between the regulatory index of mobile money and the quality dimension of financial inclusion (including the improvement of individuals' financial behavior) (Fig. 8).

Overall conclusions on hypothesis testing. The analysis shows that mobile money fintech significantly influences the first step toward financial inclusion, i.e. access to financial products and services. The first hypothesis (H1) is confirmed: The availability and access to mobile payment Fintech in Africa significantly promote the access dimension of financial inclusion in the region.

However, considering the second hypothesis (H2), there is no strong correlation between the use of mobile money fintech and the second dimension of financial inclusion, which is the use of financial products and services in general. This finding

Table 6 Descriptive statistics for mobile money regulatory index and financial behavior

Mobile money regulatory Index		Financial behavior	
Mean	78,87,857	Mean	11,37,618
Standard error	1,669,736	Standard error	1,264,832
Median	78,3	Median	9,131,216
Mode	77,6	Mode	6,438,993
Standard deviation	9,878,291	Standard deviation	7,48,285
Sample variance	97,58,063	Sample variance	55,99,304
Kurtosis	5,610,473	Kurtosis	1,662,273
Skewness	−1,56,985	Skewness	1,360,674
Range	55,75	Range	32,53,235
Minimum	41,75	Minimum	1,903,529
Maximum	97,5	Maximum	34,43,588
Count	35	Count	35

Table 7 Regression statistics for mobile money regulatory index and financial behavior

Regression statistics	
Multiple R	0,128,599
R square	0,016,538
Adjusted R square	−0,01,326
Standard error	7,532,313
Observations	35

Table 8 Some innovative Fintech business models in Africa

Sector	Business models	Examples of contributions
Mobile payment	Mobile money: services allowing users to store money and make payment transactions via a mobile device	**M-Pesa** (Kenya): inducing financial inclusion by facilitating peer-to-peer transfers, paying bills and purchasing goods (Lipa Na M-Pesa), and even saving and providing microcredit (M-shwari)
Banking and loans	White label solutions (partnership)	**Jumo** (South Africa): platform-as-a- Service that creates predictive data products and models for financial service providers, telephone companies, and even Uber with its service Jumo Drive

(continued)

Table 8 (continued)

Sector	Business models	Examples of contributions
	Banking solutions (direct loan)	**Lidya** (Nigeria): offers mobile savings and credit products directly through mobile-first technology. Its competitors are Tala and Branch (California)
Insurance (InsurTech)	Microinsurance services paid exclusively by telephone	**Bima** (Sweden): provides microinsurance, health, and financial education services to low-income and/or underserved consumers in emerging markets
Other services (Investment, cooperatives, cryptocurrency, e-commerce)	Tontine services: a group of people pool goods or services that are given to each in turn	**MaTontine** (Senegal): helping to financially include actors of the informal economy in Africa
	Crowdfunding: a platform that enables project owners to raise money from the general public	**Flowingo** (Morocco): contributing to the development and strengthening of a collaborative economy

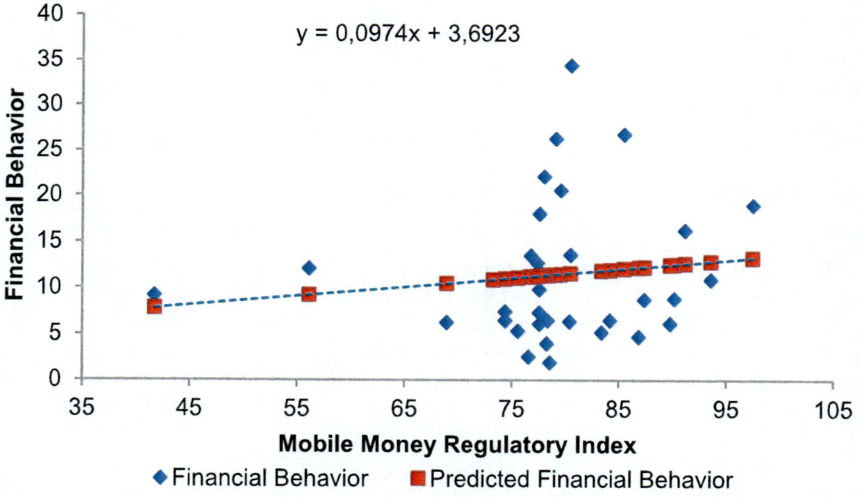

Fig. 8 Linear regression curve between the money moving index and financial behaviors. *Source* Authors' analysis

suggests that there is a need to work on how mobile money products and services can significantly contribute to financial inclusion.

Based on the indicators and the sample of countries investigated, the results of the test of the hypothesis (H3) indicate that the quality of services offered by mobile money fintech is not correlated with the quality dimension of financial inclusion of African populations.

Overall, all stakeholders in the mobile money fintech industry should work to ensure that their financial inclusion vocation does not stop at the initial stage, which is access to financial products and services, but that the usage and quality dimensions of financial inclusion are significantly impacted by the ecosystem.

4.2 *Contribution of Fintech to Financial Inclusion Through Their Innovative Business Models*

The advent and democratization of information and communication technologies have favored the birth of new business models in every sector and every region of the world. In Africa, as in most developing or emerging economies, these solutions have been able to adapt to local specificities to offer products and services that are both cheaper and highly adapted to the needs of populations and businesses. Key innovations in different sectors take advantage of geographic trends and some of the shortcomings of pre-established financial institutions. The use of advanced technologies and the collection of interesting data are also drivers for developing these innovations. Overall, Fintech's contributions are manifold. Fintech companies focus on customer/user experience. Among other things, they allow the opening of accounts using numeric identifiers. With electronic payments, both individuals and companies can open current accounts. Mobile telephony also facilitates access to financial services, even in remote areas, notably through the use of Unstructured Supplementary Service Data (USSD). With all the important data collected, Fintech can offer products and services that are more tailored to customer needs. There are also unique initiatives such as Ushahidi (a crowdsourcing platform created in 2008 in Kenya) that have proven to be very useful for the population. Through their business models and service offerings that are quite disruptive within the continent, three main sectors dominate the Fintech ecosystem in Africa: the payments sector, the credit granting and banking solutions sector, and, finally, the insurance sector.

5 Conclusion and Discussions

Financial inclusion is a real passion for governments and is supported by several financial and development organizations worldwide. It covers all dimensions of the economic life of nations.

Given the socio-economic situation prevailing in Africa and particularly in the sub-Saharan region, this concept has more meaning and scope because it contains hopes and promises. For the economies of these countries, which are classified as low- and middle-income countries, developing or emerging countries, financial inclusion is a vector for inclusive economic development, financial stability and integrity, and a guarantee of reducing inequalities, whether gender or socio-spatial. Driven by technology and innovation, the Fintech address multiple issues *vis-à-vis* individuals, companies, and economies at large. For populations, Fintech provides real access to formal financial products and services, which is a first step and a starting point for their financial inclusion. For businesses, financial inclusion is a triple win: profitability, competitiveness, and growth. Economies can therefore benefit from the virtuous circle created by financial inclusion.

Mobile payment or mobile money Fintech, as demonstrated by this paper through hypothesis testing, can significantly contribute to financial inclusion by the availability and accessibility of their products and services. However, the dimensions of adoption or use are less impacting, and the study shows that the quality of the products and services offered by Fintech does not influence the quality dimension of financial inclusion as a whole.

The limitations of this work include the fact that correlation does not mean causality. Thus, certain interpretations in a causal logic are limited. Also, the Impact dimension of financial inclusion could not be subjected to hypothesis testing due to the lack of necessary data, which must be field survey data as well. All in all, this paper has made it possible to identify areas for improvement on which the various stakeholders should focus to achieve quality financial inclusion that is more useful for both populations and nations as a whole.

Fintech has experienced two decades of great excitement around the world and particularly in emerging markets. These entities have disrupted the pre-established players and, above all, changed the approaches and ways of doing things in the entire financial ecosystem. Fintech has undoubtedly not finished revealing its secrets. They are still full of enormous potential, which is being revealed as they go along. However, they also face many challenges that need to be considered to make the most of this Fintech. Indeed, several Fintech initiatives have emerged, and with them, new risk exposures. They can open a virtual door to harmful activities and businesses as they develop. So, as Fintech technologies and innovations progress, the capabilities and resources of hackers—the other side of the coin, the dark side—also increase. Highly organized cybersecurity attacks are on the rise [34].

In some cases, they can affect the macroeconomic stability of nations. These risks are real challenges that need to be taken into consideration. Wehinger [35] raises the issue of the regulatory risks that Fintech poses. Indeed, technology generally allows to cross traditional national boundaries. With Fintech, national borders are often less relevant. This calls for regulators to take a closer look at these aspects. The sovereignty of the country is even at stake, especially in terms of financial stability and integrity. States need to set up legal jurisdictions, customer data protection bodies, and tax systems suited to this new context. Coordination and consultation between regulatory authorities, financial institutions, and Fintech must be strengthened.

In addition to the security challenge, there is also a cultural challenge, which Nolan [36] addresses. This challenge is revealed in the apprehensions of individuals toward these players and in certain obstacles that traditional financial institutions place against fintech firms, notably for commercial reasons. However, countries risk living in technological complacency without innovation, which would eventually lead to obsolescence and a loss of competitiveness concerning other countries. It is, therefore, necessary for countries to mobilize their entire local ecosystem to catch up or maintain their competitive advantage. This also involves improving the business climate. African fintech companies are also confronted with the low level of financial education of local populations and the constraints of the digital divide and digital exclusion. All these challenges naturally limit the adoption of Fintech by certain segments of society. They are sometimes confronted with a lack of financial resources (especially the difficulty of accessing financing at their launch but also during their growth phase), material resources (especially the lack of some basic infrastructure for the full deployment of Fintech), and even human resources (especially the lack of educational systems that promote innovation and creativity but also encourage entrepreneurship and integrate digital technology in its various components).

All in all, one of the emergencies in the financial ecosystem concerns its regulatory aspect. It is, therefore, necessary to accelerate the deployment of regulatory frameworks conducive to the development of Fintech and in line with international standards, particularly about the fight against money laundering and terrorist financing. The regulatory framework can even begin with establishing regulatory sandboxes. Different countries such as Tunisia and some East African countries have launched regulatory sandboxes that allow testing several innovations in the Fintech sector. There is also a need for concerted initiatives between the Fintech themselves for greater efficiency, especially mobile payment Fintech, which faces interoperability challenges.

There are several prospects for Fintech in Africa. Indeed, the continent is experiencing rapid growth in its population, which is mostly young: 65% of the population is under 35 years old, and this youth is compatible with technophilia and innovation. In addition, the continent has the fastest-growing middle class in the world, with the continent's middle-class population expected to reach 1.1 billion or 42% of its total population by 2060 [37]. The growing penetration of mobile telephony and the Internet and the continued adoption of Fintech in different countries are all interesting prospects for their development on the continent in the coming years. Strong adoption and reliance on Fintech solutions are expected in the future. WhatsApp and Facebook Messenger have a combined user base of 172 million in Africa [38]. The business media (Quartz Africa) reported that Facebook was in discussions with regulators to allow users in several African countries to transact via messaging, opening up new opportunities for payment services without forgetting the underlying risks of competition, data protection, and even state sovereignty.

More cooperation rather than confrontation is expected between Fintech and the traditional players, namely the banks, through initiatives such as open banking, whereby free access to bank data, transactions, and other financial data from banks is offered to fintech firms and other players through the use of application programming interfaces (APIs), which will contribute to the development of these entities.

Regulators and all stakeholders are mobilizing to join efforts to reduce the circulation of cash (costing various governments about 1.5% of GDP) through mobile payment, the use of QR codes and contactless payments, and even payment by bank cards (which remains under-utilized in Africa). This is intended to reduce transaction costs and further contribute to the financial inclusion of populations. However, risks must always be taken into account to ensure a bright future for Fintech with greater efficiency and scalability in all areas in which they operate.

Concerted and integrated initiatives within the continent can also boost the dynamics of the Fintech sector, particularly the African Continental Free Trade Area (AfCFTA) treaty that came into force in May 2019 and whose effective implementation is eagerly awaited. This trade agreement can create a single African market, largely free of tariffs, which will open up cross-border trade to an unprecedented extent. Eventually, the free movement of people and a single currency could be possible, which augurs well for the even more harmonious development of the Fintech nuggets. They will be able to operate with more customers and fewer barriers.

Three fundamental questions remain. Should financial inclusion be achieved at all costs? What is the real impact of financial inclusion on people and businesses? Furthermore, in terms of Fintech, what strategies for African fintech companies are resilient and highly competitive internationally?

References

1. Frost, J. (2020). The economic forces driving fintech adoption across countries. BIS Working Papers 838, Bank for International Settlements.
2. Fleischmann, S., & Sasse, E. (2018). L'Afrique, terre promise du numérique, ARTE. https://www.youtube.com/watch?v=FYsnZ5D3koA.
3. PitchBook. (2018). Emerging Tech Research: Fintech. https://pitchbook.com/news/reports.
4. The World Bank. (2018). https://www.worldbank.org/en/news/press-release/2018/04/19/financial-inclusion-on-the-rise-but-gaps-remain-global-findex-database-shows.
5. Schueffel, P. (2016). Taming the beast: A scientific definition of fintech. *Journal of Innovation Management, 4*(4), 32–54. https://doi.org/10.24840/2183-0606_004.004_0004
6. KPMG, CB Insights. (2016). The Pulse of Fintech, 2015 in Review: Global Analysis of Fintech Venture Funding. https://assets.kpmg/content/dam/kpmg/pdf/2016/03/the-pulse-of-fintech.pdf.
7. Arner, D. W., & Barberis, J. N., Buckley, R. P. (2015, October 1). The evolution of Fintech: A new post-crisis paradigm?. University of Hong Kong Faculty of Law Research Paper No. 2015/047, UNSW Law Research Paper No. 2016-62. Available at SSRN: https://ssrn.com/abstract=2676553 or http://dx.doi.org/10.2139/ssrn.2676553.
8. Skinner, C. (2018). Digital human: The fourth revolution of humanity includes everyone. John Wiley & Sons. ISBN 10: 1119511852/ISBN 13: 9781119511854.
9. PwC, Casablanca Finance City. (2018). Digitalizing Africa: The rise of Fintech Companies. https://casablancafinancecity.com/wp-content/uploads/2019/05/CFC-Africa-Insights_FinTech-Report.pdf.
10. FT Partners. (2020). Annual Fintech Almanac: Global Financing and M&A Statistics. https://ftpartners.docsend.com/view/5bfsfn5.

11. Weetracker. (2019). Decoding venture investments in Africa. https://weetracker.com.
12. The Partech Africa team. (2020). Africa Tech Venture Capital Report. https://partechpartners.com.
13. We Are Social, Hootsuite. (2019). The global state of digital in 2019. https://wearesocial.com.
14. GSMA. (2020). State Of The Industry Reports (SOTIR): The Mobile Economy, GSMA Intelligence. https://www.gsmaintelligence.com/data/.
15. Bower, J. L., & Christensen, C. M. (1995). Disruptive technologies: Catching the wave. *Journal of Product Innovation Management*. https://doi.org/10.1016/0024-6301(95)91075-1
16. Dru, J. M. (1996). *Disruption: Overturning conventions and shaking up the marketplace* (vol. 1). John Wiley & Sons Incorporated. ISBN: 9780471165651/ISBN-10: 0471165654.
17. Christensen, C. M. (1997). *The innovator's dilemma: When new technologies cause great firms to fail* (p. 1997). Harvard Business School Press.
18. Schumpeter, J. A. (1942). *Capitalism, socialism and democracy*. Taylor & Francis e-Library. ISBN 0-203-26611-0/ISBN 0-415-10762-8.
19. Lepore, J. (2014). *The disruption machine: What the gospel of innovation gets wrong*. The New Yorker.
20. Stiegler, B. (2019). *The age of disruption: Technology and madness in computational capitalism*. ISBN-13: 978-1509529278/ISBN-10: 1509529276.
21. Schrodinger, E. (1944). What is life? The Physical Aspect of the Living Cell. http://www.whatislife.ie/downloads/What-is-Life.pdf.
22. The World Bank. (2018). Financial Inclusion Overview. https://www.worldbank.org/en/topic/financialinclusion/overview#1.
23. Dupas, P., & Robinson, J. (2013). Savings constraints and microenterprise development: Evidence from a field experiment in Kenya. *American Economic Journal: Applied Economics, 5*(1), 163–192. https://doi.org/10.1257/app.5.1.163
24. Brune, L., Giné, X., Goldberg, J., & Yang, D. (2016). Facilitating savings for agriculture: Field experimental evidence from Malawi. *Economic Development and Cultural Change, 64*(2), 187–220. https://doi.org/10.1086/684014
25. Demirgüç-Kunt, A., & Singer, D. (2017). Financial inclusion and inclusive growth: A review of recent empirical evidence. World Bank Policy Research Working Paper, (8040).
26. The World Bank Global Findex Report. (2017). Global Findex Database 2017: Measuring Financial Inclusion and the Fintech Revolution. https://globalfindex.worldbank.org/sites/globalfindex/files/2018-08/Global%20Findex%20Database.xlsx.
27. Mader, P. (2018). Contesting financial inclusion. *Development and Change, 49*(2), 461–483. https://doi.org/10.1111/dech.12368
28. Foster, R. N. (1986). *Innovation: The attacker's advantage. Summit Books*. ISBN-10: 0671622501/ISBN-13: 9780671622503.
29. Lee, D. K. C., & Low, L. (2018). *Inclusive fintech: Blockchain, cryptocurrency and ICO*. World Scientific.
30. Zalan, T., & Toufaily, E. (2017). The promise of fintech in emerging markets: Not as disruptive. *Contemporary Economics, 11*(4), 415–431. https://doi.org/10.5709/ce.1897-9254.253
31. Buckley, R.P., Arner, D.W., Zetzsche, D.A., & Selga, E. (2019). The dark side of digital financial transformation: The new risks of FinTech and the rise of TechRisk. *UNSW Law Research Paper* (19–89).
32. Bae, K., Han, D., & Sohn, H. (2012). Importance of access to finance in reducing income inequality and poverty level. *International Review of Public Administration, 17*(1), 55–77. https://doi.org/10.1080/12264431.2012.10805217
33. The International Monetary Fund Financial Access Surveys . (2020). https://data.imf.org/?sk=E5DCAB7E-A5CA-4892-A6EA-598B5463A34C.
34. Callen-Naviglia, J., & James, J. (2018). Fintech, Regtech and the Importance of Cybersecurity. *Issues in Information Systems, 19*(3), 220–225. https://doi.org/10.48009/3_iis_2018_220-225.
35. Wehinger, G. (2012). The financial industry in the new regulatory landscape. *OECD Journal Financial Market Trends, 2011*(2), 30–30. https://doi.org/10.1787/fmt-2011-5k9cswmzqp7d

36. Nolan, S. (2009). Kick-starting a customer-centric revolution. *Manager: British Journal of Administrative Management, 65*, 28–30.
37. African Development Bank. (2014). Tracking Africa's Progress in Figures. https://www.afdb.org/fileadmin/uploads/afdb/Documents/Publications/Tracking_Africa%E2%80%99s_Progress_in_Figures.pdf.
38. Quartz Africa. (2018). WhatsApp is the most popular messaging app in Africa. https://qz.com/africa/1206935/whatsapp-is-the-most-popular-messaging-app-in-africa/.

Technology Platforms and Inclusivity

The Impact of Social Commercial Platforms on Entrepreneurship in a Developing Country

Nada Mallah Boustani and May Merhej Sayegh

Abstract The development of electronic commerce has led to the emergence of social commerce combining commercial and social activities by using web 2.0 and increasing the social presence of online environment. This study is proposing a set of variables related to the social aspect of social commerce platform and its impact on customer purchase behavior to enhance entrepreneurship. Toward this end, a survey is distributed to 289 Lebanese social commerce platform shoppers. The findings indicate that social presence, social value, loyalty and satisfaction have a positive impact to enhance purchase from SCP and increase sales. Our findings suggest that gender differences are significant when it is about purchase frequency and types of products.

Keywords Social commerce platform · Entrepreneurship · Social presence · Social value · Satisfaction · Loyalty

1 Introduction and Context of the Study

The Internet has opened for businesses and organizations a new way of communication with other businesses and peers [1]. Electronic commerce and Social commerce have contributed largely to the economic growth [2]. Companies were obliged to develop their online stores and they were focused on their online websites to stay competitive and keep their customers in favor of their competitors. The development of the web enabled organizations to access new markets using the electronic commerce [3, 4]. E-commerce is described as buying and selling through the Internet and is considered as an innovation that can enhance companies' entrepreneurial characteristics in developed and developing countries [5, 6]. E-commerce was considered a channel of Marketing for many firms to realize income [7] and it is considered a

N. M. Boustani · M. M. Sayegh (✉)
Saint-Joseph University, Beirut, Lebanon
e-mail: may.merhejsayegh@usj.edu.lb

N. M. Boustani
e-mail: nada.mallahboustany@usj.edu.lb

C. Dal Zotto et al. (eds.), *Smart Technologies for Organizations*, Lecture Notes in Information Systems and Organisation 60, https://doi.org/10.1007/978-3-031-24775-0_5

benefit for small and medium sized enterprises [8] which will be noticed in a later stage. Social media networks such as Facebook, WhatsApp, Twitter, Skype and other channels have evolved to offer companies spaces through which they can sell their products and this is known as social commerce [6, 9]. To gain access to social platforms, social commerce uses social media and benefits from commercial transactions to enhance the quality of a closer relationship between vendors and customers, increasing sales and encouraging loyalty to the business [10].

Nowadays, Social media is becoming a part of people's life where they can share their daily activities and tap markets and communities not reached before. Marketers are using social media to influence their customers and target new markets [11, 12]. Currently it is possible to use this activity in free social commerce platforms (Facebook, Instagram, etc.) that can give firms competitive advantage over big corporation to realize revenue if they focus on customers' attitudes toward online shopping from social commerce platforms.

This paper provides information on the overview of customers' attitudes toward online shopping, for entrepreneurs (companies) to use for the development of their businesses and optimization of their online stores and their social commerce platforms in emerging countries. That said entrepreneurs can build a selling strategy for their clients in the online environment using social commerce platform that connects buyers and sellers [13] and that can be able to generate income.

The objective of the present study is to test a probable influence of the sociodemographic characteristics of e-shoppers such as gender on participants' perceptions of social commerce platform. Gender always has a significant role toward buyer's habit and behavior [14–17]. Gender differences are based on various factors such as perception, attitudes, motivation, preferences, perceived risk and satisfaction, which influence consumer intention toward buying from social commerce platform [17].

In addition, within the social commerce context, other factors can in turn impact the perceived usefulness of the website such as social presence [18].

Furthermore, increased social presence can in turn affect other factors such as perceived usefulness of the website [18].

Therefore, this paper dives into the investigation of the probable influence of the sociodemographic factors on their buying intention through social commerce platform to improve and also focus on the impact of their social presence and values on satisfaction levels.

Thus, the study aims to address these shortcomings by offering several potential contributions. Firstly, by showing the impact of gender differences on consumers' purchase intention within social commerce platforms [17, 19–21].

It proposes a model which is then tested using an online questionnaire administered to a sample of 289 Facebook and Instagram users who have purchased from these types of social commerce platforms. Collected data is then analyzed based on the Anova technique, using IBM SPSS 20.

The research is divided into three major sections. The first section presents the theoretical development, including definitions of social commerce to be succeeded by the model and its hypotheses. The second section is about the research methodology results, theoretical and practical contributions, conclusion, limitations and the possible direction of future research.

The research questions applied in this study were as follows:

RQ1 In what way sociodemographic factors could influence the participants' perceptions about their social commerce platform customers' satisfaction?

RQ2 Is there significant positive relationship between the level of social presence, the social value, the loyalty and the level of social commerce platform customers' satisfaction that affects purchase intentions for entrepreneurs?

2 Theoretical Background

2.1 Social Commerce Platforms and Entrepreneurship

Rapid advances in the information system and the emergence of the Internet in recent years have brought new opportunities to business such as electronic and social commerce [9]. Social media and online communities have provided new opportunities to consumers [22]. Consumers use social media (Facebook, LinkedIn, Instagram,...) to engage in interaction with companies and other consumers, and companies use new channels via social networks to offer and sell their products; this is called social commerce [9]. Content is generated from online interactions via social platforms where members have facilitated information sharing globally and quickly [10, 23]. Social networks enabled companies to reach a greater number of consumers efficiently than traditional retail outlets [10, 12]. By using social media as a platform for making sale, firms can communicate with their consumers: They offer campaigns and reviews, they can interact and engage with them and create value by building better relationships [24–26]. Social platforms give consumers the opportunities to know more about their product before the actual purchase, it is a sharing place where they can take part in group discussion, forums and communities [27, 28]. Small and medium enterprises use e-commerce and many other applications such as social commerce to interact with their customers actively in social platforms such as online communities [29]. So these platforms moved to the stage of generating revenues for enterprises [30, 31], where consumer's role is much participatory than ever before and where social commerce is more than electronic commerce [27]. Firms are increasing their contributions advices to their customers on online buying and selling [27] via channels like Instagram and Facebook, in order to develop their business. These social networks are also challenging the way entrepreneurs and managers do business with the huge opportunities that these platforms offer [32, 33]. While the definition of entrepreneurship in Marketing is relevant to develop a set of processes for creating, communicating and delivering value to consumers, online social platforms and social networks are used for managing customer relationships in ways that help organization and its stakeholders by using innovative channels such as web or mobile marketing using the Internet for improving marketing efficiency and doing business [33–36]. Big Firms and entrepreneurs are nowadays aware of the effectiveness and of the huge opportunities offered by these platforms [32]. Social commerce platforms can offer firms and entrepreneurs more possibilities to develop their marketing and selling actions and reinforce their visibility in the market. These outcomes have motivated companies to develop social commerce as an essential part of their marketing communications plans [37].

Social commerce platform is considered one of the online marketing and commerce channel [38]. Using social commerce platforms, customer interactions are facilitated and participation is developed in ways that drive measurable business value that can be promoted [39].

3 Data Analysis

3.1 Data Collection

This is a descriptive cross-sectional study, which was carried out on a non-probabilistic sample of 289 participants residing in Lebanon. The questionnaires were applied using google forms link through social media. The data collection was carried out between February and March 2020.

3.2 Statistical Analysis

The exploratory analysis of the data was executed through basic descriptive statistical tools. Data centralization was done using Microsoft Excel 2013 and statistical processing was performed using IBM SPSS Statistics v.20 (IBM Corp., Armonk, NY, USA). The mean values obtained for each group were compared using the Student's t-test for independent samples. The use of parametric tests was possible because the conditions were verified, including the normality of the distribution. Additionally, the Pearson correlation coefficients were calculated for evaluation of the relations between some variables at study. For all statistical analyses the level of significance considered was 5% ($p < 0.05$).

A seven-point Likert scale was used to measure the participants' opinions regarding the different variables: 1 (totally disagree), 2 (disagree), 3 (somehow disagree), 4 (indifferent or undecided), 5 (somehow agree), 6 (agree) and 7 (strongly agree).

3.3 *Results and Findings*

3.3.1 Results of RQ1

Details of respondents' characteristics are presented in Table 1.

The survey was filled out by 289 participants: 176 (60.9%) female and 113 (39.1%) male. The majority of respondents were in the category 21–25 years (54%) age group, and the second category was 15–20 years (14.5%) age group. On employment, 130 participants (45%) were students, 116 participants (40.1%) were full-time employees, while the rest were classified as unemployed or self-employed.

About 51.6% of participants have 9–13 years of experience in internet usage, while 19.4% have 14–19 years of experience, 11% have 4–8 years of experience,

Table 1 Details of respondents' characteristics

		Frequency	Percent
Gender	Male	113	39.1
	Female	176	60.9
Age	15–20 years	42	14.5
	21–25 years	156	54.0
	26–30 years	32	11.1
	31–35 years	20	6.9
	36–40 years	10	3.5
	More than 41 years	29	10.0
Employment	Full-time employee	116	40.1
	Student	130	45.0
	Unemployed	25	8.7
	Self employed	18	6.2
Types of products purchased using social commerce platforms	Nothing	41	14.2
	Clothes and shoes	94	32.5
	Accessories	62	21.5
	Food and supplements	6	2.1
	Books	14	4.8
	Electronic devices	33	11.4
	Makeup and perfumes	39	13.5
Purchasing frequency	Never	41	14.2
	1–3 times	220	76.1
	4–6 times	10	3.5
	7–9 times	3	1.0
	More than 10 times	15	5.2
	Total	289	100.0

11.4% have 20–25 years of experience and 0.7% more than 26 years in using the Internet.

About 76.1% of the respondents have 1–3 times experience of online purchasing, 14.2% of the respondents don't have any experience in online purchasing, while the rest have more than 4 times.

The category of clothes and shoes is the most purchased type of goods (32.5%), followed by accessories (21.5%).

3.3.2 RQ2 Reliability

In order to answer the second research question, the authors started by defining each variable, then testing the reliability of the constructs items in order to control any gender differences that might be noted in the frequency of purchases or in the types of goods that might affect the types of products entrepreneurs offer on SCP.

Variable 1: Satisfaction

Satisfaction on SCP is defined as an important factor building and retaining a consumer's loyalty with the intention to repurchase the product or to reuse the service in the commercial contexts [40–45]. It is referring to user's positive emotional state about using a service or buying a product from a social commerce site when comparing experience with prior purchases [43, 46]. This state influences their purchase intention and creates a strong satisfaction toward the sites [44]. Thus, when users' satisfaction toward social commerce platform is high, it can foster their purchase intention from these platforms.

The satisfaction variable was measured using 3 items questions

1. I am satisfied with my decision on using the social commerce platform for shopping.
2. My choice to use the social commerce platform for shopping is a wise one.
3. My experience with using the social commerce platform for shopping is satisfactory.

Variable 2: Loyalty

Loyalty is defined as the behavioral intention of repeating purchase for the same product or service which can lead to increased revenues [47–49]. Customer loyalty is considered to be an important factor for vendors in e-commerce to maintain business continuity [50]. Many researches and studies have demonstrated that satisfaction was found to affect positively loyalty [51–53]. A positive relationship has been observed between overall satisfaction and attitudinal loyalty [54, 55]. Oliver [56]found that true loyalty is found when there is an embedded social network.

Loyalty on SCP was measured using 6 items questions

1. I like using social commerce platforms.
2. To me, social commerce platforms offer the best retail stores to do business with.

3. Shopping using social commerce platforms is my favorite.
4. I don't consider switching to other channels than social commerce platforms when shopping online.
5. As long as the present service offered on social commerce platforms continues, I doubt that I would switch to another channel.
6. I will continue to use social commerce platforms whenever I need to make a purchase.

Variable 3: Social Presence

Social presence is defined as the sense of "warmth" and sociability of a presence on a website [57, 58], it is a perception of social self-concept when using social commerce sites [59]. The presence of different interactional tools in social commerce websites enhances the social presence such as reviews, comments, likes, pictures, videos [58] and so enables human interactions and communication and derived sociability and sensitivity [60–62]. Social presence on websites can affect other factors such as perceived usefulness, trust, enjoyment [18]. Increased verified social presence in cyber space can affect positively relationships with the creation of results, such as satisfaction which can enhance their purchase intention [45, 63, 64].

Social presence was measured using 7 items questions

1. There is a sense of human contact in the social commerce platform.
2. There is a sense of sociability in the social commerce platform.
3. There is a sense of human warmth in the social commerce platform.
4. There is a sense of human sensitivity in this social commerce platform.
5. There are many other buyers who are interested in using the social commerce platform.
6. There are many other buyers sharing information regarding products offered on the social commerce platform.
7. There are many buyers who have bought the products offered on the social commerce platform.

Variable 4: Social Value

Social value is defined as the perception of social self-esteem derived from using social commerce platforms and sites [59, 65]. Users of social commerce sites and platforms share their experiences and communicate with other users about products and services which will enhance their sense of self-identification [66]. They evaluate it by hedonic and utilitarian values (enjoyment, performance) and by social value [45, 49]. Sharing experience and communicating enable users to increase their social value on social commerce platform and strengthen their level of satisfaction toward social media and enhance purchase intention [49, 67].

Some factors are considered determinants of the customer's loyalty to buy a service or a product. One of them is the social value as it may influence individual satisfaction creating a good feeling about a product or a service [68].

Social value was measured using 4 items questions

Table 2 The results of the Cronbach alpha

Reliability statistics		
	Cronbach's alpha	Number of items
Satisfaction	0.887	3
Loyalty	0.918	6
Social presence	0.887	7
Social value	0.916	4

1. Using the social commerce platform for shopping would help me to feel acceptable.
2. Using the social commerce platform for shopping would make a good impression on others.
3. Using the social commerce platform for shopping would improve the way I am perceived.
4. Using the social commerce platform would give its owner social approval.

The authors tested the reliability of the items in the four variables for the responses of the Lebanese participants. The results of the Cronbach alpha are presented in Table 2, and most values are close to 0.88 or higher, which means that there is internal consistency within each of the variables tested. Additionally, it was observed that the constructs with highest alpha are the loyalty to the social commercial platform ($\alpha = 0.918$) followed by social value ($\alpha = 0.916$), whereas the construct with lowest and equal internal consistency were the satisfaction and social value ($\alpha = 0.887$).

3.3.3 RQ2 and Gender Differences

Differences in the types of goods purchased through SCP (Table 3)

In the sample studied it was noted that from the list of products purchased from social commercial platforms, that (32.5%) of the participants prefer buying clothes and shoes followed by (21.5%) for accessories, (13.5%) for makeup and perfumes

Table 3 Frequencies of purchase in different types of goods purchased through SCP

List of products purchased from SCP	Frequency	Percent	Cumulative percent
Nothing	41	14.2	14.2
Clothes and shoes	94	32.5	46.7
Accessories	62	21.5	68.2
Food and supplements	6	2.1	70.2
Books	14	4.8	75.1
Electronic devices	33	11.4	86.5
Makeup and perfumes	39	13.5	100.0
Total	289	100.0	

and (11.4%) for electronic devices. Books and food don't represent a high percentage of purchasing from Lebanese participants through their online social commerce platform.

Furthermore, in order to answer the second part of RQ2, related to gender differences. Gender is defined as a socioeconomic user characteristic which has been commonly employed and tested in the field of IT acceptance, thus of online shopping behavior [19–21].

Some studies mentioned that gender is a significant factor in online consumer behavior [17] and there are gender differences in the attitude toward online purchase [15]. Studies explain existence of gender differences in e-commerce [69]. Women are more emotional than men and because online shopping is lacking face-to-face communication, it can deter them from purchasing [70]. In the same time women prefer and enjoy physical shopping more than men do. They like to feel the product they want to purchase [71]. According to Hasan (2010), the e-commerce model is more suitable for males. They can save time, efforts and it can be monetarily beneficial [17]. To repurchase from a website, men need to find informative and descriptive details about products [72] (Table 4).

Major gender differences in the types of goods and products purchased from the social commercial platforms were noted since men prefer purchasing electronic devices, foods and supplements, books, accessories and they don't even purchase makeup or perfumes; whereas women prefer buying shoes and clothes and makeup and perfumes and they don't even purchase any of foods and supplements. These results confirm the findings of [17, 19–21].

Among the population tested, the authors noted that first, 14.2% of the sample never purchase anything from the social commercial platform, the highest percentage 76.1% is for a frequency purchase of 1 to 3 times followed by persons who purchase more than 10 times 5.2%. For the frequency of the purchase using social commercial platforms there were no gender differences noted as shown in Table 5 where all percentages are almost similar and there is gender equality (Table 6).

3.3.4 Relationship Between the Four Variable—Results of RQ3

In order to answer research question 3, the authors investigated the possible gender differences as to the types of variables studied, and the results are presented in Table 7. In addition, the authors tested the level of correlation between these four variables as shown in Fig. 1.

The results of the t-tests show that no gender differences were found. Women's means are higher than men's means for the four variables studied. The highest mean is for the satisfaction, followed by the loyalty then social presence. The lowest mean is for the social value which is less than 4 for both men and women concluding that they are not influenced by the social value perceived on the social commercial platform.

For satisfaction, it is observed that women attribute higher importance (score = 4.5795 ± 1.2981) and are more satisfied when purchasing from SCP as compared to

Table 4 Percentages of purchase in different types of goods purchased through SCP by gender

			Gender		Total
			Male	Female	
List the type of product you purchase using social commerce platforms	Nothing	Count	18	23	41
		% within List the type of product you purchase using social commerce platforms	43.9%	56.1%	100.0%
		% within gender	15.9%	13.1%	14.2%
	Clothes and shoes	Count	29	65	94
		% within List the type of product you purchase using social commerce platforms	30.9%	69.1%	100.0%
		% within gender	25.7%	36.9%	32.5%
	Accessories	Count	29	33	62
		% within List the type of product you purchase using social commerce platforms	46.8%	53.2%	100.0%
		% within gender	**25.7%**	18.8%	21.5%
	Food and supplements	Count	6	0	6
		% within List the type of product you purchase using social commerce platforms	100.0%	0.0%	100.0%
		% within gender	**5.3%**	0.0%	2.1%
	Books	Count	8	6	14
		% within List the type of product you purchase using social commerce platforms	57.1%	42.9%	100.0%
		% within gender	**7.1%**	3.4%	4.8%
	Electronic devices	Count	23	10	33
		% within List the type of product you purchase using social commerce platforms	69.7%	30.3%	100.0%
		% within gender	**20.4%**	5.7%	11.4%

(continued)

Table 4 (continued)

			Gender		Total
			Male	Female	
	Makeup and perfumes	Count	0	39	39
		% within List the type of product you purchase using social commerce platforms	0.0%	100.0%	100.0%
		% within gender	0.0%	22.2%	13.5%
Total		Count	113	176	289
		% within List the type of product you purchase using social commerce platforms	39.1%	60.9%	100.0%
		% within gender	100.0%	100.0%	100.0%

Table 5 Frequency purchase differences

		Frequency	Percent	Cumulative percent
Valid	Never	41	14.2	14.2
	1–3 times	220	76.1	90.3
	4–6 times	10	3.5	93.8
	7–9 times	3	1.0	94.8
	More than 10 times	15	5.2	100.0
	Total	289	100.0	

Table 6 Purchasing frequency

			Gender		Total
			Male	Female	
Purchasing frequency	Never	Count	18	23	41
		% within purchasing frequency	43.9%	56.1%	100.0%
		% within gender	15.9%	13.1%	14.2%
	1–3 times	Count	83	137	220
		% within purchasing frequency	37.7%	62.3%	100.0%
		% within gender	73.5%	77.8%	76.1%
	4–6 times	Count	4	6	10
		% within purchasing frequency	40.0%	60.0%	100.0%
		% within gender	3.5%	3.4%	3.5%

(continued)

Table 6 (continued)

			Gender		Total
			Male	Female	
	7–9 times	Count	2	1	3
		% within purchasing frequency	66.7%	33.3%	100.0%
		% within gender	1.8%	0.6%	1.0%
	More than 10 times	Count	6	9	15
		% within purchasing frequency	40.0%	60.0%	100.0%
		% within gender	5.3%	5.1%	5.2%
Total		Count	113	176	289
		% within purchasing frequency	39.1%	60.9%	100.0%
		% within gender	100.0%	100.0%	100.0%

Table 7 Means

Group statistics				T-test	
Gender		Mean[(1)]	Std. Deviation	t	*p-value*
Social value on SCP	Female	**3.8466**	1.53968	0.361[(3)]	*0.718*
	Male	3.7876	1.22074		
Social presence on SCP	Female	**4.4034**	1.21504	0.44[(3)]	*0.66*
	Male	4.3451	1.01568		
Loyalty to my SCP	Female	**4.3466**	1.35195	0.718[(2)]	*0.473*
	Male	4.2301	1.33637		
Satisfaction of SCP	Female	**4.5795**	1.29810	1.34[(2)]	*0.181*
	Male	4.3717	1.26910		

[(1)]Scale for social value and social presence, loyalty and satisfaction: values from 1 to 7; values less than 4 is for "no influence" and values exceeding 4 for "influence"
[(2)]Equal variances assumed according to Levene's Test for Equality of Variances
[(3)]Equal variances not assumed according to Levene's Test for Equality of Variances

men (score = 4.3717 ± 1.2691), and this can also be attributed to the types of goods women tend to purchase that fits more their needs more strongly to men.

The research model presented in Fig. 1 was tested in terms of the strength of the linear regressions between the sets of individual variables, by performing a regression matrix with Pearson coefficients, as presented in Table 8. As was expected, there was a very strong correlation between the loyalty to the social commercial platform and the level of satisfaction of the participants being the highest coefficient (0.807). All other relations were weaker but positively correlated and higher than 0.5 as the regression coefficients show.

Fig. 1 Correlation between four variables

Table 8 Regression coefficients

	Loyalty to my SCP	Social presence on SCP	Social value on SCP	Satisfaction of SCP
Loyalty to my SCP	1			
Social presence on SCP	0.607	1		
Social value on SCP	0.596	0.522	1	
Satisfaction of SCP	**0.807**	0.525	0.564	1

4 Conclusion, Limitations and Future Research Venues

This research allegedly concludes that entrepreneurs can increase their sales and customers' satisfaction through the use of social commercial platform that has been demonstrated as satisfying the users' needs triggering an escalation in their loyalties. On the other hand, the perception of social presence and social value when using the social commercial platform isn't very much appreciated by Lebanese users, especially when it is related to special types of goods. Hence, the findings indicate that social presence, social value, loyalty and satisfaction have a positive impact on SCP purchasing improvement and sales increase. The findings also suggest that gender differences are significant when related and compared to frequencies of purchasing and types of products.

Regardless, in the future to target a deeper analysis on customer purchasing behavior trends affecting entrepreneurship improvement, other demographic characteristics such as level of education and living areas (Rural vs. Urban) could also be explored further, in addition to the introduction of more variables related to trust of SCP, quality of products and services provided through these innovative platforms.

Appendix A

Measurement Scales Used

Construct	Items	Source
Social value	Using the social commerce platform (Instagram, WhatsApp, …) for shopping would help me to feel acceptable	Adapted from [45]
	Using the social commerce platform (Instagram, WhatsApp, …) for shopping would make a good impression on others	Adapted from [45]
	Using the social commerce platform (Instagram, WhatsApp, …) for shopping would improve the way I am perceived	Adapted from [45]
	Using the social commerce platform (Instagram, WhatsApp, …) would give its owner social approval	Adapted [45]
Social presence	There is a sense of human contact in the social commerce platform (Instagram, WhatsApp, …)	Adapted from [73, 74]
	There is a sense of human warmth in the social commerce platform (Instagram, WhatsApp, …)	Adapted from [73, 74]
	There is a sense of human sensitivity in this social commerce platform (Instagram, WhatsApp, …)	Adapted from [73, 74]
	There are many other buyers who feel interested using the social commerce platform (Instagram, WhatsApp, …)	Adapted from [73, 74]
	There are many other buyers sharing information regarding product offered on the social commerce platform (Instagram, WhatsApp, …)	Adapted from [73, 74]
	There are many buyers who have bought the products offered on the social commerce platform (Instagram, WhatsApp, …)	Adapted from [73, 74]
Loyalty	I like using social commerce platforms (Instagram, WhatsApp, …)	Adapted from [75]
	To me, social commerce platforms (Instagram, WhatsApp, …) offer the best retail stores to do business with	Adapted from [75]

(continued)

(continued)

Construct	Items	Source
	Shopping using social commerce platforms (Instagram, WhatsApp, …) is my favorite	Adapted from [75]
	I don't consider switching to other channels than social commerce platforms (Instagram, WhatsApp, …) when shopping online	Adapted from [75]
	As long as the present service offered on social commerce platforms (Instagram, WhatsApp, …) continues, I doubt that I would switch to another channel	Adapted from [75]
	I will continue to use social commerce platforms (Instagram, WhatsApp, …) whenever I need to make a purchase	Adapted from [75]
Satisfaction	I am satisfied with my decision on using the social commerce platform (Instagram, WhatsApp, …) for shopping	Adapted from [45]
	My choice to use the social commerce platform (Instagram, WhatsApp, …) for shopping is a wise one	Adapted from [45]
	my experience with using the social commerce platform (Instagram, WhatsApp, …) for shopping is satisfactory	Adapted from [45]

References

1. Tekin, M., İnce, H., Etlioğlu, M., Koyuncuoğlu, Ö., & Tekin, E. (2018). A study about affecting factors of development of e-commerce. In N. Durakbasa & M. Gencyilmaz (Eds.), *Proceedings of the International Symposium for Production Research 2018. ISPR 2018*. Cham: Springer.
2. Qu, W. G., Pinsoneault, A., Tomiuk, D., Wang, S., & Liu, Y. (2015). The impacts of social trust on open and closed B2B e-commerce: A Europe-based study. *Information & Management, 52*(2), 151–159.
3. Laudon, K. C., & Traver, C. (2016). *E-commerce: Business, technology*. Society.
4. Gibreel, O., Al Otaibi, D., & Altmann, J. (2017). Social commerce development in emerging markets. *Electronic Commerce Research and Applications, 27* (12).
5. Alyoubi, A. (2015). E-commerce in developing countries and how to develop them during the introduction of modern systems. *Procedia Computer Science., 65*(12), 479–483.
6. Shemi, A., & Procter, Chris. (2018). Ecommerce and entrepreneurship in SMEs : the case of myBot. *Journal of Small Business and Enterprise Development; 25*(3).
7. Minculete, G. (2013). Issues regarding electronic commerce and e-marketing. In International scientific conference "strategies xxi", suppl. Technologies—military applications, simulations and resources (vol. 2, pp. 26–32). "Carol I". Bucharest: National Defence University.
8. Kartiwi, M. & Hussin, H. & Suhaimi, M. A., & Razi, M. J. M., & Amin, M. R. (2018). Impact of external factors on determining E-commerce benefits among SMEs in Malaysia. *Journal of Global Entrepreneurship Research, 8*.

9. Hajli, N. (2015). Social commerce constructs and consumer's intention to buy. *International Journal of Information Management,35*(2), 183–191.
10. Hajli, N. (2014). The role of social support on relationship quality and social commerce. *Technological Forecasting and Social Change, 87*, 17–27.
11. Hoffman, D. and Novak, T., (2012), Why Do People Use Social Media? Empirical Findings and a New Theoretical Framework for Social Media Goal Pursuit.
12. Zhou, L., Zhang, P., & Zimmermann, H.-D. (2013). Social commerce research: An integrated view. *Electronic Commerce Research and Applications, 12*(2), 61–68.
13. Turban, E., King, D., Lee, J. K., Liang, T. P., & Turban, D. C. (2015). E-commerce: Mechanisms, platforms, and tools. In: *Electronic Commerce. Springer Texts in Business and Economics* (pp. 51–99). Cham: Springer.
14. Fan, Y.-W., Miao, Y. (2012). Effect of electronic word-of-mouth on consumer purchase intention: The perspective of gender differences. *IJEBM, 10*, 175–181.
15. Hasan, B. (2010). Exploring gender differences in online shopping attitude. *Computers in Human Behavior, 4*(26).
16. Seock, Y. K., & Sauls, N. (2008). Hispanic consumers' shopping orientation and apparel retail store evaluation criteria: An analysis of age and gender differences. *Journal of Fashion Marketing and Management: An International Journal, 12*(4), 469–486.
17. Lim, P. L., & Yazdanifard, R. (2014). Does gender play a role in online consumer behavior? *Global Journal of Management And Business Research, 14*(7).
18. Gefen, D., & Straub, D. (2003). Managing user trust in B2C e-services. *E-Service Journal,2*(2), 7–24.
19. Gefen, D., & Straub, D. (1997). Gender differences in the perception and use of e-mail: An extension to the technology acceptance model. *MIS Quarterly, 21*(4), 389–400.
20. Venkatesh, V., Morris, M., Davis, G., & Davis, F. (2003). *MIS Quarterly, 27*(3), 425–478.
21. Bigne, E., Ruiz, C., & Sanz, S. (2005). The impact of internet user shopping patterns and demographics on consumer mobile buying behavior. *Journal of Electronic Commerce Research, 6*(3).
22. Gruzd, A., Staves, K., & Wilk, A. (2011). *Tenure and promotion in the age of online social media.* Wiley Online Library.
23. McLure Wasko, M., & Faraj, S. (2005). Why should i share? Examining social capital and knowledge contribution in electronic networks of practice. *MIS Quarterly, 29*(1), 35–57. Special Issue on Information Technologies and Knowledge Management.
24. De Valck, K., Van Bruggen, G. H., & Wierenga, B. (2009). Virtual communities: A marketing perspective. *Decision Support Systems, 47*(3), 185–203.
25. Long, M. C. (2011). Beyond the press release: Social media as a tool for consumer engagement. In H.S. Noor Al-Deen & J.A. Hendricks (Eds.), *Social media: Usage and impact* (pp. 145–159). Lanham, ML: Lexington Books.
26. Gillin, P. (2009). *Secrets of social media marketing.* Quill Driver Books.
27. Liang, T.-P., & Turban, E. (2011). Introduction to the special issue social commerce: A research framework for social commerce. *International Journal of Electronic Commerce, 16*(2), 5–14.
28. Wang, X., Yu, C., & Wei, Y. (2012). Social media peer communication and impacts on purchase intentions: A consumer socialization framework. *Journal of Interactive Marketing, 26*, 198–208.
29. Hajli, M. N. (2014). Social commerce constructs for strategic marketing: An emperical study. *Journal of Strategic Marketing.*
30. Wang, C., & Zhang, P. (2012). The evolution of social commerce: The people, management, technology and information dimensions. *Communications of the Association for Information Systems, 32*, 1–23.
31. Barnes, N. (2014). Social commerce emerges as big brands position themselves to turn 'follows' 'likes' and 'pins' into sales, marketing management. Association Annual Spring Conference Proceedings, Chicago, IL, March 26–28, pp. 8–13.
32. Mangold, W. G., & Faulds, D. J. (2009). Social media: The new hybrid element of the promotion mix. *Business Horizons, 52*(4), 357–365.

33. Nakara, W., Benmoussa, F., & Jaouen, A. (2012). Entrepreneurship and social media marketing: evidence from French small business. *International Journal of Entrepreneurship and Small Business, 16*(4), 386–406.
34. Kraus, S., Harms, K., & Fink, M. (2010). Entrepreneurial marketing: Moving beyond marketing in new ventures. *International Journal of Entrepreneurship and Innovation Management, 11*(1), 19–34.
35. Postman, J. (2009). *SocialCorp: Social media goes corporate*. New Riders.
36. Bulearca, M., & Bulearca, S. (2010). Twitter: A viable marketing tool for SMEs. *Global Business and Management Research: An International Journal, 2*(4), 296–309.
37. Berthon, P., Pitt, L., Plangger, K., & Shapiro, D. (2012). Marketing meets Web 2.0, social media, and creative consumers: Implications for international marketing strategy. *Business Horizons, 55*(3), 261–271.
38. Zhang, K. Z., & Benyoucef, M. (2016). Consumer behavior in social commerce: A literature review. *Decision Support Systems, 86*, 95–108.
39. Shin, D. (2013). User experience in social commerce: In friends we trust. *Behaviour & Information Technology, 32*(1), 52–67.
40. Anderson, E., & Sullivan, M. (1993). The antecedents and consequences of customer satisfaction for firms. *Marketing Science, 12*(2), 125–143.
41. Oliver, R. (1980). A cognitive model of the antecedents and consequences of satisfaction decisions. *Journal of Marketing Research, 42*(4), 460–469.
42. Oliver, R. (1993). Cognitive, affective and attribute bases of the satisfaction response. *Journal of Consumer Research*, 418–430.
43. Hsu, C.-L., & Lin, J.C.-C. (2016). Effect of perceived value and social influences on mobile app stickiness and in-app purchase intention. *Technological Forecasting and Social Change, 108*, 42–53.
44. Zhang, H., Lu, Y. B., Gupta, S., & Gao, P. (2015). Understanding group-buying websites continuance an extension of expectation confirmation model. *Internet Research, 25*(5), 767–793.
45. Gan, C., & Wang, W. (2017). The influence of perceived value on purchase intention in social commerce context. *Internet Research, 27*.
46. Oliver, R. L. (1981). Measurement and evaluation of satisfaction processes in retail settings. *Journal of Retailing, 57*(3), 25–48.
47. Oliver, R. L. (1997). *Satisfaction: A behavioral perspective on the consumer*. Irwin/McGraw-Hill, New York, NY.
48. Cronin, J. J., Jr., Brady, M. K., & Hult, G. T. M. (2000). Assessing the effects of quality, value, and customer satisfaction on consumer behavioral intentions in service environments. *Journal of Retailing, 76*(2), 193–218.
49. Gallarza, M., Saura, I., & Arteaga, F. (2013). The quality-value-satisfaction-loyalty chain: Relationships and impacts. *Tourism Review, 68*, 3–20.
50. Srinivasan, K., Rolph, A., & , Kishore, P. (2002). Customer loyalty in E-commerce: An exploration of its antecedents and consequences. *Journal of Retailing., 78*, 41–50.
51. Dube, L. M., & Maute, M. F. (1998). Defensive strategies for managing satisfaction and loyalty in the service industry. *Psychology and Marketing, 15*(8), 775–791.
52. Alonso, S. (2000). *"The antecedents and consequences of customer loyalty: The roles of customer satisfaction and consumer trust-commitment", unpublished doctoral dissertation*. University of Texas-Pan America.
53. Garcia del los Salmones, M., Perez, A., & del Bosque, A. (2009). The social role of financial companies as a determinant of consumer behavior. *International Journal of Bank Marketing, 27*(6), 467–485.
54. Pollack, B. L. (2009). Linking the hierarchical service quality model to customer satisfaction and loyalty. *Journal of Services Marketing, 23*(1), 42–50.
55. Bodet, G. (2008). Customer satisfaction and loyalty in service: Two concepts, four constructs, several relationships. *Journal of Retailing and Consumer Services, 15*(3), 156–162.
56. Oliver, R. L. (1999). Whence consumer loyalty. *Journal of Marketing, 63*, 33–44.

57. Gefen, D., & Straub, D. (2003). Managing user trust in B2C e-services. *E-Service Journal, 2*(2), 7–24.
58. Hajli, N., Sims, J., Zadeh, A. H., & Richard, M. O. (2017). A social commerce investigation of the role of trust in a social networking site on purchase intentions. *Journal of Business Research, 71*, 133–141. https://doi.org/10.1016/j.jbusres.2016.10.004
59. Sweeney, J. C., & Soutar, G. N. (2001). Consumer perceived value: The development of a multiple item scale. *Journal of Retailing, 77*(2), 203–220.
60. Hassanein, K., & Head, M. (2005). The impact of infusing social presence in the web interface: An investigation across product types. *International Journal of Electronic Commerce, 10*(2), 31–55.
61. Piller, F. T., & Walcher, D. (2006). Toolkits for idea competitions: a novel method to integrate users in new product development. *R&D Management, 36*(3), 307–318.
62. Kumar, N., & Benbasat, I. (2006). Research note: The influence of recommendations and consumer reviews on evaluations of websites. *Information Systems Research, 17*, 425–439. https://doi.org/10.1287/isre.1060.0107.
63. Hackman, M. Z., & Walker, K. B. (1990). Instructional communication in the televised classroom: the effects of system design and teacher immediacy on student learning and satisfaction. *Communication Education, 39*(3), 196–206.
64. Park, C., & Kang, B. G. (2003). Factors influencing on trust toward e-commerce by consumer experience of online buying. *Information Systems Review, 5*, 81–95.
65. Rintamäki, T., Kanto, A., Kuusela, H., & Spence, M. T. (2006). Decomposing the value of department store shopping into utilitarian, hedonic and social dimensions: Evidence from Finland. *International Journal of Retail & Distribution Management, 34*(1), 6–24.
66. Analysys. (2016). Report on mobile social commerce in China (2016). www.analysys.cn/view/report/detail.html?columnId=8&articleId=1000026. Retrieved 30 May 2016.
67. Hu, T., Kettinger, W. J., & Poston, R. S. (2015). The effect of online social value on satisfaction and continued use of social media. *European Journal of Information Systems, 24*(4), 391–410.
68. Moliner, M., Sánchez-García, J., Rodríguez-Artola, R., & Callarisa, F. (2007). Perceived relationship quality and post-purchase perceived value: An integrative framework. *European Journal of Marketing., 41*, 1392–1422. https://doi.org/10.1108/03090560710821233
69. Zhou, L., Dai, L., & Zhang, D. (2007). Online shopping acceptance model—A critical survey of consumer factors in online shopping. *Journal of Electronic Commerce Research, 8*(1), 41–62.
70. Dittmar, H., Long, K., & Meek, R. (2004). *Buying on the internet: Gender differences in on-line and conventional buying motivations*. Springer.
71. Cho, J. (2004). 'Likelihood to abort an online transaction: Influences from cognitive evaluations, attitudes, and behavioral variables.' *Information & Management, 41*(7), 827–838.
72. James, R. (2013). Student feature male and female attitudes to online shopping. *Women in Society, 5*(6).
73. Gefen, D., & Straub, D. (2004). Consumer trust in B2C e-Commerce and the importance of social presence: experiments in e-Products and e-Services. *Omega, 32*(6), 407–424.
74. Lu, B., Fan, W., & Zhou, M. (2016). Social presence, trust, and social commerce purchase intention: An empirical research. *Computers in Human Behavior, 56*, 225–237.
75. Srinivasan, S., Anderson, R., Ponnavolu, K. (2002). Customer loyalty in e-commerce: an exploration of its antecedents and consequences. *Journal of Retailing, 78*(1), 41–50.

Business Performance Management Platform for SMEs

Pedro Pimenta, Célio Carvalho, and Joaquim P. Silva

Abstract Business performance management systems are common in large enterprises and play a key role in their management. However, just a few small and medium-sized enterprises (SMEs) have implemented performance management. The obstacles to the implementation of performance management in SMEs have been identified in many studies, but solutions to overcome them are still lacking. This project proposes the development of a modular and adaptable business performance management solution, using a low-code application development platform. This solution can be implemented by SMEs in the short term, at low cost, and through a simplified process.

Keywords Business performance management · Small and medium enterprises · Low-code platform

1 Introduction

Business performance management (BPM) is not a recent research topic but continues to raise the interest of organizations operating in an increasingly complex and dynamic market environment. Enterprise managers face daily important decisions that require quick responses. Therefore, they need continuous access to useful and up-to-date information in order to monitor the performance of enterprises [1]. There are numerous tools designed to implement Business Intelligence (BI) software solutions that provide data to support managers' decisions. However, these tools have high licensing and implementation costs. Small and Medium Enterprises

P. Pimenta · C. Carvalho
Instituto Politécnico do Cávado e do Ave, 4750-810 Barcelos, Portugal
e-mail: a10206@alunos.ipca.pt

C. Carvalho
e-mail: cdcarvalho@ipca.pt

J. P. Silva (✉)
2Ai—School of Technology, IPCA, Barcelos, Portugal
e-mail: jpsilva@ipca.pt

C. Dal Zotto et al. (eds.), *Smart Technologies for Organizations*, Lecture Notes in Information Systems and Organisation 60, https://doi.org/10.1007/978-3-031-24775-0_6

(SMEs) have neither the financial resources nor the technical staff to implement these types of software solutions [2]. Therefore, it is necessary to develop a generic tool that can meet the BI requirements shared by SMEs through the integration of data from the most used ERP software packages. BPM is deployed through a balanced and dynamic system capable of supporting the decision-making process through information collection, processing, and analysis [3]. Enterprises need to measure continuously the business performance to manage the accomplishment of the business objectives.

SMEs make up more than 99% of OECD non-financial enterprises and account for more than 65% of employment and value generated [4]. BPM systems are usually implemented using BI tools and play an important role in the enterprises' competitiveness. The implementation of a BPM system offers SME managers a tool for monitoring the implementation of the strategy. However, there are many barriers to the implementation of BPM systems in SMEs. Several studies highlight the inability of SMEs to use generic BPM tools due to their size, structure, procedural complexity, and cost [2, 5]. The cost of implementing a BI solution is the first determinant for the implementation of BI in SMEs. Weak employee skills and limited financial resources are also high barriers to implement BI solutions in SMEs. BI systems require constant development and adaptation to the new enterprise's challenges [6]. Currently, BI solutions do not offer the usability, simplicity of integration, and ease of implementation required for mass adoption by SMEs [7]. This project aims at the development of an adaptable BPM solution, using a low-code application development platform, and based on the Balanced Scorecard (BSC).

The following section presents the literature review on BPM models, the BSC, and performance measurement. Section 3 presents the architecture of the development platform for BPM solutions, identifies the objectives to be achieved, and the features that will be implemented in the BPM platform, explaining how these can help SME managers. Section 4 identifies the expected results and key benefits that will be provided by the solution.

2 Related Work

To properly support decision-making, BPM systems should be balanced and dynamic. Therefore, they should combine an appropriate set of indicators, continuously monitor the internal and external context and adjust, where and when necessary, its objectives and priorities. The work of Kaplan and Norton [8, 9] and several other researchers in the 1990s served as the foundation for the BPM research performed in the following decade. The most relevant models and frameworks for implementing BPM systems in organizations were developed in those two decades [10].

There are a broad set of obstacles to the implementation of generic BPM systems on SMEs, namely [2]: lack of human resources availability, limited management skills, insufficient financial means to acquire software and consulting, a reactive approach due to lack of strategic planning, informal or poorly defined processes, and a wrong or non-existent perception of the benefits of implementing BPM systems. Over the last decades, many models have been proposed to help SMEs overcome these obstacles in the implementation of BPM systems.

2.1 Business Performance Management Models

The first BPM models were proposed in the context of large enterprises. The proposals for models adjusted to the context of SMEs only emerged in the late 1990s and early 2000s, some adapted from existing ones and others developed specifically for SMEs [10]. In the study conducted by Garengo et al. [2], five common characteristics were identified for projects to implement BPM systems in SMEs, namely: the difficulty in involving SMEs due to lack of interest or availability; incorrect use or inexistence of a BPM model; application of the model in a non-integrated way and without a holistic approach; informal and unplanned implementation process; and limited resources for data gathering and analysis. In this study, the authors analyzed eight models in the main dimensions that characterize the BPM models: strategic alignment, strategic development, stakeholder focus, balancing, dynamic adaptation, process orientation, detail/depth and scope, causality between results and determinants in performance measurement, and clarity and simplicity. Among the models studied, BSC has become the most widely known and used [10]. It was introduced as a model to support the conversion of the strategy into action. Although strategic alignment and development are typical dimensions of the models developed for large enterprises, these dimensions are also important for SMEs as long as they are properly adjusted to their needs. SMEs should include these dimensions in their BPM system, while continuing to focus on operational aspects [11].

2.2 Balanced Scorecard Applied to SMEs

The BSC model proposed by Kaplan and Norton [8] proposes a holistic vision for the implementation of a BPM system. The key idea is that models based on financial indicators do not encompass the complexity of an organization's performance, just as an organization's performance cannot be measured by operational indicators alone. Therefore, the model must be balanced and based on various types of metrics: financial and non-financial, short and long term, internal and external. In addition, it combines various perspectives: financial, customer, internal processes, and learning and growth. Although initially designed for large enterprises, BSC has been successfully applied in many SMEs. Recent research studies concluded that the adoption of the BSC has a positive impact on business performance [11, 12]. To implement a BPM system, the BSC model obliges organizations to define a solid strategy, set objectives, and make actions consistent with the strategy and objectives. Frequent change in strategy is the major cause of failure of the BSC in SMEs [13].

For the successful implementation of the BSC in an SME, the role of the enterprise's information system is fundamental, as it will have to make available the data used to calculate the indicators used to measure business performance.

2.3 Enterprise Performance Measurement

Performance measurement is a central point for the implementation of BPM systems, as it is the result of measurement that supports the decision-making processes at the strategic, tactical, or operational level. According to the BSC model, the enterprise's vision and strategy must be converted into specific objectives from each of the model's perspectives. For each objective, one or more measures must be defined, and targets must be established. The measures can be obtained from financial and non-financial data. The information systems that support the business processes of enterprises, e.g. ERP, CRM, SCM, etc., are the main source of data. BPM systems can also integrate external data or elements provided directly by the users to get measures and calculate performance indicators, i.e., a quantitative or qualitative indicator that reflects the state or the progress of the enterprise, the business unit, or the individual [14]. To evaluate the performance of a business unit or process, a set of key performance indicators (KPIs) is generally defined, which can be calculated directly from the source data or from the combination of several other KPIs. Each KPI is associated with a target, against which the value of the indicator is compared, thus allowing a quantitative and immediate assessment of performance [14]. Based on these principles, KPIs can be defined to measure each enterprise process, for example, the performance of the sales department can be measured by the combination of visits to customers, percentage of sales proposals accepted, and the invoicing volume. The data can be gathered from different data sources, such as enterprise's transactional systems, Excel sheets, or from manual direct users input into the BPM system.

Despite the considerable research in performance models, there are still many difficulties in the implementation of the BPM systems in SMEs. Although there is evidence that use of KPIs in combination with the BSC model is suitable for measuring performance in enterprises, these methods require in-depth knowledge, and their implementation can be complex and lengthy. SMEs often lack the necessary finance and knowledge on how to implement this approach [15].

3 Business Performance Management Platform

This section presents the architecture, objectives, and features of the BPM platform for SMEs. The project will be implemented on the OMNIA cloud-based platform provided by Primavera BSS (https://omnialowcode.com/), making use of a low-code application development model.

3.1 Architecture and Development Approach

According to Gartner, a low-code application platform supports the rapid development, implementation, execution, and management of applications in a single development stage, using high-level declarative programming abstractions such as model-driven and metadata-based programming languages. They support the development of user interfaces, business logic and data services, and improve productivity at the expense of vendor portability compared to conventional application platforms. Gartner predicts that, by 2024, low-code application development will account for over 65% of application development [16].

OMNIA supports the rapid development of solutions, enabling data integration from business software applications (ERP, CRM, SCM, etc.) and facilitating the evolution of the software applications to keep them dynamically aligned with the business. OMNIA platform can also be used to gather data directly from end-users whenever the data is not available in the business software applications. To overcome the obstacles to the adoption of performance management by SMEs, in particular, the lack of financial resources, the reduced availability of top management staff, and the limitations of management skills, we will adopt an innovative template-based approach to the development of BPM solutions. Each template offers a set of ready-to-use basic features, from data source connectors to the KPIs processing and visualization.

Just as ERP solutions are based on generic modules that can be shaped to fit the characteristics and operation of each enterprise, the solution to be developed within this project will be based on templates that meet the typical needs of SMEs and can be adjusted to each enterprise. As a result of this approach, the cost and time of implementation of a BPM solution can be significantly reduced, increasing the perception of SMEs about the objectives and benefits of the BPM systems.

The first templates to be developed will cover the financial and customer perspectives. The first phase of the project will just connect to the ERP software provided by Primavera BSS. The BSC model will be used to identify the features and KPIs to be included in the first templates. We will perform a literature review to identify the sales area indicators used in BSC implementation projects in SMEs, categorize the objectives and select the KPIs. Subsequently, the objectives and KPIs will be prioritised through interviews with a group of SME managers and cross-checked with measures that are available in the ERP software or can be obtained from direct user input. Figure 1 presents the structure of the template to be developed on OMNIA. New data connectors can be developed to integrate data from any business software application. The settings layer allows to adjust the template to the specific characteristics of the enterprise in the development of BPM solution. The template layer corresponds to all KPI management, objectives, and respective categories. The objectives will be detailed in KPIs and these will be defined according to the measures obtained from external data sources or by user input.

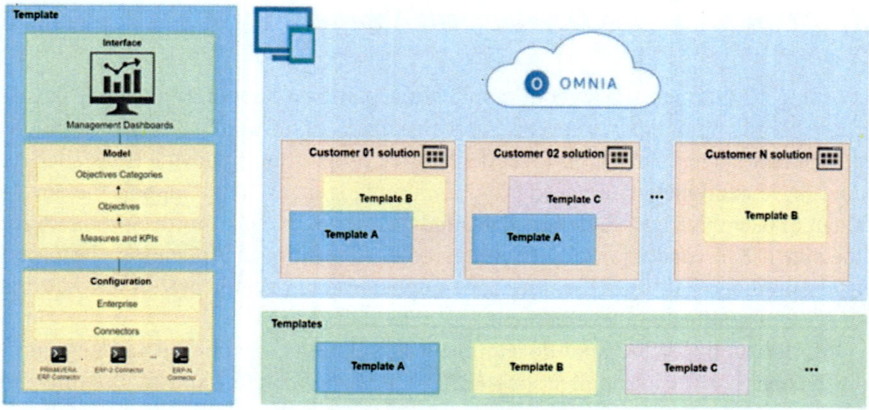

Fig. 1 Structure of OMNIA template (left) and solution overall architecture (right)

The top layer corresponds to the interface that will allow the interaction with data dashboards and management reports. The interface can also be used to implement data gathering processes, for example, the sales objectives per enterprise or employee. Each template can be replicated for multiple customers and customized to their specific needs. In addition, new templates can be developed, based on generic templates, for supporting specific needs of a set of enterprises or market niche. OMNIA allows to include, at any time, new templates in the solutions that are being deployed by customers. After importing the template structure into the solution, we just need to configure and feed it with data. The template will include most of the enterprise needs and can be easily adapted and extensible. This architecture brings great agility in the implementation of solutions.

3.2 *Objectives and Features*

The project democratizes SMEs' access to decision support instruments. For example, in the sales area, we can identify the following common objectives: invoicing volume, number of customers achieved, number of scheduled visits, number of leads generated, and customer satisfaction index. The standardization of the solution using templates allows the removal of the main obstacles to the adoption of BPM systems by SMEs. Even if not specifically developed, it can be adjusted to each SME and offer the main benefits of BPM, a fundamental tool for business management in an increasingly competitive market. Considering this general objective, several objectives have been outlined for the solution to be developed: (a) the platform must support the adding of new data connectors; (b) the templates should be instantiated with the enterprise's data, including data sources; (c) the instantiation of the template should allow adding categories, configuring KPI's and objectives, defining users and access profiles; (d) an application may combine several templates; and (e) the interface layer should be highly and easily customizable in terms of layout and visualization elements.

The objectives management should allow the registration, editing, removal, and monitoring of objectives. Initially, four categories are considered, corresponding to the existing BSC perspectives, that can integrate many subcategories. Each subcategory may have a different importance by assigning a weight. Objectives are associated with a category or subcategory and mapped in KPIs. The objectives can be global to the enterprise, associated with a unit/department or be assigned to one or more people. For each target, a weight and an expected value must be defined. SME managers will be able to implement Frolick and Ariyachandra's BPM framework [17]: select the objectives and plan the targets, according to the desired strategy, monitor and analyze the results, so they can take the appropriate corrective actions.

4 Expected Results

The implementation of this project will provide a set of generic templates for the implementation of BPM systems in the sales department of SMEs in a simple and fast manner and adjusted to its reality. The BPM system implemented in this way will be able to combine several interfaces, namely, dashboards with KPIs, reports, and data input forms. After the implementation of this project which, according to our conviction, will materialize the proposed objectives, it is planned to develop more templates to support the SMEs in the other BSC perspectives. After many decades of research and the successful implementation of BPM systems in SMEs, the adoption rate of BPM by SMEs remains very low. The implementation of BPM in SMEs could have a very positive impact on the economy. With this project, major obstacles to the implementation of BPM systems in SMEs will be reduced due the template-based standardization and low-code. The key expected benefits of the proposed BPM Platform are: (a) extremely short delivery time; (b) possibility of phased implementation of BPM by areas with immediate ROI; (c) low cost for implementing and running the solution; (d) no need of deep knowledge about BPM models and methods; and (e) fast perception about the benefits of BPM systems.

References

1. Neely, A. (1999). The performance measurement revolution: Why now and what next?. https://doi.org/10.1108/01443579910247437.
2. Garengo, P., Biazzo, S., & Bititci, U. S. (2005). Performance measurement systems in SMEs: A review for a research agenda. *International Journal of Management Reviews, 7*, 25–47. https://doi.org/10.1111/j.1468-2370.2005.00105.x.
3. Neely, A. D., Adams, C., & Kennerley, M. (2002). *The performance prism: The scorecard for measuring and managing business success*. Financial Times/Prentice Hall.

4. OECD. (2017). *Small, medium, strong. Trends in SME performance and business conditions.* OECD Publishing. https://doi.org/10.1787/9789264275683-en.
5. Smith, M. H., & Smith, D. (2007). Implementing strategically aligned performance measurement in small firms. *International Journal of Production Economics, 106*, 393–408. https://doi.org/10.1016/j.ijpe.2006.07.011.
6. Olszak, C. M., & Ziemba, E. (2012). Critical success factors for implementing business intelligence systems in small and medium enterprises on the example of Upper Silesia, Poland. *Interdisciplinary Journal of Information, Knowledge, and Management, 7*, 129–150. https://doi.org/10.28945/1584.
7. Llave, M. R. (2017). Business intelligence and analytics in small and medium-sized enterprises: A systematic literature review. *Procedia Computer Science, 121*, 194–205. https://doi.org/10.1016/j.procs.2017.11.027.
8. Kaplan, R. S., & Norton, D. P. (1992). The balanced scorecard–measures that drive performance. *Harvard Business Review, 70*, 71–79.
9. Kaplan, R. S., & Norton, D. P. (1996). Using the balanced scorecard as a strategic management system. *Harvard Business Review* 75–85.
10. Taticchi, P., Tonelli, F., & Cagnazzo, L. (2010). Performance measurement and management: A literature review and a research agenda. https://doi.org/10.1108/13683041011027418.
11. Heinicke, A. (2018). Performance measurement systems in small and medium-sized enterprises and family firms: A systematic literature review. *Journal of Management Control, 28*, 457–502. https://doi.org/10.1007/s00187-017-0254-9.
12. Malagueño, R., Lopez-Valeiras, E., & Gomez-Conde, J. (2018). Balanced scorecard in SMEs: Effects on innovation and financial performance. *Small Business Economics, 51*, 221–244. https://doi.org/10.1007/s11187-017-9921-3.
13. Rompho, N. (2011). Why the balanced scorecard fails in SMEs: A case study. *International Journal of Business and Management, 6*. https://doi.org/10.5539/ijbm.v6n11p39.
14. Popova, V., & Sharpanskykh, A. (2010). Modeling organizational performance indicators. *Information Systems, 35*, 505–527. https://doi.org/10.1016/j.is.2009.12.001.
15. Janickova, N., & Zizlavsky, O. (2019). Key performance indicators and the balanced scorecard approach in small and medium-sized enterprises: A literature review. In *17th International Conference on Perspectives of Business and Entrepreneurship Development in Digital Transformation of Corporate Business*, Brno, Czech Republic (pp. 55–64).
16. Vincent, P., Lijima, K., Driver, M., Wong, J., & Natis, Y. (2019). Magic quadrant for enterprise low-code application platforms.
17. Frolick, M. N., & Ariyachandra, T. R. (2006). Business performance management: One truth. https://doi.org/10.1201/1078.10580530/45769.23.1.20061201/91771.5.

Breaking the Monopoly of Facebook? Toward the Building Blocks of a Unified Model: Review

Puzant Balozian and Samia Curdy

Abstract A growing number of Social Network Sites (SNS) users are raising questions on the privacy and security of their information on the popular platforms. In recent years, various SNS have experienced several system crashes causing distress for a majority of users, as well as some high-profile scandals of security and privacy breaches. Subsequently, current users of these SNS are contemplating switching to less popular alternatives that may have a more reliable network and system. In the extant literature there is a gap in explaining this switching phenomenon with a comprehensive model. This paper addresses this gap with a literature review. The end result is a unified model of antecedents for switching from a social media to another, along with developing several relational propositions. Thus, this study responds to an important gap in the SNS literature, and guides future research to solve a contemporary problem in our world.

Keywords Social network sites · SNS switching · SNS monopoly · Review

1 Introduction

This study defines social networking sites (SNS) as digital platforms that enable users to share information about themselves as well as retrieve information on other users. Social media has become a crucial element in today's society because it connects millions of users and enables them to interact with each other. Social media also represents a communication and interaction medium not only for users but also for businesses and institutions enabling them to reach out to customers and engage with them. Social media is essentially an online platform where users share and interact with each other. In this study, robust online social platforms like Facebook, Instagram, Twitter, WhatsApp, and YouTube will be referred to as Big Tech SNS as

P. Balozian (✉)
James Madison University, Harrisonburg, USA
e-mail: balozipx@jmu.edu

S. Curdy
Lebanese American University, Byblos, Lebanon

C. Dal Zotto et al. (eds.), *Smart Technologies for Organizations*, Lecture Notes in Information Systems and Organisation 60, https://doi.org/10.1007/978-3-031-24775-0_7

they are the platforms with the highest number of users. These Big Tech SNS are pioneers in the industry and will be compared to alternative smaller SNS (Alt Tech SNS) companies that are claiming to offer better privacy and are seemingly the most promising alternatives to these Big Tech SNS.

Facebook is the most well-known of all SNS because of the sheer number of its features [1]. Facebook's attributes have made it one of the most attractive platforms for users and hackers alike. Currently in the world, there are nearly 2.4 billion Facebook users, around 330 million Twitter users, and approximately 2 billion YouTube users [2]. Furthermore, messaging platform WhatsApp has about 1.5 billion active users as of 2020 [3]. Social networking sites (SNS) are consistently exposed to data breaches and privacy scandals since SNS contain some of the largest databases for consumer information, with huge ramifications of security and privacy concerns for users. Meanwhile other alternative smaller tech companies are boasting with better security and privacy features, demanding and posing a more competitive SNS market. The scope and purpose of this study is to find out the antecedents that push the users to switch from Big Tech to Alt Tech. The extant literature lacks an explanation of this switching phenomenon with a comprehensive synthesized model. This paper addresses this gap with a systematic literature review. The end result is a unified model of antecedents for switching social media, along with developing several relational propositions. Such theory building and proposition advancing studies are common in the extant information systems literature (See [4–8]). The security aspect of organizations and SNS companies are outside the scope of this study in order to keep the volume of articles manageable and for the sake of keeping the topic of this study focused (for the recent existing security literature and reviews see [9–14]). In the next paragraphs and with a special emphasis on Facebook, we further highlight the importance of our study.

One of the most infamous Facebook data breaches occurred in September 2018 when 50 million users' private information was left exposed [15]. Mark Zuckerberg, the founder of Facebook, testified that Cambridge Analytica stole the personal data of the users that were subjected to this breach [16]. This Facebook incident did not only leave members of Facebook (FB) vulnerable, but it also provided hackers with the access to any accounts linked to a member's FB account (i.e. AirBnb and Tinder) [17]. The complexity and connectedness of today's digital environment enables hackers to single-handedly compromise a plethora of data through the compromise of a single platform.

The Cambridge Analytica scandal set users off in a struggle to comprehend FB's integrity and the level of security in which their data is being stored and protected. Eventually, users started to think that the only way to protect their data would be to delete their accounts and discontinue Facebook usage. A survey conducted by Leswing [18], with a sample size of 1,000 Americans, was taken shortly after the FB scandal in 2018 with results to prove that the breach had an effect on users: 17% of respondents said they deleted the Facebook app from their phone over privacy concerns and 9% reported deleting their Facebook account altogether. However, the survey only provided insight into users' initial reactions to the scandal and no further information was provided whether these users eventually re-downloaded and/or reactivated their FB accounts.

Within the same week of the scandal, the hashtag, "#DeleteFacebook" [17] started trending and "within a two-hour period…it was mentioned 40,398 times" [16]. It not only raised awareness on the Cambridge Analytica scandal, but it also initiated users into becoming more cautious of the information they share, and it also led some to delete their accounts. A number of celebrities and public figures were involved in popularizing the trending hashtag and raising awareness as they continued to abandon FB. Among the public figures that tweeted the trending hashtag were WhatsApp co-founder Brian Acton [19], Apple co-founder Steve Wozniak [20], and entrepreneur Elon Musk [21]. Following Steve Wozniak's tweet were singer Cher, actors Will Ferrell and Jim Carrey, and former talk show host Rosie O'Donnell claiming FB's privacy practices as unsafe and unethical [20].

On another hand, FB's monopolization of the market leaves users with little to no intention to discontinue use as it has become a main form of communication. However, there are emerging FB alternatives that users might consider switching to, claiming to have stronger security and privacy over users' sensitive information. These FB alternatives function similarly to FB with features that allow users to post pictures and make online connections. These alternatives include but are not limited to: Diaspora, Ello, Vero, Mastodon, and Raftr [22]. Despite the availability of these alternatives, FB still remains to be the preferred online platform arguably because of the extent of connectedness of users [23].

In this study, we synthesize a growing literature on SNS switching, in order to better understand this phenomenon and come up with a unified model. The model could be a stepping stone to summarize the research till date, in order to explore other dimensions related to SNS switching. The remainder of this paper will be divided as follows: the emerging themes are discussed in the literature review followed by the comprehensive unified model of the antecedents of Big Tech SNS switching toward lesser-known alternatives. Seven propositions are developed that highlight the relations among the constructs. Finally, we underscore our theoretical and practical contributions and we direct future research toward dimensions that would enable researchers to better understand why Facebook is still a monopoly and how to enhance a healthier SNS environment.

2 Literature Review Methodology

In order to narrow the search process, this study employed the databases known to be inclusive of many of the business and social science disciplines. These databases are as follows: ABI Inform, Business Source Complete, Informs Pub Online, and Science Direct. After refining the databases, the keyword searches that this manuscript included were interchanged between both the title and abstract sections. The keywords were as follows: social networking sites (SNS), SNS continuance, SNS continuance intention, SNS discontinuance, privacy, privacy concerns, social media, risks, self-disclosure, antecedents, and later on after the initial reading of the titles, the keywords social media technology (SMT) and technology invasiveness (TI) included in the search. The keyword Facebook (FB) was also employed due to the relatively high frequency of SNS studies conducted that used Facebook as the platform to test users' behavior and attitude toward SNS.

Since SNS usage continuance is a fairly new topic, the research was limited to the period from 2010 to 2019 inclusive with a focus on articles in the senior basket of eight journals in information systems. The adoption of the narrow 10-year timeframe in the literature review was justified by the relative novelty of the data scandals on SNS. In the analysis and synthesis of this study we also included articles from prior to 2010 and articles within the timeframe but not related to SNS (e.g. general privacy papers), in order to enrich cross discipline pollination of constructs. This exercise greatly enhanced the comprehensiveness of the unified model.

Finally, this study considered only empirical articles that tested structural models. All the conceptual papers were left outside the scope, although we made use of review papers on privacy (e.g. [4]). We excluded the conceptual papers for two reasons: to keep the scope of the literature review more manageable since the privacy literature is voluminous and to build a unified model based on empirically proven models and constructs. The level of analysis in the literature result section is not organizational, but the societal level. Nevertheless, organizational level articles were screened for relevant constructs (or adaptation of constructs) to the research at hand. The unit of analysis is the individual user. Based on the reading of the titles of hundreds of articles from online databases we short listed the titles of articles of which we wanted to read the abstract. We read the abstract of dozens of articles that resulted in identifying 18 manuscripts. In order to keep the study focused yet comprehensive, the references of the articles used for this research were further scanned in search of any studies that were relevant to this article. We identified 18 articles based on their content and relevancy and tested constructs related to SNS switching. The eighteen articles are marked by an asterisk "*" in the references.

3 Literature Analysis and Results

This section of the paper identifies the emerging trends and relationships gathered throughout the research that lend further understanding into SNS privacy behavioral change. The conceptual relationships that emerged throughout the research consisted of various controls that affected member attitudes and privacy-related aspects. The following paragraphs analyze the seven emerging relationships from the synthesis of our systematic review. These are: (1) the effects of attitude toward change on SNS privacy behavioral change, (2) the impact of switching stressors on attitude toward change, (3) the impact of privacy concerns on attitude toward change, (4) the impact of perceived control on attitude toward SNS change, (5) the impact of social influence of dissenters on attitude toward change (i.e. discontinuing, switching), (6) the impact of privacy awareness on privacy concerns, and (7) the impact of past bad experience on privacy concerns.

3.1 *Effects of Attitude Toward Change on SNS Privacy Behavior Change*

Throughout this manuscript, the different relationships that collectively effect SNS privacy behavior change have been studied and hypothesized. In this context, attitude refers to a user's positive evaluative effect about using a social networking site [24]. Throughout the users' SNS experience, an attitude toward the platform will eventually be formulated and either a positive or negative judgment will be formed. The value added from SNS forms an important incentive to encourage a more positive attitude toward SNS while the lack of value added may encourage a more negative attitude toward SNS. Attributes that affect a user's judgment and can be seen as an added value are part of user experience. User experience can range from the application's user interface which contributes to ease of use and how accessible basic functions are to privacy-related incidents.

The most pivotal issue that users face is the fear and distrust regarding their personal privacy [25, 26]. Ad-free browsing experiences and transparency in regard to information privacy are factors that aid in a more enjoyable experience and will instill a positive outlook on SNS rather than a negative outlook. However, users may become victim to cyber risks as SNS popularity increases. This is likely to create adverse attitudes in users toward SNS and discourage them from regularly using SNS [24]. Alternatively, a user might misjudge the integrity of an SNS through the provider's behavior. Contributing factors to a user's misjudgment include the extent to which a provider collects personal information about a user's location, activity, and usage behavior [27, 28]. The more distressed a user feels toward the privacy of their information, the more likely it is to affect their attitude toward change. The degree to which user attitude is affected ultimately determines the likelihood that the user will leave Big Tech SNS for Alt Tech SNS. Thus, this study proposes the following:

Proposition 1 Attitude toward SNS change has a positive effect on SNS behavior change.

3.2 *Impact of Switch Stressors on Attitude Toward SNS Change*

When users have the ability to switch SNS platforms from Big Tech to Alt Tech, the behavioral response is not immediate, and users have to consider the cost and effort of switching [29]. The latter are then considered to be switching stressors perceived by the user that aid in the behavioral response to be taken. Furthermore, techno-invasion was used to describe social media's invasive nature which is said to be a main stressor on the user, which in turn increases a user's stress levels [30].

A main instigator of stress in users is the switching costs that a user senses, eventually leading to SNS exhaustion. The first switching cost is the transition cost which involves the stress of having to learn and adapt to new SNS technologies and situations, along with the search costs in looking for SNS alternatives [31]. When a user wants to adopt an ALT Tech SNS technology, the ability to grasp the controls and benefits from using social media is crucial. If a user has already been a long-time loyal user of a popular social media platform like FB, the effort of having to learn to use a new Alt Tech SNS becomes an insurmountable task which deters the user from moving away from Big Tech.

The second switching stressor is represented by sunk costs. This cost alludes to the psychological commitment to keep using specific SNS because of the effort it took a user to learn the technology and the time it took the individual to incorporate the use of SNS into their daily lives [31]. In the event of considering switching platforms, the users might find themselves having to use several different online platforms to compensate for what one platform was offering them, which would then cause the users to reassess SNS switching.

A contributor in SNS switching are the FB alternatives available to the user and the ease of switching to less popular alternatives. Among the aforementioned FB alternatives, Vero is the most similar to FB and is a fast-growing FB alternative with nearly 3 million members [32]. Vero shares similar features to FB, but it also differentiates from it—claiming an ad-free experience and no user data mining. Although Vero developers collect user statistics this feature is turned off by default and if the user decides to turn it on, they have full access to the information being collected [22]. Diaspora is another FB alternative with approximately 50,000 new users joining in the past six months by the time of the writing of this manuscript [22]. Diaspora prides itself on its decentralization and is tailored more for the tech savvy user with the network consisting of many networks known as pods. Instead of user information being stored centrally, they are stored in pods and the infrastructure is distributed by the users themselves. If the user has good technical know-how, they can operate their own pods, which essentially function as servers. This means that users have full control over the privacy of their data [22]. What might deter a user from committing to a switching from Big Tech SNS to Alt Tech SNS is the user interface of the SNS. User interface and design impact the use of a system [33]. It is the designer's role to make the design of the alternative social media platforms similar to that of the current monopolistic ones so that Alt Tech SNS closely resemble Big Tech SNS in design and features. This minimizes the learning curve between the two platforms and increases the likelihood of switching and committing.

Although these FB alternatives do not seem to be as popular as FB, the user count is steadily increasing in the aforementioned alternate platforms. The steady growth indicates that users are becoming more interested in information privacy and might be developing privacy concerns from allocating too much of their information on FB and prefer to switch to a more transparent online platform.

The feeling of having to use too many alternatives to interact at the same degree is the third type of switching stressor known as replacement overload [31]. Replacement overload is also said to create an unfavorable psychological perception in terms of switching-exhaustion. Switching-exhaustion emphasizes how integral FB has become to a user since the platform offers all services on one platform with the suitability of having an already existing friend network. Thus, this manuscript proposes the following:

Proposition 2 Switch stressors have a negative effect on attitude toward SNS change.

3.3 Impact of Privacy Concerns on Attitude Toward SNS Change

The degree to which the users are seeing their digital information being collected, shared, and processed in a fair way defines privacy concerns [34]. The increase of digitalized personal information and social media's popularity poses a new threat for users in terms of their information privacy. Though information sharing through online media platforms has existed since the Internet, it is the increase in data on these social platforms that is leaving users concerned with the possible misuse of their information.

To better understand internet privacy concerns (IPC), Lowry et al. [34], Benamati et al. [35], and Bansal et al. [25] introduced the four main dimensions of IPC that are as follows: collection, errors, secondary use, and improper access. *Collection* has been uniformly defined as the concern surrounding the collection of sensitive information that is being collected and stored in online databases [25, 34–36]. Thorough research shows that collection is one of the most important dimensions of information privacy and is a "core component of privacy concerns" [36]. Users on the Internet want to feel like the information that they are providing these online websites with is relevant to their online transactions. Privacy concerns stem from instances where users find themselves needing to provide more than the minimum information required in order to complete a specific transaction. The intrusive nature of online social platforms makes individuals becoming aware that they are providing too much of sensitive information.

Moving on to the second IPC dimension, *errors* were collectively defined as the inadequacy of data collectors to properly protect an individual's private information [25, 34–36]. When setting up a profile, users will provide personal information about themselves and entrust the organization to make sure that the data being collected is double-checked for accuracy. However, often an organization will not check the authenticity of an online profile—this gives users the ability to create online profiles under aliases and fake celebrity accounts. However, user identity is now being confirmed with the placement of a blue check mark next to the username of public figures and celebrities across all SNS platforms [37]. The lack of auditing information inserted to create a user's profile as well as checking for accuracy of accounts is considered as an error and can spark privacy concerns in a user. Moreover, it is important that organizations check for information accuracy because for example, a user's password can be changed by a third party and the account holder will not be notified of this change until after the password has been changed. It is the organization's responsibility to always check that any changes in personal information were modified by the user and not by a third party, creating errors in the information and therefore, sparking privacy concerns in a user.

Lowry et al. [34] found that the third construct of IPC is unauthorized *secondary use* which is a user's concern about the possibility of their private information being used for secondary reasons and without authorization. Unauthorized secondary use of information is a main factor in prompting user privacy concerns. Oftentimes, online media platforms will sell private user information as a main source of revenue without the user's consent. This not only applies to the personal information a user has provided to online platforms, but it is also the tracked activity of a user. Tracked activity can refer to all the actions a user makes online which include but are not limited to search history, private chats, and even data submitted in FB quizzes. The information exchanged in private chats can range from trivial messages to significant information about a person's health.

The last common dimension of IPC is *improper access* which refers to the degree in which an individual's private information is readily available to unauthorized personnel [34]. This dimension refers more to the information on an organizational level and how stringent a company's security is over the control of user data. Unencrypted passwords of accounts could be stored in company databases and accessible to employees of the organization whether the employee is authorized or unauthorized to view this information. Furthermore, occurrences of security-related incidents that a social media user might be exposed to could trigger their awareness of the possibility of improper access to their personal data.

In addition to the four aforementioned dimensions of internet privacy concern, Hong and Thong [36] found that control and awareness were also dimensions of IPC, whereby control refers to the perception of control that a user has over his/her personal information collected by websites. Awareness is how aware a user believes online websites are pertaining to information privacy practices [36]. The researchers found that consumers consider the gain or loss of control crucial to their interaction with websites [36] and need to feel as if they control the decisions taken in regard to their data.

While Hong and Thong [36] refer to control mainly in the context of websites, the dimensions of SNS privacy concerns appear to be more sensitive. Even though websites collect information about the user, Big Tech SNS like FB go further and can collect user's private messages, status updates, comments, and any form of communicated text to infer personality traits. This helps Big Tech SNS to improve their product for users and advertisers and creates a platform where ads become better suited for the user. The patent of the software responsible for this type of psychological profiling was gained by Facebook in 2012 [38]. In the case of Big Tech SNS, user information that is being collected and sold to third parties is sold without the user's consent. The lack of transparency that Big Tech SNS exhibit gives users the idea that they do not have control over their information and that elicits privacy concerns [17]. As a consequence of the above, we formulate the following proposition:

Proposition 3 Privacy concerns have a positive effect on attitude toward SNS change.

3.4 Impact of Perceived Control on Attitude Toward SNS Change

Perceived control is frequently used to describe the individual's perception level of control such as self-efficacy and how an SNS allows the member to control use of information through privacy settings [39]. Through perceived control, a user's trust either increases or decreases based on the experience. Furthermore, the data that is being collected is said to be used by the organization in order to better create business strategies for the consumer. While some organizations have been benefiting from selling private user information, other companies have remained ethical and chosen not to share sensitive data and in return this has given users a sense of control over their information [39].

Lately, governmental bodies intervened to increase user perceived control. The General Data Protection Regulation (GDPR) is a law in the European Union and the European Economic Area that addresses transfer of information between companies and individuals. The recent surge of FB data privacy notifications and the privacy settings that are no longer turned on by default are impositions made by GDPR [40]. GDPR regulations have increased user perceived control and can be seen as a major contributor in affecting users' attitude toward SNS change, thus a decisive element for either switching to SNS or sticking to Big Tech SNS. Although the regulations that have been imposed on Big Tech SNS seem to be keeping users from switching, it is still undetermined if these Big Tech SNS are being forced to comply or are willingly complying with these regulations. Perceived control can be seen to have significant effects in reducing a user's privacy concerns as well as being considered to be the most notable factor affecting privacy concerns within the SNS context [40].

The term self-efficacy can be defined as a person's ability to perform tasks with the skills that they possess and can influence an individual's behavior on what actions to take [41]. Self-efficacy is also considered to assimilate control and give users the perception of having the ability to control their sensitive information. Kim et al. [41] divided self-efficacy into two types, technology self-efficacy and information self-efficacy; where the former refers to the adeptness a user possesses in using the SNS while the latter refers to an individual's beliefs in their ability to share valuable information with others. With social media being a relatively new concept, technology self-efficacy can affect a user's perception of control because of the learning curve a user faces in operating SNS. If a member of SNS doubts their ability to use social media technologies, it can become a critical barrier in continued usage and the feeling of loss of control. In order to gain some control, members of online social networks should be trained on the SNS platform whether formally or informally. For example, the two-step verification is a feature known to many users but not used by all. It is available on Big Tech SNS and users can activate it in order to add extra barriers to their account and provide users with a stronger sense of control over their accounts.

Moreover, perceived networking assistance is also found to provide users with a sense of control over their personal network and networking activities through a manageable personal network with the help of SNS [24]. Searching for old and new friends by education, location, or workplace is an example of an SNS tool that helps members manage their personal networks. Moreover, being able to make new connections with other users online through shared similar interests and backgrounds is especially convenient on Big Tech SNS because of its connectedness. The manageability of personal online networks takes part in shaping members' attitudes toward SNS. Thus, this manuscript proposes that:

Proposition 4 Perceived control has a negative effect on attitude toward SNS change.

3.5 Impact of Social Influence Dissenters on SNS Attitude Toward Change

Social influence concerns the psychological needs of a user and the pressure to conform to specific expectations from users' reference groups. Reference groups refer to a user's online and offline social network that is comprised of friends, family, and coworkers who are important to the individual. For example, messages a user receives from a member of their reference groups will influence their attitude toward and perception of expected participation on online social networks [24]. The expectations imposed on a user through their reference group will affect a user's attitude toward whether or not the imposition is perceived positively or negatively. For example, when social influence is favorable, users will internalize these beliefs and develop a similar attitude toward online social networks.

A positive attitude toward online social networks will add value to social networking for the user and will encourage a user to remain active on online social networks. Moreover, users may consider social networking sites as unfavorable when their reference group starts to believe the same. What is still unapparent in research is the effect of the critical mass on users and what amount of a user's personal network should do a mass exodus in order for the user (and others) to follow.

Similarly, the source of social influence is not only comprised of a user's personal network, but it is also from influencers like politicians, entrepreneurs, prominent figures in the technology field, and celebrities. Tech titan and co-founder of Microsoft, Bill Gates, has limited how much tech his kids consume and according to several sources, did not allow his kids to have phones until the age of fourteen [42]. Billionaire Mark Cuban also manages his kids' screen time and has restricted the use of all technology after specific times [42]. When users witness major tech giants practicing caution on Big Tech SNS, it is likely to affect users' activity on SNS.

This social influence that is brought upon by social influencers and dissenters has been defined as subjective norm by Loiacono and McCoy [30]. Subjective norm is the influence that a social network has on a user's decision on whether or not to perform a behavior. Thus, this manuscript proposes that:

Proposition 5 Social influence dissenters have a positive effect on the attitude toward SNS change.

3.6 Impact of Privacy Awareness on Privacy Concerns

Part of what makes social media desirable is the ability to connect with the people you know; a person's constant desire to be present and aware in their online community is pertinent to social media's survival and growth. Not only users enjoy sharing and interacting with their network, but users will also feel more encouraged to share and communicate on social media if they feel the same is being reciprocated [34]. However, it can be argued that users who practice caution over their data, and know the repercussions of over-sharing information online, tend to exhibit less desire for online awareness and more privacy awareness [34]. Over time, a user adds more personal information on his/her social media. This creates a more or less complete profile of the user. His/her likes, dislikes, interests, religious and political preferences, and psychological make-up become available to even strangers. Frequent posting behavior makes an individual easily discoverable in comparison to users who do not share the same desire to post and have provided limited information about themselves on their online profiles.

In light of the extant literature, Lowry et al. [34] uses "awareness" in the context of social awareness and while social awareness is relevant to SNS usage, it is the level of privacy awareness that impacts Big Tech SNS usage. Privacy awareness pertains to the level of knowledge that a user has in terms of their online privacy and how a user's privacy awareness can impact privacy concerns. However, due to recent regulations set by GDPR, Big Tech SNS are obliged to notify users of data privacy-related settings available on platforms which lends users a stronger sense of awareness over their online privacy [40], without totally minimizing the fact that users are basically bound to accept SNS conditions if they want to use their platforms. Otherwise they will remain isolated from their social media friends and feel de-platformed.

Corresponding to the aforementioned, internet, computer or privacy self-efficacy, and privacy knowledge are defined as knowledge [43]. Generally, it is the people who know more about information privacy and the risks pertaining to sharing your information online that have higher privacy concerns. In addition, an individual's privacy awareness is formed by both their privacy exposure in the media as well as any prior experiences [35]. Based on the findings of this study, privacy awareness not only stems from personal experience, but it can also come about from having heard or read stories through the media. Thus, this manuscript proposes the following:

Proposition 6 Privacy awareness on SNS privacy and security has a positive effect on the privacy concerns of users.

3.7 Impact of Past Bad Experience on Privacy Concerns

Privacy concerns can be termed as a person's belief that the information they are sharing online is at risk [44]. Prior to any negative experiences on these social networking sites (SNS), users are often unaware of the threat that Big Tech SNS pose in terms of their data vulnerability. In some instances, users are aware of these threats but do not perceive Big Tech SNS as posing any high-level threats as they have not been exposed to any prior negative experiences to elicit such a reaction. It is often assumed that far-reaching global companies like FB would have the utmost tight-knit security systems with well-protected databases. This assumption creates a sense of security in users who have entrusted platforms with sensitive data and therefore have no reason to question a company's integrity. However, it is not until users experience security breaches and fraudulent activities that they become more aware of the risks they are in [24]. The robustness of social media platforms like Facebook is always vulnerable to security breaches and has been victim to online privacy invasions.

Users could also be made aware of a privacy breach through notifications that they receive from FB; advising them to change their passwords or to make sure that double-verification is activated. Constant exposure to notifications about data privacy and headlining articles on data breaches could evoke a user's curiosity in terms of the vulnerability of their data as well as encourage users to take initiative into switching to smaller alternatives in which they feel their data is more protected in comparison to Big Tech SNS. After all, the hackers are after the big techs for greater impact.

Social media operators should focus on improving their privacy protection technology in order to compensate for a user's prior negative experiences [44]. Although being informed of these breaches through media outlets is not a direct form of privacy invasion, it is sufficient enough to raise questions and concerns regarding the aforementioned. However, it still remains that people who have been subjected to negative experiences tend to be wearier of the information that they are sharing on social media as well as the actions they are taking [45]. Thus, this manuscript poses the following:

Proposition 7 Past bad experiences of security or privacy breaches have a positive effect on the privacy concerns of SNS users.

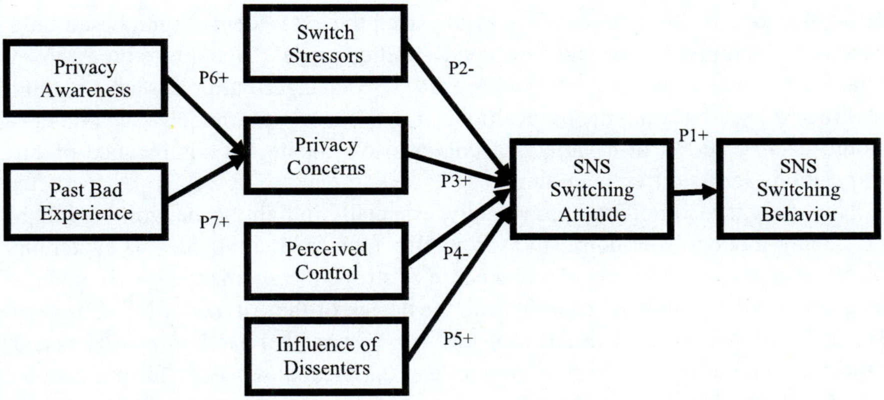

Fig. 1 The unified research model

4 The Unified Research Model

As shown in Fig. 1, the research model includes the following constructs: past bad experiences, privacy awareness, social influence of dissenters, privacy concerns, perceived control, switch stressors, benefits, attitude toward change, and SNS privacy behavior change.

The unified model builds on the literature review of the previous sections. It shows that Privacy Awareness and Past Bad Experiences of security and privacy breaches are positively related to increased Privacy Concerns. Consequently, Privacy Concerns and Influence of Dissenters have positive relationships on heightened Attitude toward SNS Switching. On the other hand, Switch Stressors and Perceived Control of data have both a negative relationships toward Attitude of SNS Switching. In turn a heightened SNS Switching Attitude leads SNS Switching Behavior.

To the best of our knowledge, this is one of the first studies to consider a unified model of SNS switching in the information privacy context. In the remaining section, we highlight our theoretical and practical contributions, before giving directions for future research.

5 Contributions, Conclusion, and Future Research

This paper presents an in-depth literature review on SNS switching. This study contributes to theory and research by developing a unified model of SNS switching antecedents along with relevant propositions of the relationships among the constructs. Review articles that synthesize literature, develop theoretical models, and advance propositions to be tested in future research abound in Information Systems discipline [5, 7, 8], including in privacy [4] and information security [6]. The extant literature shows there is paucity in articles that combine all antecedents in SNS switching, thus we comprehensively synthesized the antecedents found based on a systematic literature review and developed a unified model that helps to break down what factors impact the user attitude toward SNS change. Future research should be directed toward testing the propositions of such theoretical models. Such models should also be tested in a variety of contexts to validate the significance of the propositions across different cultures.

This study is also important practically, especially in light of the world community facing monopolistic domination from Big Tech SNS companies. If switching platforms and leaving Facebook are a cause of stress, the alternative social media's designers have a significant role to make the design of the Alt Tech SNS similar to Big Tech SNS to minimize the stress on the users. On another hand, campaigners and influencers can have an important role to lead mass exoduses away from monopolistic Big Tech SNS. If a user switches to a new platform and does not find his or her friends there, he/she will likely come back or reactivate the old platform. This also invites future research to investigate by using focus groups, interviews, and qualitative analysis to shed light on what might cause or trigger a mass exodus. These directions of research could not only enrich literature and theory, but also change our SNS experiences and our world for the better.

References

1. *Fox, J., & Moreland, J. J. (2015). The dark side of social networking sites: An exploration of the relational and psychological stressors associated with Facebook use and affordances. *Computers in Human Behavior, 45*, 168–176.
2. Clement, J. (2020). *Global social media ranking 2019*. https://www.statista.com/statistics/272014/global-social-networks-ranked-by-number-of-users/.
3. Davies, W. (2020). *What's wrong with Whatsapp*. https://www.theguardian.com/technology/2020/jul/02/whatsapp-groups-conspiracy-theories-disinformation-democracy.
4. Bélanger, F., & Crossler, R. E. (2011). Privacy in the digital age: A review of information privacy research in information systems. *MIS Quarterly*, 1017–1041.
5. Cram, W. A., Brohman, K., & Gallupe, R. B. (2016). Information systems control: A review and framework for emerging information systems processes. *Journal of the Association for Information Systems, 17*(4), 216–266.
6. *Cram, W. A., Proudfoot, J. G., & D'arcy, J. (2017). Organizational information security policies: A review and research framework. *European Journal of Information Systems, 26*(6), 605–641.
7. Li, S. S., & Karahanna, E. (2015). Online recommendation systems in a B2c e-commerce context: A review and future directions. *Journal of the Association for Information Systems, 16*(2), 72–107.
8. Lim, S. Y., Jarvenpaa, S. L., & Lanham, H. J. (2015). Barriers to interorganizational knowledge transfer in post-hospital care transitions: Review and directions for information systems research. *Journal of Management Information Systems, 32*(3), 48–74.
9. Balozian, P., & Leidner, D. (2016). Is security menace: When security creates insecurity. In *Proceedings of International Conference on Information Systems*.
10. Balozian, P., & Leidner, D. (2017). Review of IS security compliance: Toward the building blocks of an IS security theory. *The Data Base for Advances in Information Systems, 48*(3), 11–43.

11. Balozian, P., & Leidner, D. (2017). The assumptions and profiles behind IT security behavior. In *Proceedings of 50th Annual Hawaii International Conference on System Sciences*.
12. Balozian, P., Leidner, D., & Warkentin, M. (2019). Managers' and employees' differing responses to security approaches. *Journal of Computer Information Systems, 59*(3), 197–210.
13. Cram, W. A., D'arcy, J., & Proudfoot, J. G. (2019). Seeing the forest and the trees: A meta-analysis of the antecedents to information security policy compliance. *MIS Quarterly, 43*(2), 525–554.
14. Moody, G. D., Siponen, M., & Pahnila, S. (2018). Toward a unified model of information security policy compliance. *MIS Quarterly, 42*(1), 285–311.
15. Lee, D. (2018). *Facebook security breach: Up to 50m accounts attacked*. https://www.bbc.com/news/technology-45686890.
16. Hsu, T. (2018). *For many Facebook users, a 'last straw' that led them to quit*. https://www.nytimes.com/2018/03/21/technology/users-abandon-facebook.html.
17. Wakefield, J. (2018). *Is leaving Facebook the only way to protect your data?* https://www.bbc.com/news/technology-43469656.
18. Leswing, K. (2018). *Nearly one in 10 Americans surveyed say they deleted their Facebook account over privacy concerns*. https://www.businessinsider.com/delete-facebook-statistics-nearly-10-percent-americans-deleted-facebook-account-study-2018-4.
19. Gerken, T. (2018). *Whatsapp co-founder says it is time to delete Facebook*. https://www.bbc.com/news/blogs-trending-43470837.
20. Fleig, S. (2018). *These high-profile users are quitting Facebook*. https://www.usatoday.com/story/tech/talkingtech/2018/04/11/these-high-profile-users-quitting-facebook/503730002/.
21. Locke, T. (2020). *Elon Musk: Delete Facebook, 'it's lame'*. https://www.cnbc.com/2020/02/10/elon-musk-delete-facebook-its-lame.html.
22. Dredge, S. (2018). *You've decided to delete Facebook but what will you replace it with?* https://www.theguardian.com/technology/2018/mar/31/youve-decided-to-delete-facebook-but-what-will-you-replace-it-with.
23. Jackson, T. (2020). *What can you use instead of Google and Facebook?* https://www.bbc.com/news/business-50460712.
24. *Chen, R., & Sharma, S. K. (2015). Learning and self-disclosure behavior on social networking sites: The case of Facebook users. *European Journal of Information Systems, 24*(1), 93–106.
25. *Bansal, G., Zahedi, F. M., & Gefen, D. (2015). The role of privacy assurance mechanisms in building trust and the moderating role of privacy concern. *European Journal of Information Systems, 24*(6), 624–644.
26. *Miltgen, C., & Smith, H. J. (2015). Exploring information privacy regulation, risks, trust, and behavior. *Information & Management, 52*(6), 741–759.
27. *Gerlach, J., Buxmann, P., & Diney, T. (2019). "They're all the same!" stereotypical thinking and systematic errors in users' privacy-related judgments about online services. *Journal of the Association for Information Systems*, 787–823.
28. *Xu, H., Teo, H. H., Tan, B. C. Y., & Agarwal, R. (2010). The role of push-pull technology in privacy calculus: The case of location-based services. *Journal of Management Information Systems, 26*(3), 137–176.
29. *Maier, C., Laumer, S., Eckhardt, A., & Weitzel, T. (2015). Giving too much social support: Social overload on social networking sites. *European Journal of Information Systems, 24*(5), 447–464.
30. *Loiacono, E., & McCoy, S. (2018). When did fun become so much work. *Information Technology & People, 31*(4), 966–983.
31. *Maier, C., Laumer, S., Weinert, C., & Weitzel, T. (2015). The effects of technostress and switching stress on discontinued use of social networking services: A study of Facebook use. *Information Systems Journal, 25*(3), 275–308.
32. Swartz, J. (2018). *Facebook Rival Vero faces a user backlash*. https://www.barrons.com/articles/facebook-rival-vero-faces-a-user-backlash-1519924496.
33. Balozian, P., & Leidner, D. (2014). Tablet image: Visibility feature designs increase tablet use. In *Proceedings of Americas Conference on Information Systems*.

34. *Lowry, P. J., Cao, J., & Everard, A. (2014). Privacy concerns versus desire for interpersonal awareness in driving the use of self-disclosure technologies: The case of instant messaging in two cultures. *Journal of Management Information Systems, 27*(4), 163–200.
35. *Benamati, J. H., Ozdemir, Z. D., & Smith, H. J. (2017). An empirical test of an antecedents—Privacy concerns—Outcomes model. *Journal of Information Science, 43*(5), 583–600.
36. *Hong, W., & Thong, J. Y. L. (2013). Internet privacy concerns: An integrated conceptualization and four empirical studies. *MIS Quarterly, 37*(1), 275–298.
37. Arshad, H. (2016). *Complete guide to social media account verification: SEJ*. https://www.searchenginejournal.com/the-complete-guide-to-social-media-account-verification/172832/.
38. Cellan-Jones, R. (2018). *Facebook explored unpicking personalities to target ads*. https://www.bbc.com/news/technology-43869911.
39. *Hajli, N., & Lin, X. (2016). Exploring the security of information sharing on social networking sites: The role of perceived control of information. *Journal of Business Ethics, 133*(1), 111–123.
40. Nadeau, M. (2019). *What is the GDPR, its requirements and facts?* https://www.csoonline.com/article/3202771/general-data-protection-regulation-gdpr-requirements-deadlines-and-facts.html.
41. *Kim, J., Lee, C., & Elias, T. (2015). Factors affecting information sharing in social networking sites amongst university students. *Online Information Review, 39*(3), 290–309.
42. Berger, S. (2018). *Tech-free dinners and no smartphones past 10 Pm—How Steve Jobs, Bill Gates and Mark Cuban limited their kids' screen time*. https://www.cnbc.com/2018/06/05/how-bill-gates-mark-cuban-and-others-limit-their-kids-tech-use.html.
43. *Miltgen, C. L., & Peyrat-Guillard, D. (2014). Cultural and generational influences on privacy concerns: A qualitative study in seven European countries. *European Journal of Information Systems, 23*(2), 103–125.
44. *Wang, L., Sun, Z., Dai, X., Zhang, Y., & Hu, H.-H. (2019). Retaining users after privacy invasions. *Information Technology & People, 32*(6), 1679–1703.
45. *Ozdemir, Z. D., Smith, H. J., & Benamati, J. H. (2017). Antecedents and outcomes of information privacy concerns in a peer context: An exploratory study. *European Journal of Information Systems, 26*(6), 642–660.

Digital Transformation, Value Creation and Sustainability

Digital Transformation and Its Contribution to Corporate Social Responsibility as a Sustainable Development Strategy Within Organizations: A Literature Review

Yassmin Diab and Imed Ben Nasr

Abstract This paper analyzes the academic literature that highlighted the role of digitalization in the implementation of sustainability strategies within organizations. It questions how digital transformation could help organizations to integrate the corporate social responsibility perspective in their structure, their value proposal, business, and management methods (ethics, values…). The main results of the literature review insist on the positive contribution of digital transformation in the implementation and success of corporate social responsibility strategies. Even though the reviewed literature does not give insights into the conditions and implementation processes underlying such positive effect, it highlights the key factors that could positively link digital transformation and the development of corporate social responsibility strategies. These results appear useful both for business practice and future research.

Keywords Digitalization · Digital transformation · Corporate social responsibility · Sustainable development strategy

1 Introduction

Everything around us nowadays is based on technology, it is no wonder that our lives revolve around a digital and automated orbit. Let us travel together back in time and remember the initial industrial revolution where they used water and steam to power mechanical production facilities, then came the second industrial revolution where electric power and division of work was utilized for mass production. After that, we all remember the third revolution in industry where the introduction of electronics and information technology enabled a breakthrough that led to the automation of

Y. Diab (✉)
Beirut Arab University, Beirut, Lebanon
e-mail: ygd272@student.bau.edu.lb

I. B. Nasr
Excelia, La Rochelle, France
e-mail: bennasri@excelia-group.com

C. Dal Zotto et al. (eds.), *Smart Technologies for Organizations*, Lecture Notes in Information Systems and Organisation 60, https://doi.org/10.1007/978-3-031-24775-0_8

production lines. This is generally referred to as the digital revolution, catalyzed by the development of semiconductors, mainframe computing (1960s), personal computing (1970s and 1980s), and the Internet (1990s).

Today, we are experiencing the fourth industrial revolution. Starting at the turn of the century, it bases on and extends the third digital revolution and it is characterized by a much more ubiquitous and mobile internet, by smaller and more powerful sensors that have become cheaper, by artificial intelligence (AI) and machine learning. Similar to other revolutions that took place before, this fourth industrial revolution has the possibility to enhance the global income and expand the standard and quality of life for all populations around the world, while implementing sustainable development goals. Technology has changed and evolved every aspect of our daily lives and made it easier, with various options available, to improve and increase the productivity of our work. We can thus say that the digital transformation is the most essential deliverable of the fourth industrial revolution.

The digital transformation taking place nowadays is leading modern organizations to switch from traditional to digital business models. Progression towards an even more digital world is underway with all industries transforming their traditional operations into digitalized business operations. This transformation process poses challenges and pushes businesses to create awareness among the people working in their organization to undergo new strategies. In order to meet the global climate goals of 2050, the transformation into a circular and thus sustainable economy needs to be accelerated. Such an acceleration is possible through a responsible digitalization. In order to become more socially responsible, a lot of organizations are utilizing their digital and technology innovations to implement strategies characterized by sustainable corporate social decisions. The key success factor for incorporating digital transformation into an effective corporate social responsibility (CSR) is leadership, together with an authentic commitment of leaders and managers within organizations to go through a sustainability-oriented process of change.

Sustainability within organizations may have many different meanings, such as taking a more ethical approach towards employees, shareholders as well as towards the social and environmental issues. Digital transformation can lead organizations to be socially responsible by supporting the development of sustainable strategies to the benefit of society and the whole economic system. The aim of this paper is to conduct a review of the existing literature to explore digital transformation, its impact on CSR and its contribution to sustainable development through the operationalization of the main CSR principles within organizations.

The paper will be structured as follows: An introduction to the history of digital transformation, review of the conceptual models of digital transformation, review of definitions of sustainable development in literature, overview of CSR within organizations, review of digital transformation and its contribution to CSR, and sustainability within organizations and finally concluding all the above.

2 Literature Review

2.1 Definition of Digital Transformation

In literature, there is not a commonly accepted definition of Digital Transformation (DT). The available definitions show significant variations around the concept of DT that mainly result from the widely spread and diverse academic backgrounds of DT scholars and practitioners. Digital transformation is a relatively new concept in the field of Information technology and business, which has attracted the attention of researchers but at the same time is lacking consensus on a unified definition. Researchers have adapted different approaches to define DT. Some tackled it from a business process point of view and others from a technological perspective. However, many researchers have conceptualized DT as a business strategy and not only as a technological process. Digital transformation is thus related to a revolution impacting the overall work environment (including strategy, management, and people) [1]. On a similar note, digital transformation is referred to as a continuous process characterized by different digital phases and going through a complete redesign to achieve new digital customer experience and enhanced digital value propositions [2].

Between 1990 and 2000, for example, digitalization enabled payments to be done online and not only physically in the store like the old traditional way. In the following decade from 2000 till 2010, smart devices started emerging and, together with the expansion of social media platforms, they transformed the retail way of how customers and businesses deal with each other. Educational courses started becoming digital, online payments such as those taking place on PayPal have increased, and online banking and commerce have taken off to become the normal way of doing business. Nowadays, thanks to new communication and mobile technologies, a massive amount of customers' data can be collected online and leveraged to advertise more effectively by tailoring advertising to fit customers' specific needs. Table 1 illustrates the chronological history of digital transformation.

Digitalization is considered as one of the primary impetus of today's development that is linked to achieving higher competitiveness by improving the economic, social, and environmental aspects in life [3]. DT is a process used to enhance the performance of organizations, by improving their use of technologies to transform customer relationships, modify internal processes, and capitalizing on new value propositions [4]. On another note, Mazzone highlighted that DT is a continuous digital evolution of the organization, a redesign of business models that introduces new process ideas on a strategic level [5]. To sum up, the different definitions of digital transformation depict a networking of businesses and customers across all value-added chain segments, which include the application of new technologies and appropriate skills and analysis to evolve business models and processes, ultimately leading to an increased performance of organizations [2]. The benefits and positive impacts of DT on business and the economy as a whole highlight that DT is a new form of change that can enable technology to transform existing business [6].

Table 1 The emergence history of digital transformation

Timeline/age	Year	History details	Change	References
Introduction	1970–1989	New emergence that led to the Internet (ARPANET, Telenet)	Analog computers switched to digital, automation of teller machines started, industrial robots emerged, etc.	Schwab [6]; Schallmo et al. [2]
Growing	1990–2005	Invention of the World Wide Web, mainstreaming of the Internet	Organizations started adopting digital channels and managed to change processes to make better use of technology	
Current	2005–2020	Web 2, social media, smartphones	The whole world is embracing digital transformation as a strategy and not a process, towards a more sustainable digital world	
Future	2020+	Imagination/virtual age (AI, machine learning, smart places, intelligent spaces)	Introduction of virtual reality, artificial intelligence and machine learning, 3D internet, progression of the concept of Internet of Things (IoTs), intelligent spaces and smart places, Big Data and augmented analytics, etc.	

Digitalization from the notion of technology alone may not yield results. However, if coupled with strong strategic leadership, technology can transform firms into digitally mature organizations, which outperform those that are not digitally mature [7]. A lot of organizations lack leadership even if they possess digital capabilities to transform their processes; this lack of leadership prevents organizations to strategize and develop in the right direction. In a McKinsey report of 2017 [8], DT is defined as a process of reinvention. It is not related to the process of digitizing a product or process, it is rather a redesign of processes or an enhancement that generates for instance new customer experience using technology platforms. It is strategy, not technology, that steers DT in the right direction. Thus, in order for organizations to gain value out of the transformation, they should adopt a carefully coordinated digital approach across four areas: first of all, organizations should grasp what is the real value of digitalization; then, they should redesign their structure and processes to target more profitable yet economical changes; third, they would need to back such changes with effective partnerships; and finally, organizations should make sure to evaluate the risk when going through this process of change. If organizations are to respect the sustainable development goals, this implies that organizations act as good citizens and not only work for generating profits. Such an approach needs to become an imperative for the DT process [8] and can only be enabled by an established strategic leadership.

Nwaiwu's literature review highlighted that there is a lack of connection or application between practice-oriented and academic-based research related to the subject of digital business transformation [4]. This implies that future research should emphasize this gap to ensure that the academic research is practiced by organizations adopting digital transformation. His research findings point out that digital transformation for organizations does not refer to the adoption of technology as this alone does not efficiently transform the organization. Further research reconfirmed that digital business transformation relates to a strategy that has to be aligned with the overall corporate strategy of an organization [4]. On the same note, a new philosophy of digital transformation highlights the appeal of digital entrepreneuring, defined by Gunasilan as "Technopreneurship," a new concept to strengthen organizations by enabling collaborative digital entrepreneurial ventures [9]. In other words, the researcher emphasized that when startup organizations collaborate with developed organizations, they capitalize on their digital transformation by combining their internal capabilities with external ones and thus maximizing their efficiency. Collaboration in digital transformation can build innovation-oriented partnerships that can extend the best capabilities of both organizations [9].

The potential of DT to achieve the sustainable development goals in the economy has been explored in literature [10]. It is an important area to look into and understand how DT can contribute to the success of SDGs from a local perspective. DT could help organizations become more efficient in adopting a sustainable strategy aligned with their overall strategy. ElMassah and Mohieldin highlighted that decentralized economies or governance structures yield successful transformation as they involve local communities to be part of the decision making. Digital transformation can contribute to the accomplishment of sustainability if complete transparency and thus free exchange of information is ensured [10]. This ensures that sustainability can be accomplished successfully locally which in turn will ensure the global success. On top of that, in order to have a successful implementation of digital transformation within organizations, there must be a change of organizational culture and thus of people's awareness working in the business [1].

Table 2 summarizes all the above-mentioned conceptualizations of digital transformation from the perspective and approach of different researchers.

Table 2 Definition of DT in literature

Author	Definition	Approach
Orbik and Zozulakova (2019)	"DT is revolutionizing all industries, changing not only business processes and models, but the work environment and the way of communication" [1]	Strategic
Javonovic et al. (2018)	"DT is the primary impetus of today's development that is linked to achieving higher competitiveness by improving the economic, social, and environmental aspects in life" [3]	Technological and strategic
Schallmo et al. (2017)	"DT framework includes the networking of actors such as businesses and customers across all value-added chain segments and the application of new technologies. As such, DT requires skills that involve the extraction and exchange of data as well as the analysis and conversion of that data into actionable information" [2]	Technological and strategic
Schwab (2016)	"DT is a new era of the development of new form of change that can enable technology to transform existing business" [6]	Strategic
Mazzone (2014)	"DT is the deliberate and on-going digital evolution of a company, business model, idea process, or methodology, both strategically and tactically" [5]	Technological and strategic
Westerman et al. (2011)	"DT—the use of technology to radically improve the performance or reach of enterprises—is becoming a hot topic for companies across the globe. Executives in all industries are using digital advances such as analytics, mobility, social media, and smart embedded devices—and improving their use of traditional technologies such as ERP—to change customer relationships, internal processes, and value propositions" [7]	Strategic

(continued)

Table 2 (continued)

Author	Definition	Approach
Gunasilan (2019)	"DT—a new concept to strengthen organizations by enabling collaborative digital entrepreneurial ventures. Entrepreneurship as a driver of the digital transformation defined as 'Technopreneurship'" [9]	Strategic

2.2 Review of the Conceptual Frameworks or Theories on Digital Transformation

This section analyzes the relevant theories that have been developed by different scholars on digital transformation and their application within the organizational context [4]. Some of these theories are relevant and empirically tested, whereas some others lack the applicability of the theory within companies.

The frameworks that looked at digital transformation from a technological and strategic perspective include the "Six Keys to Success" framework proposed by Kavadia et al. (2016), "The Digital Reinvention" framework by Berman et al. (2016), the "Digital Transformation Model" proposed by Matt et al. (2015), the "Technology Acceptance Model (TAM)" proposed in 1989 by Davis, and the "Unified Theory of Acceptance and Use of Technology (UTAUT) model" proposed by Venkatesh et al. (2003). Table 3 summarizes the findings of the relevant frameworks discussed in digital transformation and presents a comparison of the selected frameworks.

The above review of the theoretical frameworks, whether their approach is academic or business oriented, helps in understanding the parameters used to support organizations during their digital transformation journey [4]. The "Six keys to Success" and the "Digital Reinvention" frameworks highlighted the dimensions that are key to success for organizations going through a digital transformation. However, these frameworks present some limitations as they could not link these dimensions to the performance of organizations. Focused on the general business model they failed to link the technology development with the organization itself. Therefore, they are considered as business tools for organizations to understand "how to" transform, but not complete in terms of achieving digital business transformation.

TAM and UTAUT frameworks looked at the technology adoption from a narrow point of view: the "personal" rather than the overall point of view. Therefore, they lack the depth to analyze how different organizations are transformed by digitalization. This confirms that digital transformation represents rather an organizational strategy, and not only an intention and behavior analysis aiming at technology adoption. Despite the above-mentioned limitations, most frameworks capture that people's technology acceptance and leadership are key factors to enable a successful digital transformation within organizations.

Table 3 Comparison of the relevant conceptual and theoretical frameworks on DT

Conceptual/theoretical framework	Brief of the conceptual/theoretical framework	Origin of the framework (business/academic)	Critical analysis/views/limitations	Scientific validation of the framework?
Six Keys to Success (Kavadia et al. (2016))	Business model that links technology and markets. Six dimensions are keys to success for an organization to implement a successful digital transformation: product/service offering, closed-loop process, asset sharing, usage-based pricing, a collaborative ecosystem, and an adaptive organization	Business	Failed to consider the managerial role for the successful implementation of DT, leadership dimension is neglected	No
Digital Reinvention Framework by Berman et al. (2016)	Organizations' strategic focus should integrate new expertise and new methods of work processes in order to achieve a successful implementation of DT within the organization	Business	The framework could be useful; however, the limitation is that it is purely theoretical and not tested within organizations	No
Digital Transformation Framework by Matt et al. (2015)	A digital transformation strategy should consider four dimensions to digitally transform an organization: use of technologies, changes in value creation, structured change, and financial aspect	Academic	Not very detailed as it lacks the central concept that integrates all the mentioned dimensions. Need to validate through more research	No

(continued)

Table 3 (continued)

Conceptual/theoretical framework	Brief of the conceptual/theoretical framework	Origin of the framework (business/academic)	Critical analysis/views/limitations	Scientific validation of the framework?
Technology Acceptance Model (TAM) by Davis (1989)	The model seems to predict the user intention and behavior towards adoption and use of technologies	Academic	There is no strategic direction that explains the adoption of technology from an organizational point of view, rather it focuses on a "person" point of view	Yes
The Unified Theory of Acceptance and Use of Technology (UTAUT) by Venkatesh et al. (2003)	UTAUT is an integration of previous theories (TAM, theory of planned behavior TPB, IDT innovation diffusion theory, social cognitive theory, and motivation model) that aims to explain user intention and user behavior. The four dimensions of the UTAUT are: effort expectancy, social influence, performance expectancy, and facilitating conditions	Academic	UTAUT correlates the social, economic, and environmental (sustainability) factors with the behavioral change of business. It suggests that businesses are susceptible to adopt and accept technology if these factors are developed	Yes

The result of this review of frameworks suggests that future research should investigate on DT by combining business and academic frameworks to achieve clear empirical evidence concerning the impact of DT on organizations.

2.3 Definition of Sustainable Development

The United Nations defined sustainable development (SD) as "development that meets the needs of the present without compromising the ability of future generations to meet their own needs" [11]. The idea behind this definition is to develop while preserving resources from depleting in the future, which explains the difference between being environmentally friendly and being sustainable. In 2015, the member countries of the UN nations approved a new plan for sustainability that should yield results by 2030. They decided on 17 goals that promote sustainability development to ensure a better and safer world for the people and planet.

Sustainability was argued in research as a source for competitive advantage [12]. The ultimate goal of SD is the long-term stability of the economy and environment; and this is only attainable through the integration and acknowledgment of economic, environmental, and social concerns throughout the decision-making process [13, 14]. The aim of the sustainable development goals is to ensure a good equilibrium between needs of the economy, environmental, and social aspects, which allows to preserve resources for the present and upcoming generations. Avoiding the over consumption of natural resources while ensuring not to exploit these resources is key to achieving a healthy and balanced environment that is as well socially efficient [15]. These goals, set by the UN and agreed upon by 193 countries, represent the starting point of a transforming process into a longer sustainable world to fight poverty and attain sustainable development within the next 15 years [16].

The three elements that are essential to the sustainable development goals can be classified as environmental, economic, and social elements. These elements may introduce complications in balancing their objective and success. The question is which aspect will take precedence and if two are conflicting, which of these elements can be successfully measured in terms of success or failure [14]. For example, an economical goal can be moving energy to non-polluting sources, which may contradict social goals as the poor will probably not be able to afford it and the burden on them will increase. Therefore, the goals that contribute efficiently to society, economy, and environment should complement and not contradict each other to reach and build sustainability in the world. Researchers stressed that for all the sustainable goals to be achieved equally and fairly, global governance systems must be designed for the effective integration of all three elements [13]. This supports the argument that digital transformation can play a major important role in achieving the implementation of sustainable development strategy. On a similar note, other research defined sustainability as a choice of selection, rather a mindset to respect the overall environment and the social goals [16]. This further supports the idea that the three elements—economic, environmental, and social—are the fundamental pillars of sustainability. On the other hand, other research argued that production and consumption (economic theory) need to transform in order to achieve sustainability [17]. Therefore, sustainability is understood as the accommodation of the economic needs, in addition to the social, and to the environmental requirements of the present time and of the future generations.

Literature had studied the relationship between sustainability and financial performance [18]. Pyka confirmed that sustainability strategy served as a support factor enhancing the effect of managerial capability on financial performance. This positive correlation confirmed that employing or adopting a digital business strategy will enhance the financial performance of the organization. Sustainability has become an important part of business operations. On another note, further research emphasized that there is a lack of attention towards the implementation of sustainability strategies within the overall business strategy of an organization [19]. Ukko et al. highlighted the factors required to ensure a successful implementation of an efficient corporate sustainability strategy. This includes having a proper organizational structure that is driven by a strong leadership. These new strategies should take into consideration cost reduction and optimization of processes by improving efficiency. Therefore, another definition that was introduced by Ukko et al. for sustainability is the fact that it is a strategy that needs to be developed by integrating the sustainable development principles mentioned in literature into business operations [19].

Moreover, the relation between sustainability and organizations was discussed by Engert and Baumgartner in the context of organizational strategies [20]. Organizations should shift their thinking from a competitive strategy to incorporate a sustainable strategy. Competitive strategies within organization definitely yield profit to shareholders, however sustainable strategies yield gains to stakeholders by increasing partnerships to improve competitiveness. Sustainability within organizations is an important topic that consist in finding a balance between generating profits while ensuring the safety of the other people and the planet. Therefore, for organizations to do so and become sustainable thinkers, they must undergo a shift in their economic or financial mindset and evolve it into a more social and economically sustainable mindset. This shift does not happen over a day or so, this is part of a continuous and long-term process of change, led by managers and leaders inside the organization and aimed at ensuring the adoption and implementation of sustainable strategies. The authors' research emphasized the conflict or contradiction that organizations face nowadays between the financial-economical view and the sustainable view [20]. They further highlighted the importance of educating leaders and managers about sustainability. Managers or leaders need to be aware that they are accountable for making sound decisions beneficial to all stakeholders. Sustainable strategies within organizations are not only based on the management approach as an economical practice but rather on a social approach (culture and people).

2.4 *Definition of Corporate Social Responsibility (CSR) Within Organizations*

When we refer to a "corporate entity," we are simply pointing to an organization or corporation which exist to generate profit and maximize this profit. However, there is an increased pressure nowadays on organizations to act as a good citizen to the society and environment [20]. The legitimization of achieving corporate success has been depicted in a lot of research as possible if organizations adopt Corporate Social Responsibility (CSR) within its organization's strategy. Therefore, the concept of Kumar of CSR was originated from the concept of organizations leaving an impact of social, economic, and environmental nature [1]. Orbik and Zozulakova stated that the problem of CSR was apparent in the 1980s where organizations in the US used monopoly practices which led European companies to lose business. This is when CSR started emerging to protect organizations from the exploitation of their practices, that can cause an economic and social cost.

Several research studies highlighted the positive impact of introducing CSR strategies and practices within organizations. Researchers addressed the necessity of CSR and digital transformation as main factors to achieve organizational competitiveness [1], including the improvement of financial performance for the organization, a better reputation for the business, improved employee motivation, enrichment of brand image, decreasing operating costs, and increased sustainability for the environment and society. This helps to differentiate an organization from its competitors. These CSR-driven benefits favor the goodwill of the organizations as well as that of our society as a whole.

On the same note, other research highlighted that organizations should act in an ethical and sensible way towards society and environment and that this is not a differentiation strategy within their organization, it is rather a normal strategy to obtain legitimacy [21]. Kumar examined the concept of Corporate Social Responsibility (CSR) as a strategic business decision that maintains a balance between the social, environmental, and economic benefits of stakeholders and shareholders. CSR was always understood as beneficial from a social point of view by doing good deeds and sponsorships to the society. It was rather a philanthropic concept. However, a proper CSR strategy should be mirroring the organizational overall strategy. It is about realizing a competitive advantage in the society that can enhance sales while cutting operational costs by improving the supply-chain both economically and socially. CSR is described as a responsibility of the shareholders and directors of the organization onwards employees, shareholders, stakeholders, and also the environment and society [21]. In 2005, the concept of CSR became more prevalent towards the concept of business. It stated the major six principles that are supposed to guide organizations in doing business through legitimate and fair practices to promote governance and ethics. Adhering to these six principles would certainly constitute the social responsibility of a company. Briefly, these six principles of CSR are listed below as follows:

1. "Businesses should have a proper conduct in line with Ethics, Transparency and Accountability."
2. "Businesses should provide goods and services that are safe and contribute to sustainability throughout their life cycle."
3. "Businesses should protect the well-being of all employees."
4. "Businesses should respect and promote human rights."
5. "Businesses should respect, protect, and make efforts to restore the environment."
6. "Businesses, when engaged in influencing public and regulatory policy, should do so in a responsible manner."

Organizations should evolve their strategy and undergo a redesign of their processes in order to re-align and adopt a sustainable culture [21]. Palma et al. pointed out that managers play a fundamental role in integrating a sustainable approach in the organization's strategy, in order to make it ecologically and socially acceptable by society. It is imperative to keep the socio-economic environment embedded within the organizational strategy [9]. The authors stressed out the importance of a proper implementation of corporate social responsibility strategies within organizations to survive the worldwide competitive struggle. They analyzed an effective approach to assess the competitive advantage of the organization by considering the practices of the organization in building its competitive strategy. These competitive indicators are the People, Profits, Planet, and Partnership. Their research confirmed that the integration of the corporate social practices of the organization in the development of sustainable strategies is essential.

Moreover, a study conducted in Bangladesh on the entrepreneur's personal social responsibility (PSR) has confirmed that the entrepreneurs' PSR has a positive effect on the longevity and credibility of the CSR of their organizations [22]. The important element in establishing a reputable CSR lies in the intangible asset of organizations which is their people or what is called the human capital [22]. Basically, if a person's behavior is based on good values and ethics and the person tends to take responsibility on a personal level towards society, this will definitely reflect positively on and improve the quality of CSR. In other words, personal social responsibility has a positive effect on CSR.

3 Digital Transformation and Its Contribution to Corporate Social Responsibility as a Sustainability Strategy Within Organizations

Sustainability concerns have forced organizations to reconsider their strategies to incorporate the social well-being of the society and the organization. In order for organizations to implement effective sustainability-focused strategies, they have to consider a process of change. This is not about product change or change of a specific service that an organization offers; it is about creating new strategies and a full transformation of the organization's business process.

Research emphasized that managers should benefit from the digital transformation by utilizing technology to evolve their internal processes and value propositions [15]. Emas focused on the implementation of digital transformation within the overall organization that also sustains their responsibility to remain socially and economically efficient towards the society. Three blocks of digital transformation including an evolvement of the organization's customer experience, operational processes, and its overall business model have stressed the importance of leadership within this transformation to accomplish an effective CSR strategy.

A lot of organizations are utilizing IT capabilities and systems to enhance customer experience by the understanding of customers and by building effective cost structures that positively affect society. For example, restaurants adjust their price structures to fit the demand of people due to weather or inventory or proximity to closing times. Other examples include insurance companies and banks that redeveloped their cost structures to be more socially acceptable. All this is possible thanks to the IT-driven business analytics they use to understand their customers better. This leads to CSR and sustainability by opening to radical reinventions in seeking new sustainable sources of income. These adjustments or the concept of building a new reinvention of the fundamental business can yield real benefits and represents an important phase to drive organizations into digital transformation.

Digital transformation or reinvention does not refer to taking an existing product and digitalizing it by introducing an e-commerce site or digitizing a customer experience [13]. A new concept or approach for organizations as a tool to support them in developing their economic, social, and environmental competitiveness in order to achieve their CSR strategies has been discussed in literature [21]. With the help of IT systems, organizations should develop specialized CSR teams to implement CSR strategies that formulate policies for their social corporate programs. These programs have clear objectives that run with the purpose of achieving social well-being and are aligned with the objectives and strategies of the overall business. The programs are led by decision makers and employees that can make this process effective within the organization [21]. In other words, these CSR programs can be linked with the development of communities, aligning with educational or health projects in the community, thus contributing to improve societies and the overall environment.

Other researchers drew some similarities and differences between the concept of business processes reengineering (BPR), which complements CSR strategies, and digital transformation [2]. Their research reconfirmed the contribution of what digital transformation can add to the development of CSR or sustainable development strategies of organizations. Their research designed a roadmap as an approach to the DT of business models. The last stage, which they called digital implementation, concerns the reengineering of an organization's strategy to redesign the digital customer experience and develop a digital value-creation network by integrating partners [2]. They proved that DT refers to a continual process and not just to technology enablement, which confirms research addressing DT as a tool to achieve CSR as a sustainability strategy within organizations [2, 21].

Other research looked into the achievement of the Sustainable Development Goals (SDGs) on a local level and how DT can affect the achievement of these goals [22]. The commitment of world leaders towards the adoption of such goals in 2015 put pressure and challenged every country to localize the SDGs. Considering these national and global efforts towards sustainable development, including the process of localizing these efforts, technology and innovation play an important role to achieve or facilitate sustainability [10]. ElMassah and Mohieldin examined several case studies of DT application in 7 countries, considering also the progress towards localizing the SDGs. Their findings confirmed that digital transformation improves the process of localizing SDGs. Big Data and e-governance can facilitate the effective implementation of sustainable development through the localization of the SDGs [10].

On the same note, similar research studied the impact of digitalization on the development of sustainability in societies and confirmed that the level of digitalization can improve the sustainable economic development on a country level [12]. Porter et al. confirmed that as the digitalization level increases, the economic development level also increases through higher competitiveness, innovativeness, and entrepreneurial activities. This is reflected in a higher GDP in the more digitalized countries. Additionally, their research concluded that the social aspect of the country is also positively influenced by the digitalization. However, the ecological environment is neglected showing a negative impact of digitalization. The authors explained that the cultural dimensions or other differences between nations may influence the process of digitalization.

Digital transformation does not only base on a technical capability, it rather depends on a managerial and operational capability that leads to a successful sustainable development in the organization [18]. This is because digital transformation implies a redesign of business processes taking into consideration the sustainability strategy, and this requires not only technical capabilities but also experienced managerial and organizational capabilities to succeed. Research showed that sustainable development is positively correlated with financial performance. Therefore, this is a confirmation that digital transformation supports and improves the sustainability development strategies of organizations by means of capable managers leading the transformation and redesigning business processes. However, research that empirically analyzes the effect of DT on sustainability within organizations is still poor [18].

Therefore, in order to achieve a successful digital transformation within organizations, a sustainable CSR strategy coupled with an effective leadership able to align the organizational strategy with a sustainable approach is needed. Corporate Leadership seems to be a critical component for creating a sustaining value in organizations. Table 4 summarizes the role of DT in CSR to ensure sustainability development within organizations along with the essential components needed to drive this transformation effectively.

Table 4 Summary of researchers' perspectives on the role of digital transformation in achieving CSR

Researchers	Perspectives on DT and CSR	Success drivers
Westerman et al. (2011)	CSR is imperative to reach a successful DT. This is done by evolving customer experience, operational processes, and the overall business model [7]	Leadership Effective CSR aligned strategy
Mazzone (2014)	DT is a new way to re-think, reinvent, redesign the organization in order to transform [5]	Strategic re-alignment/reprocess/redesign/CSR strategy
Schallmo et al. (2017)	DT is a continual process. Business model innovation is required to set a roadmap for successful digital transformation [2]	Roadmap of DT for business models: digital reality (value-added analysis of customer requirements), digital ambition, digital potential (enablers + best practices), digital fit (strategies, CSR, customer orientation), and digital implementation (redesign of business model)
Kumar (2017)	Developing CSR as a strategy by utilizing the technology systems [21]	CSR as a strategic responsibility for DT/leadership/personal corporate responsibility
Ukko et al. (2019)	Digital transformation is not a technical capability but rather a managerial and operational capability that leads to sustainability performance within the organization [18]	Sustainability strategy positively correlated to organizational financial performance/strong leadership
Gunasilan (2019)	DT is linked with digital entrepreneurship referred to as "Technopreneurship" [9]	Internal organization capabilities, innovation partnerships
ElMassah (2020)	Successful DT contributes to localization of sustainable development goals in the economy [22]	Organizations should build their own sustainable strategies and goals to contribute to the overall transformation

4 Conclusion

The review of the various conceptual and theoretical frameworks relevant to digital transformation within organizations has depicted a lack of adequate research towards the implementation of sustainability strategy within the organization's business strategy, or what is referred to as the Corporate Social Responsibility (CSR) and its contribution to digital transformation. The theoretical frameworks of TAM and UTAUT have addressed the technology adoption and user behavior intention towards technology from a "person" point of view and not from an "organizational" point of view. Technology has developed rapidly in today's environment where industries and organizations are highly impacted by this transformation. Countries around the world as well as major industries have noticed that their natural resources have been exploited in order to keep up with the rapid economic growth that is happening all around the world. Therefore, organizations should benefit from the digital transformation by utilizing technology as a strategy to evolve their internal processes and enhance their value propositions in order to become more sustainable socially and economically.

This paper analyzed the business and theoretical frameworks and looked into the existing academic literature to look for a logic and introduce a research-based analysis to find a proper fit of corporate social and digital transformation. CSR and digital transformation are relatively new concepts that are reshaping the relationship between technology and organizational strategy. Corporate Digital Responsibility (CDR) is the new approach of organizations that reinforces the corporate social responsibility by adapting to the impact of technological and digital transformation. DT can serve as a framework for understanding how the digital reinvention or evolution can also evolve towards a sustainable development. It impacts the sustainability aspects of doing business. DT is the primary step in helping leaders in the economy and in the organizations achieve a better digital sustainable world with un-depleting resources for the future.

Today everything is dependent on the capabilities of technology. In business nowadays, the major revolution is to switch from traditional to digital business models to maximize competitiveness. Digital transformation is one of the important developments in the business world. Its contribution to organizations for achieving a successful and sustainable development depends on the managerial capabilities and not only on the technical capabilities of organizations. It is important that the overall organizational strategies ensure proper alignment with its technical capability and includes the knowledge and competencies to integrate and redesign their business processes by taking into consideration the sustainability strategy or what we call the corporate social responsibility strategy of organizations. DT processes need experienced managerial and organizational capability to succeed.

Therefore, organizations should benefit from digital transformation if they evolve their strategy and transform their processes by showing CDR and implementing a Corporate Social Responsibility strategy that calls for a digital but sustainable culture. In order to accomplish that, managers and leaders in organizations play a fundamental role in transforming the organization into a sustainable entity. Leadership is an essential organizational trait for the success of such transformation, as it is critical for creating and sustaining the appropriate values within the organization. Thus, the success of DT depends not only on the IT capabilities, but also and especially on the right combination of other factors such as proper managerial and leadership skills as well as organizational thinking and commitment within the organization. For future research, it is important to explore the notion of corporate leadership and its relation to the success of digital transformation implementation. There is not sufficient academic studies linking DT and leadership. It is advisable to research these dimensions further to explore the relationship between digital transformation success and leadership effect on CSR within organizations. The limitation in this literature review is that most of the research studies looked at and concentrated only on qualitative studies and did not apply any qualitative analysis which could result in more significant information that is related to the topic. Therefore, for future research, we could know more about the digitalization transformation process and successful implementation to learn more in depth about successful strategies that organizations follow to encourage digital transformation implementation and know more about what kind of real experiences they have passed through to implement DT.

References

1. Orbik, Z., & Zozuľaková, V. (2019). Corporate social and digital responsibility. *Management Systems in Production Engineering, 27*, 79–83.
2. Schallmo, D., Williams, C., & Boardman, L. (2017). Digital transformation of business models—Best practice, enablers, and roadmap. *International Journal of Innovation Management, 21*, 1740014.
3. (2018). Digitalization and society's sustainable development—Measures and implications. *Zbornik radova Ekonomskog fakulteta u Rijeci: časopis za ekonomsku teoriju i praksu/Proceedings of Rijeka Faculty of Economics: Journal of Economics and Business, 36*.
4. Nwaiwu, F. (2018). Review and comparison of conceptual frameworks on digital business transformation. *Journal of Competitiveness, 10*, 86–100.
5. Mazzone, D. (2014). *Digital or death: Digital transformation—The only choice for business to survive smash and conquer* (1st ed.).
6. Arroyo Lazo, M. A. (2018). Schwab, Klaus. The fourth industrial revolution. Ginebra: World Economic Forum, 2016, 172 pp. *Economía, 41*, 194–197.
7. (2011). *Digital business transformation and strategy: What do we know so far?* https://www.researchgate.net/publication/322340970_Digital_Business_Transformation_and_Strategy_What_Do_We_Know_So_Far.
8. (2017). *Digital reinvention: Unlocking THE 'HOW'*. McKinsey. https://www.mckinsey.com/business-functions/mckinsey-digital/our-insights/digital-reinvention-unlocking-the-how.
9. Gunasilan, U. (2019). Entrepreneurship as a driver of the digital transformation. *International Review of Management and Marketing, 9*, 23–29.
10. ElMassah, S., & Mohieldin, M. (2020). Digital transformation and localizing the sustainable development goals (SDGs). *Ecological Economics, 169*, 106490.
11. (1987). *Report of the world commission on environment and development: Our common future.* https://digitallibrary.un.org/record/139811.
12. Porter, M. E., & van Linde, C. (1995). Toward a new conception of the environment-competitiveness relationship. *Journal of Economic Perspectives, 9*, 97–118.
13. Dernbach. (2003). Achieving sustainable development: The centrality and multiple facets of integrated decision making. *Indiana Journal of Global Legal Studies, 10*, 247.
14. Stoddart, H. (2011). A pocket guide to sustainable development governance. *Commonwealth Secretariat.*
15. Emas, R. (2015). *The concept of sustainable development: Definition and defining principles.* https://sustainabledevelopment.un.org/content/documents/5839GSDR%202015_SD_concept_definiton_rev.pdf.
16. Freyling, V. The importance of all sustainable development goals (SDG).
17. Harris, J. M. (2004). Global environmental challenges of the twenty-first century: Resources, consumption, and sustainable solutions. *Ecological Economics, 50*, 315–316.

18. Pyka, A. (2017). Dedicated innovation systems to support the transformation towards sustainability: Creating income opportunities and employment in the knowledge-based digital bioeconomy. *Journal of Open Innovation: Technology, Market, and Complexity, 3.*
19. Ukko, J., Nasiri, M., Saunila, M., & Rantala, T. (2019). Sustainability strategy as a moderator in the relationship between digital business strategy and financial performance. *Journal of Cleaner Production, 236*, 117626.
20. Engert, S., & Baumgartner, R. J. (2016). Corporate sustainability strategy—Bridging the gap between formulation and implementation. *Journal of Cleaner Production, 113*, 822–834.
21. Palma, L. C., Pedrozo, E. Á., & Alves, N. B. (2019). Sustentabilidade, organizações e formação de gestores: Uma pesquisa exploratória em cursos De administração no Rio Grande do Sul. *Revista de Administração da UFSM, 11*, 1324–1343.
22. Kumar, S. (2017). Position and technology of corporate social responsibility in company management. *International Journal of Recent Research Aspects, 4*(3), 126–131.

The Role of Emerging Technologies in Shaping the Value Innovation Aptitude Toward Competitive Advantage and Sustainability

Mohammed A. Hajar, Daing Nasir Ibrahim, Mohd Ridzuan Darun, and Mohammed A. Al-Sharafi

Abstract Maintaining competitive advantage and long-term success has been very challenging for firms, particularly with the current highly competitive business environment. The scale for successful innovation has been more complicated and directly linked with the customers' value proposition. In this context, companies must adapt to a dynamic process of value innovation to open up new market spaces, make the competition irrelevant, and create valuable intangible resources, including customer satisfaction and loyalty. The main purpose of this paper is to provide a new perspective to achieve competitive advantage and Sustainability through the logic of value innovation and within the context of the Resource-Based View (RBV) of the firm. This paper aims to spotlight the role of emerging technologies in shaping the new business strategy of market transformation to value-creating intangible assets instead of physical, tangible ones. The study presents three test cases of strategic moves of value innovation in digital companies. It shows the significance of emerging technologies in creating intangible resources and competencies and their role in achieving competitive advantage and Sustainability. Thus, this study encourages firms to focus on value innovation and suggests a conceptual framework to achieve competitive advantage and Sustainability by promoting customer satisfaction and loyalty.

Keywords Emerging technologies · Value innovation · Intangible resources · Competitive advantage · Sustainability

M. A. Hajar (✉) · D. N. Ibrahim · M. R. Darun
Faculty of Industrial Management, Universiti Malaysia Pahang, Lebuhraya Tun Razak, 26300 Kuantan, Pahang, Malaysia
e-mail: eng.mohammed.hajar@gmail.com

M. A. Al-Sharafi
Department of Business Analytics, Sunway University, 47500 Bandar Sunway, Selangor, Malaysia
e-mail: mohamed.a.alsharafi@gmail.com

C. Dal Zotto et al. (eds.), *Smart Technologies for Organizations*, Lecture Notes in Information Systems and Organisation 60, https://doi.org/10.1007/978-3-031-24775-0_9

1 Introduction

The nature of innovation itself is getting more sophisticated as the market becomes more mature and global. Enterprises keep shifting their innovation strategies to improve their competitive position and adapt to the market's rapid dynamic changes. The rapid evolution in the business environment helped in developing innovative knowledge encompassing a wide range of domains and perspectives. Since the 1930s, when Schumpeter defined innovation as a source of economic and technological change, the managing aspects of generating, recognizing, and implementing successful innovation processes are getting prevalent in strategic management literature [1–3]. Although Schumpeterian focused on technological innovation, innovation management remains the buzzword that indicates the transformation of new ideas and business opportunities outputs into values and, thus, market success [4–6].

Value innovation concentrates on providing a breakthrough in value for customers and firms by enabling business differentiation, reducing the relevance of competition, and evolving new uncontested market space [7–9]. The logic of value innovation tends to drive organizations to break out the bloody value-cost trade-off competition by focusing on the creation of a quantum leap in value instead of scattering resources and capabilities trying to beat the existing competition [7]. Therefore, firms have started shifting their strategies toward intangible breakthrough value innovation to provide customers' desired value and achieve high profitable growth [4, 10, 11]. For instance, companies like Facebook, Uber, YouTube, Amazon, and Airbnb, with the help of the digital revolution and merging technologies like the Internet of things (IoT), big data analytics, artificial intelligence (AI), blockchain, and machine learning, managed to compete on superior value delivery to the customer as a basis for competitive advantage and profitable growth. By applying a value innovation logic they made the competition irrelevant and created uncontested market spaces and new industries [12, 13].

Emerging technologies are science-based innovations that have the potential to create a new industry or transform an existing one [11, 14]. They significantly affect a firm's strategies and performance by shifting value chains, digitization of goods, and the locus of innovation. In this context, the emerging technologies associated with innovation strategies represent the progressive development of competitive advantage [15, 16].

This paper intends to shed light on the logic of value innovation and its significance in achieving Sustainability. Particularly, this study links the value innovation notion of the Blue Ocean Strategy and the Resource-Based View (RBV) theory of the firm with the emerging technologies perspective to attain superior competitive advantage and Sustainability. In a way, this paper encourages organizations to transform their business strategies from the physical, tangible assets-focused type of innovation to value breakthrough intangible assets-focused innovation to create a quantum leap in value, improve customer value, enhance customer satisfaction, and promote customer loyalty. This should be achieved by utilizing emerging technologies, ultimately resulting in superior competitive advantage and Sustainability. Thus, the paper answers the research question, "How can emerging technologies shape value innovation to create a competitive advantage and achieve sustainability?". In this context, this paper employs a qualitative research approach to discuss three value innovation case studies in digital companies that use technology to create new value in business models, customer experiences, and the internal capabilities that support its core operations. Furthermore, the analytical case studies examined the firms' behavior in enabling customer value by focusing on intangible resources and capabilities. Moreover, this study contributes to the strategic management literature by suggesting a conceptual framework to enhance customer satisfaction and promote customer loyalty as intangible resources to achieve a competitive advantage and hence, Sustainability.

The remaining part of this article is organized in the following way: a review of the related literature is presented in Sect. 2, focusing on value innovation, RBV, and value innovation within the RBV context. Then, Sect. 3 presents the value innovation-related case studies, while Sect. 4 discusses the paper outcomes and suggests a conceptual framework to achieve business sustainability through the value innovation approach. Finally, the implications of the paper are presented in Sect. 5, while the conclusions, hindrances encountered during the study, and recommendations for future studies are presented in Sect. 6.

2 Related Literature

2.1 Value Innovation

Value innovation (VI) refers to "the parallel drive towards differentiation and low cost to create a quantum leap in value for both customers and the firm" [17]. The logic of VI was conceptualized by Chan and Renée based on the ex-post study of over 150 cases from 30 industries and the business lunch of about 100 companies worldwide [17, 18]. According to Kim and Mauborgne [7], VI does not necessarily occur in developing new technology or being first to market rather than being able to link innovation to value.

Value innovation concentrates on providing a breakthrough in value for customers and firms through enabling business differentiation, reducing the relevance of the competition, and evolving new uncongested marketspace [7, 17, 19]. The VI logic goes beyond just an innovation; it is an approach that covers the whole activities of the firm to break out the value-cost trade-off by reshaping an industry's competitive capability to create new offerings while eradicating and minimizing the factors of competitive interest to the industry [17, 20–22]. Different from the competitive strategic approach, which strives toward increasing the market share in an already existing competitive market space, the VI strategy drives toward creating a new uncongested marketspace with novel demands and better profitability [7].

The concept of VI has been enriched with the Blue Ocean Strategy framework that comprises a set of effective analytical tools like "Strategy Canvas, Four Actions Framework, and Six Paths Framework" to define and reconstruct market competition factors and boundaries. This is to create new value innovation elements that offer an entirely new experience to customers [18, 21]. However, VI can only be achieved when innovators are able to align innovation with utility, price, and cost [21]. In addition, VI is a dynamic process in which companies need to continuously monitor their value curves for Sustainability and continuous growth [23].

Dillon et al. [24] identified two main drivers of VI in providing the most important customers with exceptional value, as well as increasing enterprise value; the sustained desire to meet customers' demand drives continuous improvement in the enterprise value. Mele [25], Mele et al. [26] presented VI from the perspective of integrating resources within the collection of actors operating within service-dominant logic. According to Mele et al. [26], VI enables higher value co-creation, which can be achieved through the integration of resources such as knowledge and skills within a network of firms. Moreover, Matthyssens et al. [27] viewed VI as a path to sustainable competitive advantage by re-conceptualizing the business model or the industry to usher in better and new customer value.

2.2 *The Resource-Based View (RBV)*

The RBV of a firm addresses the means of explaining competitive advantage. It provides the theoretical model to analyze and interpret firms' capabilities and resources for achieving sustainable competitive advantage and, in return, superior performance and growth [28–30]. The RBV focus goes beyond cost reduction or efficient resource utilization and also encompasses the "concept of difficult-to-imitate attributes of the firm" as the foundation to gain competitive advantage and better performance [30]. Thus, the RBV discusses the relationships that exist between resources, strategy, performance, as well as a competitive advantage and firm-level processes [31].

In the RBV context, firms' resources have been classified into tangible (TR) and intangible (IR) resources. The TR refers to the physical resources with a set of values that can be liquidated easily; this includes resources like cash, machinery, inventory, land, or buildings. On the other hand, IR is the long-term assets that do not exist in physical form and add value to the business; these include knowledge, skills, experiences, innovation, intellectual property, firm's reputation (customer satisfaction), brand name (customer loyalty), and organizational procedures. Despite the importance of all resources to determine firm success, RBV literature reveals the key role of IR as "conferring a sustainable competitive advantage and superior performance" [30, 32, 33].

As per Barney [34], "a firm can only gain a sustainable competitive advantage if its resources have specific attributes of being valuable, imperfectly substitutable, rare, and imperfectly imitable". Furthermore, Fahy [35] evaluated a "resource-based model of global sustainable competitive advantage" and suggested the need for resources to drive value creation and be robust against duplicative efforts in order to gain sustained competitive advantage. Wang and Lo [28], Clulow et al. [29] discussed value creation more broadly with respect to customer-focus perspective, explaining the firm's key intangible resources in creating customer value. Moreover, Khan et al. [33] empirically investigated the influence of investment in key intangible resources and capabilities in Pakistan and concluded the significant role of intangible resources and capabilities in enhancing sustainable competitive advantage and firm's performance.

2.3 Value Innovation Within RBV Context

Value innovation has been defined in both innovation and RBV literature as an essential factor for firm's long-term success. For instance, while Kim and Mauborgne [23] defined it as the cornerstone of any innovation process, Wang and Lo [28] introduced value innovation as the motive of customer-focused performance, providing an efficient and smooth flow of customer satisfaction experiences. As per Matthyssens et al. [27], "the only way to escape cut-throat competition and sustain competitive advantage is through launching new value concepts and continuously re-invent the way customer value is created and delivered".

In the RBV context, VI is the creativity in "identifying, developing, deploying, combining and coordinating the capabilities and resources of a firm to be inimitable, valuable, rare, and non-substitutable in order to develop sustainable competitive advantage and long-term superior performance" [30, 32]. A strong parallel has been observed between the RBV reasoning and the Blue Ocean Strategy when focusing on the VI attributes as a key for superior performance, competitive advantage, and sustainable growth. For instance, VI, in the customer-focused view of RBV, plays a key role in supporting, improving, and redefining the quality and price of the attributes that drive customer-related performance and enhance customer value, customer satisfaction and loyalty [28]. Similarly, the Blue Ocean Strategy focused on the utility, price, and cost position attributes and their alliance with innovation activities to achieve VI [21].

3 The Role of Emerging Technologies in Shaping Value Innovation

The strategic moves of value innovators have been remarkable in creating new marketspaces and industries since the early days. Lately, the digital revolution and the growing awareness of strategic management knowledge aided in rapidly transforming the global market [12, 13]. This has led to the appearance and high rise of new corporations like Facebook, Uber, YouTube, Amazon, Huawei, and Airbnb which managed to achieve business differentiation, make the competition irrelevant, and create new marketspaces and industries based on emerging technologies and intangible breakthroughs of value innovation logics. On the contrary, big companies like Nokia, Kodak, Blackberry, Xerox, and Yahoo failed to cope with the market transformation to offer new and valuable experience to customers. This section highlights three value innovation cases that successfully achieved competitive advantage and profitable growth by creating new and superior customer value. The case studies have been selected based on the strategic moves of successful digital companies that managed to adopt emerging technologies to create value innovation, and hence, competitive advantage and long-term success. The study employed analytical case studies methodology as to examine the firm's behavior toward enabling customer value by focusing on intangible resources and capabilities. Westgren and Zering [36] considered case studies a useful analytical tool for studying organizational behavior and business strategy of firms and industries in transition. Thus, three case studies were analyzed based on secondary data of annual reports, review articles, news articles, and companies' websites.

3.1 Facebook

Facebook was first launched in 2004 by Mark Zuckerberg and his university colleagues Dustin Moskovitz, Chris Hughes, and Eduardo Saverin as a platform to allow the creation and maintenance of social ties among university students [37]. Now, Facebook has grown to the status of one among the five largest companies in the world in terms of market value as per statista.com in 2018. According to Statista.com, by June 2021, Facebook is the 3rd most visited global website with daily average active users of 1.9 billion.

Facebook showed great success in establishing new market space by providing a quantum leap in value and offering an entirely new experience to customers compared to other social network websites like chats, blogs, and emails. The emerging technologies, like Internet of things (IoT), big data, and artificial intelligence of social network, assisted Facebook to break off the competition and create an intangible value innovation. According to Hart et al. [38], Facebook "plays an important role in the process of forming and managing social capital by improving self-esteem and low life satisfaction, as well as crystallizing relationships that may remain short-lived". In this context, Facebook provided an intangible breakthrough value innovation that enhanced customer satisfaction and loyalty by offering new user experience in terms of positive emotions. Sas et al. [39] empirically investigated Facebook success and suggested three drivers of positive emotional values that attract and retain Facebook users as "prevalent memorable experiences recorded between friends that engender feelings of connectedness and relive the good old times; participants' engagement in public performance which experiences entertainment; and positive events that made public and people start to derive additional emotional benefits".

Despite the presence of strong competitors like Myspace and Twitter, Facebook sustains superior competitive advantage through pursuing the dynamic and continuous process of value innovation. For instance, Facebook evolved to keep offering customers a leap in value beyond social networking. Thus, it becomes an efficient medium for teaching, news, company forums, discussion groups, communication channels, commercial purposes, online sales, and so forth. Furthermore, Facebook established powerful B2B businesses particularly in marketing by taking advantage of the huge number of users (2.38 billion), efficient data analysis competencies, and third-party applications to provide superior value for business partners.

3.2 *Uber*

As a multinational transportation network company, Uber Technologies provides a range of logistics and transportation services such as "ride service hailing, bicycle sharing, peer-to-peer ridesharing, and a food delivery system". Since its foundation in 2009, Uber has grown rapidly by providing unprecedented value in the taxi industry [40]. By 2019, Uber was the largest ride-hailing company with over $110 million globally, 69% passenger transport market share, and 25% food delivery market share (statista.com).

The innovative business model of Uber, which leveraged on the emerging technologies of machine learning and blockchain [41], managed to sustain a competitive advantage by re-conceptualizing the Taxi industry, and thus, changing the whole value network [42]. In other words, Uber managed to obtain a value innovation breakthrough for customers and itself by providing business differentiation and low-cost services based on the integration of intangible resources or dynamic capabilities. In fact, Uber succeeded without major financial resources and steely commitment. As the largest worldwide taxi company, Uber has no taxicab assets as the drivers are independent contractors using their own cars to drive on behalf of Uber.

The remarkable delivery of superior customer value, which Uber competes on, played a key role in attaining competitive advantage and profitable growth. According to Cramer and Krueger [40], "Four factors likely to contribute to the higher utilization rate of UberX drivers are: (i) Uber's more efficient driver-passenger matching technology; (ii) Uber's larger scale which supports faster matches; (iii) inefficient taxi regulations; and (iv) Uber's flexible labor supply model and surge pricing which more closely matches supply with demand throughout the day".

3.3 *Airbnb*

Airbnb is an "online marketplace and hospitality service brokerage company founded in 2008 which currently provides access to more than 6 million unique places in its accommodation marketplace in approximately 100,000 cities and 191 countries around the globe" (Airbnb.com). Like Uber, Airbnb witnessed tremendous growth through enabling business differentiation and low-cost services by using intangible resources integration and a peer-to-peer business model to offer a brand-new experience to customers.

With the efficient use of emerging technologies like cloud computing, big data, and Geographic Information System (GIS), Airbnb was able to achieve value innovation through reconstructing the industry's boundaries, tapping uncongested marketspace and offering alternative benefits to customers, such as "cost-savings, household amenities, and more authentic local experiences". According to Guttentag et al. [43], tourists were highly attracted to Airbnb owing to its "experiential and practical attributes, and the motivating factors of interaction, home benefits, novelty, sharing an economy ethos, and local authenticity".

Furthermore, Airbnb invested in creating intangible resources to sustain competitive advantage, including trust, customer satisfaction, and loyalty. Ert et al. [44] marketed trustworthiness and reputation as imperative conditions for Airbnb transactions which were achieved through online review-score and previous experience.

4 Discussion

Despite the rapidly growing technological evolution and its enormous impact on market transformation, customer value is still the keyword for a firm's success and survival. Technological innovation without value tends to be a technology-driven or market pioneering that customers may not accept, thus turning into a failure like Motorola Iridium for example [21]. In this context, value innovation relates to providing the most important customer with exceptional value and increasing the value of the enterprise, whereas the sustained desire to meet customers' satisfaction drives a continuous increase in the value of the enterprise [24]. Thus, innovation has become a creative tool for identifying, developing, deploying, combining, and coordinating a firm's resources and capabilities, including technology, skills, and experience to create exceptional value, and thus, developing sustainable competitive advantage and long-term superior performance. As discussed in the previous cases, value innovation did not occur due to technological breakthrough in using Internet or mobile applications, which can simply be designed by fresh graduates or a freelance team. It rather occurred thanks to highly creative strategical thinking that allowed to utilize the intangible resources of the available Internet technology and smartphones aligned with skills and knowledge to provide unprecedented customer value.

The Innovation and RBV literature intensively highlighted the importance of customer satisfaction and loyalty to achieve superior performance and sustainable competitive advantage [28, 29, 32, 33, 45, 46]. Wang and Lo [28], Clulow et al. [29], Kamasak [32], Khan et al. [33], Simon and Luc [45], Bellingkrodt and Wallenburg [46] addressed the influence of customer-focused and customer-oriented business principles on customer satisfaction and concluded that both Innovation and RBV have a strong positive effect on customer satisfaction and loyalty, and hence, competitive

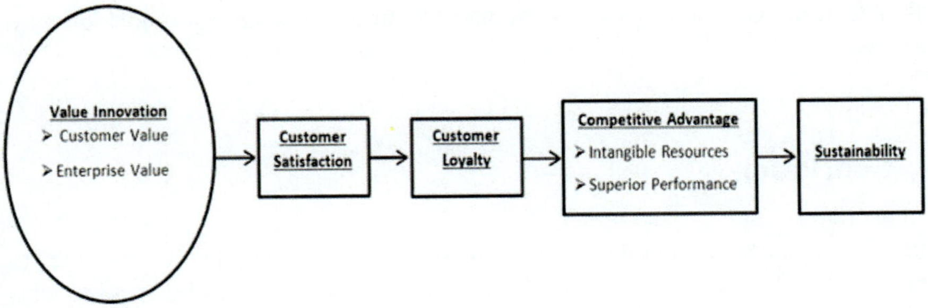

Fig. 1 Value innovation proposed conceptual framework

advantage and Sustainability. Therefore, value innovation helps to deliver superior customer value that leads to higher customer satisfaction, increased sales and productivity, and greater customer loyalty, whereas more innovation and development activities and higher employees' satisfaction ultimately result in sustainable competitive advantage [28, 32, 33, 46]. This study proposes a conceptual framework to approach Sustainability through the development of value innovation that enhances customer satisfaction, promotes customer loyalty, and creates a competitive advantage (Fig. 1).

5 Implications

The business environment is getting more challenging, especially with globalization helping to share innovation knowledge, secure resources, and increase the dominancy of large international companies. Simple innovation is no longer enough to maintain profitability and sustain a competitive advantage. The rising power of new ambitious players in the global economy, such as China and India, with surplus financial and technological resources, broke out the wont of monopoly or long-term industrial property dominance. Thus, the dynamic business differentiation, achieved by providing superior customer value that is able to build up reputation (customer satisfaction) and brand loyalty (customer loyalty), is imperative for a firm's success and long-term survival. It is highly recommended for entrepreneurs to pursue the logic of value innovation to focus on creating a leap in value and making the competition irrelevant rather than focusing on beating the existing competition. Besides that, innovators are required to pay more attention to the significance of emerging technologies to create intangible resources and competencies, and their key role in achieving competitive advantage and Sustainability.

Accordingly, the significance of the concept of value innovation goes beyond providing a quantum leap in value, but also plays a major role in creating intangible resources, like knowledge, skills, experience, customer satisfaction, and loyalty, which are critical for achieving competitive advantage and Sustainability. Therefore, the dynamic process of value innovation leads to the creation of more intangible valuable resources, and hence, sustainable competitive advantage and long-term success.

6 Conclusion

The rapid dynamic change in the business environment encouraged companies to improve their competitive positions by adopting more innovative business models in order to sustain and survive. The concept of business differentiation and low cost has become the concern of many organizations to gain a competitive advantage. This paper sheds light on the logic of value innovation and its significance to enhance customer satisfaction, promote customer loyalty, and thus, attain superior competitive advantage and achieve Sustainability. Nevertheless, this study encouraged innovators to focus on emerging technologies and intangible resources in establishing their value innovation strategies in order to improve their value and that of their customers, make the competition irrelevant and open new market space. The present paper adopted a qualitative method to investigate the strategic moves of some digital companies including Facebook, Uber Technologies, and Airbnb to enable customer value, break out the competition, and create uncongested marketspaces based on intangible resources and capabilities. In conclusion, the current paper proposed a conceptual framework to build competitive advantage and approach Sustainability by embracing value innovation strategies.

This paper has some limitations with regard to the measurements and analysis as no empirical data were collected to verify outcomes. Likewise, the conducted case studies relied on analyzing the strategic moves and were not supported by market analysis nor empirical analysis. On the contrary, future researchers are recommended to use empirical methods to further investigate value innovation and its influence on competitive advantage and Sustainability. Also, we encourage further researchers to conduct empirical studies to deeply investigate the factors enabling the success of innovative companies.

Acknowledgements We gratefully thank the Universiti Malaysia Pahang for their endless support that greatly assisted the research.

References

1. Schumpeter, J. (1934). *The theory of economic development*. Harvard University Press.
2. Schumpeter, J. (1939). *A theoretical, historical and statistical analysis of the capitalist process, business cycles*. McGraw-Hill.
3. Malerba, F., & McKelvey, M. (2020). Knowledge-intensive innovative entrepreneurship integrating Schumpeter, evolutionary economics, and innovation systems. *Small Business Economics, 54*(2), 503–522.

4. Lusch, R. F., & Nambisan, S. (2015). Service innovation: A service-dominant logic perspective. *MIS Quarterly, 39*(1).
5. Liao, S.-H., Fei, W.-C., & Liu, C.-T. (2008). Relationships between knowledge inertia, organizational learning and organization innovation. *Technovation, 28*(4), 183–195.
6. Tidd, J. (2001). Innovation management in context: Environment, organization and performance. *International Journal of Management Reviews, 3*(3), 169–183.
7. Kim, C., & Mauborgne, R. (1999). Strategy, value innovation, and the knowledge economy. *MIT Sloan Management Review, 40*(3), 41.
8. Hajar, M. A., Ibrahim, D. N., Ridzuan, M., & Al-Sharafi, M. A. (2020). Value innovation activities in the wireless telecommunications service sector: A case study of the Malaysian market. *Journal of Global Business Insights, 5*(1), 57–72. https://doi.org/10.5038/2640-6489.5.1.1125.
9. Hajar, M. A., et al. (2022). The effect of value innovation in the superior performance and sustainable growth of telecommunications sector: Mediation effect of customer satisfaction and loyalty. *Sustainability, 14*(10), 6342. https://www.mdpi.com/2071-1050/14/10/6342.
10. Hajar, M. A., Ibrahim, D. N., & Al-Sharafi, M. A. (2018). *Value innovation in the Malaysian telecommunications service industry: Case study*. Presented at the International Conference of Reliable Information and Communication Technology, Putrajaya, Malaysia.
11. Hajar, M. A., Ibrahim, D. N., Darun, M. R., & Al-Sharafi, M. A. (2020). Improving mobile value-added services (MVAS) in Yemen mobile companies. *International Journal of Information Technology and Language Studies, 4*(2).
12. Warner, K. S., & Wäger, M. (2019). Building dynamic capabilities for digital transformation: An ongoing process of strategic renewal. *Long Range Planning, 52*(3), 326–349.
13. Brynjolfsson, E., & McAfee, A. (2011). *Race against the machine: How the digital revolution is accelerating innovation, driving productivity, and irreversibly transforming employment and the economy*. Brynjolfsson and McAfee.
14. Srinivasan, R. (2008). Sources, characteristics and effects of emerging technologies: Research opportunities in innovation. *Industrial Marketing Management, 37*(6), 633–640.
15. Soares, O. D., da Cruz, A. M., Pereira, G. C., Soares, I. M., & Reis, A. J. (1997). *Innovation and technology—Strategies and policies*. Springer Science & Business Media.
16. Al-Sharafi, M. A., Arshah, R. A., Abu-Shanab, E. A., & Alajmi, Q. (2019). The effect of sustained use of cloud-based business services on organizations' performance: Evidence from SMEs in Malaysia. In *2019 5th International Conference on Information Management (ICIM)* (pp. 285–291). IEEE.
17. Kim, C., & Mauborgne, R. (1997). Value innovation: The strategic logic of high growth. *Harvard Business Review, 82*(7–8), 172–180.
18. Wubben, E. F., Düsseldorf, S., & Batterink, M. H. (2012). Finding uncontested markets for European fruit and vegetables through applying the Blue Ocean Strategy. *British Food Journal, 114*(2), 248–271.
19. Loosemore, M., & Richard, J. (2015). Valuing innovation in construction and infrastructure: Getting clients past a lowest price mentality. *Engineering, Construction and Architectural Management, 22*(1), 38–53.
20. Kim, C., & Mauborgne, R. (1999). Creating new market space. *Harvard Business Review, 77*(1), 83–93.
21. Kim, C., & Mauborgne, R. (2005). *Blue ocean strategy: How to create uncontested market space and make competition irrelevant*. Harvard Business School Press.
22. Hajar, M. A., Alkahtani, A. A., Ibrahim, D. N., Darun, M. R., Al-Sharafi, M. A., & Tiong, S. K. (2021). The approach of value innovation towards superior performance, competitive advantage, and sustainable growth: A systematic literature review. *Sustainability, 13*(18), 10131.
23. Kim, C., & Mauborgne, R. (2005). Value innovation: A leap into the blue ocean. *Journal of Business Strategy, 26*(4), 22–28.
24. Dillon, T. A., Lee, R. K., & Matheson, D. (2005). Value innovation: Passport to wealth creation. *Research-Technology Management, 48*(2), 22–36.

25. Mele, C. (2009). Value innovation in B2B: Learning, creativity, and the provision of solutions within Service-Dominant Logic. *Journal of Customer Behaviour, 8*(3), 199–220.
26. Mele, C., Russo Spena, T., & Colurcio, M. (2010). Co-creating value innovation through resource integration. *International Journal of Quality and Service Sciences, 2*(1), 60–78.
27. Matthyssens, P., Vandenbempt, K., & Berghman, L. (2006). Value innovation in business markets: Breaking the industry recipe. *Industrial Marketing Management, 35*(6), 751–761.
28. Wang, Y., & Lo, H.-P. (2004). Customer-focused performance and its key resource-based determinants: An integrated framework. *Competitiveness Review: An International Business Journal, 14*(1/2), 34–59.
29. Clulow, V., Barry, C., & Gerstman, J. (2007). The resource-based view and value: The customer-based view of the firm. *Journal of European Industrial Training, 31*(1), 19–35.
30. Madhani, P. M. (2009). Resource based view (RBV) of competitive advantages: Importance, issues and implications. *KHOJ Journal of Indian Management Research and Practices, 1*(2), 2–12.
31. Habbershon, T. G., & Williams, M. L. (1999). A resource-based framework for assessing the strategic advantages of family firms. *Family Business Review, 12*(1), 1–25.
32. Kamasak, R. (2015). Determinants of innovation performance: A resource-based study. *Procedia-Social and Behavioral Sciences, 195*, 1330–1337.
33. Khan, S. Z., Yang, Q., & Waheed, A. (2019). Investment in intangible resources and capabilities spurs sustainable competitive advantage and firm performance. *Corporate Social Responsibility and Environmental Management, 26*(2), 285–295.
34. Barney, J. (1991). Firm resources and sustained competitive advantage. *Journal of Management, 17*(1), 99–120. https://doi.org/10.1177/014920639101700108.
35. Fahy, J. (2002). A resource-based analysis of sustainable competitive advantage in a global environment. *International Business Review, 11*(1), 57–77.
36. Westgren, R., & Zering, K. (1998). Case study research methods for firm and market research. *Agribusiness: An International Journal, 14*(5), 415–423.
37. Ross, C., Orr, E. S., Sisic, M., Arseneault, J. M., Simmering, M. G., & Orr, R. R. (2009). Personality and motivations associated with Facebook use. *Computers in Human Behavior, 25*(2), 578–586.
38. Hart, J., Ridley, C., Taher, F., Sas, C., & Dix, A. (2008). Exploring the facebook experience: A new approach to usability. In *Proceedings of the 5th Nordic Conference on Human-Computer Interaction: Building Bridges* (pp. 471–474). ACM.
39. Sas, C., Dix, A., Hart, J., & Su, R. (2009). Dramaturgical capitalization of positive emotions: The answer for Facebook success?. In *Proceedings of the 23rd British HCI Group Annual Conference on People and Computers: Celebrating People and Technology* (pp. 120–129). British Computer Society.
40. Cramer, J., & Krueger, A. B. (2016). Disruptive change in the taxi business: The case of Uber. *American Economic Review, 106*(5), 177–182.
41. Bashir, M., Yousaf, A., & Verma, R. (2016). Disruptive business model innovation: How a tech firm is changing the traditional taxi service industry. *Indian Journal of Marketing, 46*(4), 49–59.
42. Antikainen, M., & Valkokari, K. (2016). A framework for sustainable circular business model innovation. *Technology Innovation Management Review, 6*(7).
43. Guttentag, D., Smith, S., Potwarka, L., & Havitz, M. (2018). Why tourists choose Airbnb: A motivation-based segmentation study. *Journal of Travel Research, 57*(3), 342–359.
44. Ert, E., Fleischer, A., & Magen, N. (2016). Trust and reputation in the sharing economy: The role of personal photos in Airbnb. *Tourism Management, 55*, 62–73.
45. Simon, A., & Luc, H. P. Y. (2012). Improving innovation and customer satisfaction through systems integration. *Industrial Management & Data Systems, 112*(7), 1026–1043.
46. Bellingkrodt, S., & Wallenburg, C. M. (2015). The role of customer relations for innovativeness and customer satisfaction: A comparison of service industries. *The International Journal of Logistics Management, 26*(2), 254–274.

Big Data Analytics: Toward Smarter Marketing Decisions in Value Creation

Inas Zein Eddine and Imed Ben Nasr

Abstract Drawing on the Big data and the marketing literature, this paper aims to provide an inclusive analysis of the Big Data (BD) and Big Data Analytics (BDA) methods, and attempts to draw attention to the remarkable contributions of BDA to marketing decisions, if linked with the building of a consolidated BD infrastructure. Primarily, at the customer level, BDA has managed to offer marketers an improved set of data-driven engagement, acquisition, and retention decisions. Additionally, at the market level, BDA has given them the capability of coming up with new product developments, as well as dynamic pricing and advertising plans that best suit their customers' needs and wants. Yet, toward value creation, the leveraging of BDA should be accompanied by the creation of an organized BD set-up that can efficiently and effectively combine physical, human, and technological efforts.

Keywords Big data · Big data analytics · Big data infrastructure · Marketing decisions · Value creation

1 Introduction

Being introduced as the oil of modern marketing, Big data Analytics (BDA) has pushed marketing researchers and practitioners to look deeply into new visualizations, and search for new understandings to book their places in the "Big data" era. Primarily, the marketing arena is undergoing critical transformations with the development of new digitalized marketing networks [1], in which the analytics brought by the emergence of Big Data (BD) is radically changing marketers' perspectives toward new tactics of customers' understandings [2]. In this vein, BDA is expected to dominate the contemporary marketing approaches, market evaluations, and customers'

I. Z. Eddine
Beirut Arab University, Beirut, Lebanon
e-mail: irz258@student.bau.edu.lb

I. B. Nasr (✉)
Excelia, La Rochelle, France
e-mail: bennasri@excelia-group.com

C. Dal Zotto et al. (eds.), *Smart Technologies for Organizations*, Lecture Notes in Information Systems and Organisation 60, https://doi.org/10.1007/978-3-031-24775-0_10

analyses [3]. Indeed, the emergence of BD has raised new discussions on the need for new policies and procedures to improve BD integration [4]. Accordingly, marketers look for new opportunities via BDA to enhance their decision-making processes, as there is an urgent need to link "Big data" with "Big impact" [2]. Nevertheless, the vision is still blurry regarding the association of specific data types with the suitable analytics' methods, and the framework needed for their appropriate implementation [5]. In addition, great expectations on BDA are faced with the alerts of considering it as a publicity leading to disappointments. In this context, and driven by the need to derive clearer customers' understandings from BD [6], marketers are heavily relying on the role of BDA as an enabler for data-driven value creation processes. Some researchers have argued that BDA is the "fourth paradigm of science" [7], a "new paradigm of knowledge assets" [8], or "the next frontier for innovation, competition, and productivity" [9]. Indeed, the potential contributions of BDA have been highlighted throughout the literature, from its impacts on the business environment [2, 10] to its role in decision-making processes and in advancing organizational functions [11]. However, the synthesized literature on BD and BDA prospects in marketing is limited, and research in this area has shown a critical gap in displaying the supporting value of BD to the marketing function in organizations [12]. Accordingly, in this paper, we extend the research by exploring and demonstrating the potentials of BDA in the marketing field; in essence, we expect to obtain an in-depth understanding of the contributions of BDA integration in marketing decisions at the customer and market levels, in association with a consolidated BD infrastructure.

Certainly, this research work contributes to the body of knowledge in various means. To the authors' best knowledge, this literature review is among the first to address the revolutionary role of BDA in the marketing field, in association with the building of BD infrastructure. Additionally, it brings researchers and practitioners a clarified view of the existing studies in this area, in which it sheds light on the importance for marketers and decision-makers to capitalize on BDA through their decision-making processes. Indeed, in this review, we argue that there is a need for a transformation in the marketing field by considering the effects of the BD era in understanding the dynamism of customers' needs and expectations.

The remainder of this paper is structured as follows. In the next section, an understanding of BD and BDA is displayed. In the third section, an overview of marketing fundamentals in the BD era, and the elements that form a strengthened BD infrastructure are presented. Then, in the fourth section, the contributions of BDA at the customer and market levels are revealed, to be then followed by the fifth section that highlights the conclusion.

2 The Understanding of Big Data and Big Data Analytics

2.1 The Definition of Big Data

Since the emergence of BD, its definitions have been masked by conceptual vagueness, failing to cover its wide range of concepts through specific explanations. The absence of a formal definition has directed research into unexplored and challenging roads [13]; however, this ambiguity has not prevented business researchers and practitioners in extending, or even disregarding prior definitions and suggesting new ones. Commonly, BD has been described implicitly through its characteristics, technological methodologies, success stories, and impacts on enterprises, business operations, and societies. In parallel, trials in explicitly describing BD have also failed, in which there has been no settlement on what common entity this term would be linked to. BD reflects a diversity of different entities encompassing, but not restricted to information assets, data collections, analytical procedures, storage expertise, and infrastructures. "Big Data refers to datasets whose size is beyond the capacity of typical database software tools to capture, store, manage, and analyze data." This definition comes as an illustration of the critical challenges that companies are confronted with, which nowadays make the dealing with huge amounts of data in traditional ways unfeasible [9].

In parallel, BD definitions have been categorized into three core groups [13], which are illustrated in Table 1. The first group of definitions covers BD characteristics, which in accordance with Laney's framework, discusses the three dimensions of data volume, velocity, and variety [14]. In this context, it is argued that, even if this framework has not managed to explicitly define BD [15], it well reflects BD's main characteristics and can be designated as the 3V's framework, representing the increasing amount of data by "Volume," the quick pace of data formation by "Velocity," and the diverse richness of data by "Variety." This framework has been later developed by other authors to incorporate two other "V"s: Value associated with data quality and Veracity associated with the need to be aware of data value [16]. Additionally, the second group of definitions is linked with the technological needs for data processing. Research has argued that the traditional analytical methods used cannot have the capacity in holding BD [17]. Similarly, it has been considered that "Big Data is less about data that is big than it is about a capacity to search, aggregate, and cross reference large data sets" [18]. The third group refers to the impact of BD. Here BD is seen as a cultural, technological, and scholarly phenomenon, where computation influence and algorithmic precision are exploited by effective technology interfaces, and large data sets are allocated through analysis, objectivity, and accuracy [18]. In this regard, BD economic impacts have been emphasized as follows "Big Data technologies describe a new generation of technologies and architectures, designed to extract an economic worth from very large volumes of a wide variety of data, by enabling high velocity capture, discovery and/or analysis" [19].

Table 1 Definitions of big data and main references

Different orientations of big data	Big data definition	References
Big data from a characteristics orientation	– 3v's model (volume, velocity, variety) – Value and veracity	[14, 15]
Big data from a technological needs orientation	– BD is less about data that is big than it is about a capacity to search, aggregate, and cross reference large data sets	[18]
Big data from an impact orientation	– BD is a cultural, technological, and scholarly phenomenon, where computation influence and algorithmic precision are exploited by effective technology interfaces, and large data sets are allocated through analysis, objectivity, and accuracy – BD technologies describe a new generation of technologies and architectures, designed to extract an economic worth from very large volumes of a wide variety of data, by enabling high velocity capture, discovery, and/or analysis	[18, 19]

2.2 *Big Data Analytics*

Many companies have the equipment and expertise to handle limited structured data, but when it comes to BD, they are stuck with capabilities creation processes to derive actionable intelligence on a timely basis. BD does not always fit into structured frameworks, it rather necessitates the replacement of traditional analytics by new types of data processing and interpretation approaches [20]. In this vein, and to capitalize on the opportunities that BD offers, BDA plays an important role in providing methods to gain better organizational and managerial perceptions through data management and analysis [21]. Indeed, BDA is defined as "a process of inspecting, differentiating and transforming big data, with the goal of identifying, mining and communicating useful information, suggesting conclusions, and helping to take accurate decisions" [22]. Thus, BDA is considered as a critical tool for enhancing productivity and revenues where it is associated with cost reductions [23]. In this manner, BDA has dramatically changed from a descriptive approach to a more inquisitive one, in which it encompasses four different types of analytics [22] as illustrated in Fig. 1 and more specifically described in the following section.

Types of Big Data Analytics

Descriptive analytics. This approach replies to companies' "what" question [24]. Indeed, as a straightforward statistical approach, descriptive analytics is deployed by nearly all companies that need to describe their database components and form

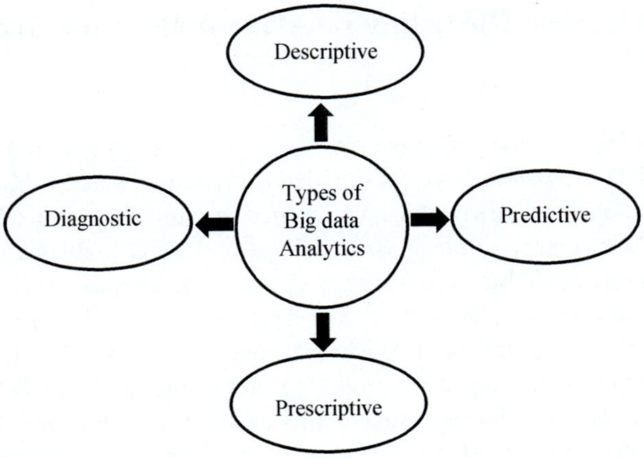

Fig. 1 Types of big data analytics

a variety of historical data arrays, including central tendency estimations, dispersion estimations, charts, sorting and sampling techniques, frequency and probability distributions.

Predictive analytics. Targeting improvements in products and services, predictive analytics entails the use of sophisticated software and research techniques for the detection of predictive variables in order to build predictive frameworks into a descriptive analysis. This kind of analysis provides a set of probabilities and solutions for the future based on careful studies of past and present data patterns. Indeed, predictive analytics is highly utilized in social media, consumer relationship data management, and sales lead scoring [25].

Prescriptive analytics. Despite the fact that some companies might not prefer such linear way of analysis, prescriptive analytics provides them with prescriptions by recommending the needed actions in real time and in a more focused way. Prescriptive analytics has the ability to shape and enhance the management of decision-making processes through a precise allocation of resources and authority, and clarified operational tactics [26].

Diagnostic analytics. Through collected data configurations, diagnostic analytics expands its support to the other types of data analytics to uncover hidden patterns, utilize past data, and discover the causes of situations and answers the "Why" question [24]. This diagnostic process is greatly efficient in forecasting and maintenance as well as in detecting errors and analyzing social media interactions.

3 An Overview of Marketing Fundamentals in the Big Data Era

By referring to the Customer Relationship Management revolution in the late 1990s, the historical data explosions have not only led to "potential benefits" for companies [6], as the latter appear to pass by three phases incorporating Data enthusiasm (Investment phase), Data disappointment (frustration), and Data realism (Reinvestment phase), as illustrated in Table 2. In this manner, and in an attempt to launch competitive digital-business tactics, gain value creation prospects, and enhance market performance, companies have deployed substantial investments in BD [2]. At first, the investment in BD sounds innovative, profitable, and competitive, where top management is sometimes trapped by some successful examples in the business sector and by some software providers or Information technology consultants. However, the more explosions in data investments take place, the more failed projects appear, inducing a second phase of data frustration after all the promises of success. This is not to say that companies' investments in BD always end up with failures, but to emphasize the need for a richer analysis and understanding of BD investments [27] to be reflected in strategies rethinking, research expanding, and renovating of data in the reinvestment phase [32].

In fact, these investments have given the rise of newly emerged digital channels; however, they have brought critical challenges for marketers such as dealing with customers' increasingly changing interactions. Primarily, "as the ecosystem of consumers becomes increasingly digitized, analytics is moving from a qualification and measurement tool to a core business decision-making capability" [28]. BD cannot generate value by itself, advanced analytical models are needed to achieve multi-channel analysis and real-time customization processes [9]. Accordingly, and as major marketing approaches have been recently developed such as marketing automation, data-centric marketing, and customer centricity and engagement, a consolidated BD infrastructure is the key toward value creation, where the marketing field is moving toward a mandatory transitional period of data-driven decisions to maintain its survival [28].

Table 2 Data cycle phases within companies, and main references

Data cycle phases	Explanation	References
Data enthusiasm (investment phase)	– Promises in profitable, innovative, and competitive investments – Success stories, software providers, and IT consultants impact on top management strategies	[6]
Data disappointment (frustration phase)	– Value destruction – Negative returns and waste of resources	[6]
Data realism (reinvestment phase)	– Strategies rethinking – Research expanding – Effective renovating of data into a sustainable performance	[32]

3.1 The Laying Foundations of BD Infrastructure Across Marketing Departments

At first sight, companies may think that all it takes is buying some software, hiring some experts, and BD implementation can be accomplished; however, it is not the case. According to Cisco, 60% of companies agree that BD enhances decision-making processes and increases their competitiveness, but only 28% point out that data is strategically providing them with an added value [29]. Hence, for companies to achieve long-term success in the BD era, a careful investment in people, systems, processes, and in the organization itself would facilitate marketing departments' work toward value creation.

People. The creation of marketing intelligence has changed marketing decision-making processes, as marketers' reach has shifted into a strategic realm with more innovations and brand positioning strategies [30]. Indeed, marketing departments have become a driving force in data integration developments, where they are switching to a more predictive approach focusing on customer and market-driven data; they are typically moving from reactive analyses to more proactive agendas-setting strategies. With the aim of efficiently exploiting BD, firms have recently started hiring and training BD scientists; as in many companies, analytical competences and marketing intelligence skills are missing. The understanding of how the marketing intelligence function develops is considered as the core of analytics success, as it starts with a supplier role by delivering what is needed and ends up with an orchestrating role embedded along marketing decisions [31]. Accordingly, this pathway of marketing intelligence would provide a valuable input for marketing departments based on accurate and clear market requirements. In parallel, teamwork approaches with an incremental analytical team building are argued to strengthen marketing departments, as long as team members of business analysts and Information technologists have a space for innovation and development, and have the needed training and education.

Systems. Large technological companies have conveyed the importance of focusing on technologies contributions for customers and not on the technological advances themselves [32]. In this sense, the marketing field is an advocate of this understanding, as it always emphasizes customer-driven systems rather than technology-driven ones. Hence, it is of a great importance to create an environment that has the ability to maintain data integration and provide an integrated data ecosystem within multiple sources, as many companies might be operating through multiple channels or even in multiple countries. Certainly, the initial response for BD integration is building up big systems and incorporating all BD analytical frameworks. However, the previous customer relationship approaches should be taken into account by companies, which need to include further examinations when integrating their data [33].

Processes. Processes mainly entail all the approaches from data input to data real-time accessibility, as well as the communication between analytical teams and the (marketing) management. Undeniably, these processes call for companies' coverage of policies implementation, as they should not be in contrast with the safe flow of data and the marketing intelligence role. In this vein, analytical teams are supposed to clearly communicate their findings to marketing departments, where they should work hand in hand among well-defined processes, fueled with relevant information and insightful reports. In parallel, and in order for processes to be more effective, analysis should always start with a business question able to combine the right focus and use of resources [34]. Accordingly, marketing departments will then guarantee the consensus of the market demands, determine the level their brand and customer analysis can compete with, as well as be able to determine whether to come up with some incremental improvements or go with major strategic changes.

Organizations. The importance of embedding a BD-oriented corporate vision is highly emphasized by the organizational literature. In order for organizations to be successful, a critical ingredient of such a vision is for skilled people to work in an enhanced environment [35]. In this vein, and in order for BDA not to become a losing strategy, enough attention should be devoted to embed BDA internally and train organizational members accordingly, as each BD solution may report a business problem [36]. This will be sustained by the establishment of a data-driven culture and the acquisition of the top management support.

Establishing a Data-driven Culture. An effective BD implementation entails the shift from the reliance on intuition or gut feeling to more data-based insights. Marketing has always been blamed in offering creative ideas without considering their effects on business performance [37]: "The marketing field is chockablock with creative thinkers, yet it's short on people who lean toward an analytic, left-brain approach to the discipline" [38]. Although marketers' approaches are initially based on scientific meta-analysis of the marketing discipline, there are still debates of whether a balance between a fact-based and intuition-based marketing might be created. In this vein, according to Evan "Rather than making decisions based on experience, people need to be convinced to trust data as well as informed recommendations" [39]. Indeed, numerical values are always trusted by executives and this is reflected by their commitment to a data-driven culture. However, human/emotional factors in decision-making processes are still needed to maintain an agile organizational culture based on facts as well as on creativity [40].

Top Management Support. Top management commitment plays a critical role in BDA integration and acceptance [41]. However, despite that the business literature agrees that top management support is a prerequisite for inducing stronger analytical functions [42], it still warns that a new set of leadership skills is needed to accompany the implementation of new strategies and analytical business frameworks. Particularly, in the BD context, this cannot come into real unless top management admits the advantages of BD, including innovation opportunities and information-based products.

4 Value Creation Through BDA Application in Marketing Decisions

Primarily, the creation of a well-designed BD infrastructure gives marketers the capability in launching new channels of decoding data, as humans are not able to analyze and interpret all the anonymous variables around [43]. Indeed, technological advances allow marketers to adopt a more reasoned approach to data analysis, as they receive every minute large and new sets of data awaiting their evaluation [44]. However, due to the variety of data sources including web analytics, mobile applications, and digital advertisements, and due to customers' interactions across different channels, marketers are getting fragmented data and distorted insights [45]. As BD has developed as a significant element for knowledge generation and the improvement of decision-making processes [46], the main concern is related to the capability to access and analyze large streams of data, and turn them into transferrable knowledge to be shared with decision-makers [47]. In this regard, capabilities in BDA can offer help in running sophisticated and advanced data mining methods and investing in data-rich environments [48].

4.1 BDA Contributions at the Customer Level

BDA is highlighted as a critical element in value creation, as it induces transparency establishment, needs detection and performance enhancement [9]. Indeed, with instantaneous data triangulation opportunities [49], the leveraging of BDA reinforces marketing competences in attaining enriched insights [50], and business superiority [51]. From a behavioral perspective, some studies have highlighted the importance of monitoring customers' interactive roles, being the stimulators for new engagement activities [52, 53]. BDA predominantly entails behavioral examinations that can detect errors before failure takes place [54]. As customers' preferences are constantly and frequently changing, now mainly seeking the figurative, cultural, and emotional aspects of companies' offerings [55], marketers are now badly needing BDA to gain clearer understandings about their customers' needs and wants [56].

Studies have debated that BD has enabled companies to come up with customized offerings across their customers' different buying channels [57–59]. In this regard, BDA is considered as a customer self-service and product personalization driver, where both are made easier through the real-time extraction of customers' perceptions and recommendations [60]. In this context, the importance of personalization in differentiating loyal customers from new ones has been highlighted [58]. Indeed, personalization gives companies the likelihood of maintaining customers' satisfaction by enhancing their role in being potential controllers of their buying practices [49]; for instance, recommendation systems represent great examples in simplifying customers' decision-making through customized suggestions that will translate their choices into an added value [61]. Additionally, based on their current or future location, customers are offered product suggestions through controlled timing. In this vein, the location-based log-in data via web pages and social networks has been used to discover user profiles and then applied to location-based targeted marketing with the help of regression and classification techniques [62]. Hence, through such targeted marketing efforts, companies gain the possibility of attaining greater returns on their marketing investments, as long as they are associated with their customers' understandings [63]. Furthermore, to attract more customers, companies need to dynamically set competitive prices [10]. Indeed, the BD-driven insights give them the opportunity to adjust their pricing decisions in harmony with their customers' spending actions [47]. Moreover, BDA has been considered as a visibility provider for customers in the supply chain [10], as well as a supplier for cohesive understandings of customers' multiple offerings from different sources [58].

4.2 BDA Contributions at the Market Level

From web to retail outlets, BD has become an avenue for novel products innovations. Gut has nothing to do with competition in the twenty-first century, where companies are shaping their products based on customers' involvement activities [64]. In fact, it is argued that customers' involvement in BD-enabled mechanisms minimizes risks associated with decision-making [56]. In parallel, the role of BDA in supporting companies to better design and execute their marketing initiatives is also highlighted [65], particularly the launching of new product developments. Leveraging BDA systems stimulates a prepared environment for nurturing, executing, and embedding innovations. Indeed, companies' capitalization on BD learning potentials offers them new opportunities to exploit value creation [66]. In this regard, some studies have proved that significant positive relationships exist among BDA, service supply chain innovation capabilities, and service supply chain performance [67]; this is in line with studies conducted on the effects of BD on firm performance thanks to new developments in products, customer relationships, and supply chain [68]. Accordingly, the consolidation of a valuable marketing scope is associated with the innovative approach brought by BD [50] and BDA transformation potential in accelerating innovative initiatives. Moreover, the importance of data in terms of pricing methodologies has been emphasized [69]. Indeed, the application of BDA offers a great opportunity for dynamic pricing strategies that can bring out enhanced price examinations and new market entries [70]. In this manner, it is argued that BDA facilitates the automation of rivals' analysis [47], in which companies are now capable of not only having competitors recognized, but also discovering the competitive product frameworks, pricing policies, and customers' feedback; in essence, this is to better understand the risks, enhance companies' pricing decisions, and bring out new forms of value. Furthermore, BDA does provide a better allocation for marketing budgets and marketing plan schemes; mainly, this is through the exact view given for sales forecasts, as well as for purchasing pattern tracking [67] and advertising campaigns. For instance, social media effects on acquisition are found partial, which will in turn stimulates the question by marketers of whether to capitalize on social media in acquisition strategies or not [71].

5 Conclusion

BDA has laid the basis of how firms could manage the daily technological updates and innovations, of how to exploit the hidden opportunities around, and most importantly of how to mirror customers' behaviors and intentions through novel initiatives. In this vein, marketers have been greatly affected by this data evolutionary disruption. Indeed, BDA has guided its customers' purchasing journeys, where it has managed to help them tune to every single customer and to reply with agile engagement, retention, and acquisition plans. Through BDA, marketers are now able to give birth to new product innovations and design strategies, to monitor their promotional tactics, to take care of their pricing strategies, and to put their place-based strategies in place. In parallel, the value creation road cannot be attained without an organized BD infrastructure, where the marketing intelligence function is transmitted over departments, systems are designed for better data management, processes are based on effective communication, and the organization is strengthened by a data-driven culture establishment and the top management support.

Accordingly, the road toward the consolidation of smarter marketing decisions and value creation processes cannot be in isolation anymore from the emergence of BDA, which is changing the rules of the game and the dynamics of how firms in general, and marketers in particular, could survive in the BD era. Yet, as companies are heavily relying on detailed and accurate customers' data, they are getting wedged with more critical questions of how much privacy customers do have as BD has taken a big part in their daily life, and how much security should be put into action as data becomes much more accessible and more unsecured pathways are on the way. Mainly, customers' data is being collected invisibly by companies without them being aware of their data collection or even usage [72]. However, it is argued that the relationship between privacy concerns and customers' sharing of data has not yielded clear results yet, as customers do not have constant behaviors; some may have uncertainties about their privacy, yet their personal information traces are everywhere. Additionally, it is debated whether companies' concerns about moral decisions and legislations will affect their path of data collection. In this vein, the need for ethical decisions has been stressed [73], as when privacy issues are formulated with moral customer-centric approaches, companies will be able to maintain their reputation and minimize customers' churns and market losses [74]. Accordingly, companies will definitely be able to confidently and safely move toward value creation in this continuously challenging era.

References

1. Moorman, C. (2016). Celebrating marketing's dirty word. *Journal of the Academy of Marketing Science, 44*, 562–564.
2. Chen, H., Chiang, R., & Storey, V. (2012). Business intelligence and analytics: From big data to big impact. *MIS Quarterly, 36*, 1165.
3. Wamba, S. F., Gunasekaran, A., Akter, S., Ren, S. J.-F., Dubey, R., & Childe, S. J. (2017). Big data analytics and firm performance: Effects of dynamic capabilities. *Journal of Business Research, 70*, 356–365.
4. Pasquale, F. (2015). *The black box society*. Harvard University Press.
5. Wedel, M., & Kannan, P. (2016). Mapping the boundaries of marketing: What needs to be known. *Journal of Marketing, 80*, 97–121.
6. Meer, D. (2013). The ABCs of analytics. *Strategy Business, 70*, 6–8.
7. Strawn, G. O. (2012). Scientific research: How many paradigms? *Educause Review, 47*, 26.
8. Hagstrom, M. (2012). High-performance analytics fuels innovation and inclusive growth: Use big data, hyperconnectivity and speed to intelligence to get true value in the digital economy. *Journal of Advanced Analytics*, 3–4.
9. Manyika, J., Chui, M., Brown, B., Bughin, J., Dobbs, R., Roxburgh, C., & Byers, A. H. *Big data: The next frontier for innovation, competition, and productivity*. https://www.mckinsey.com/business-functions/mckinsey-digital/our-insights/big-data-the-next-frontier-for-innovation.
10. Akter, S., & Wamba, S. F. (2016). Big data analytics in e-commerce: A systematic review and agenda for future research. *Electronic Markets, 26*, 173–194.
11. Waller, M. A., & Fawcett, S. E. (2013). Data science, predictive analytics, and big data: A revolution that will transform supply chain design and management. *Journal of Business Logistics, 34*, 77–84.
12. Pantano, E., Giglio, S., & Dennis, C. (2019). Making sense of consumers' tweets. *International Journal of Retail & Distribution Management, 47*, 915–927.
13. De Mauro, A., Greco, M., & Grimaldi, M. *What is big data? A consensual definition and a review of key research topics*. https://www.researchgate.net/publication/265775800_What_is_Big_Data_A_Consensual_Definition_and_a_Review_of_Key_Research_Topics.
14. Laney, D. (2001). 3D data management: Controlling data volume, velocity and variety. *META Group, 949*, 1–4.
15. Zikopoulos, P. (2012). *Understanding big data*. McGrawHill.
16. IBM. *Diagnostic analytics 101: Why did it happen?* https://www.ibm.com/blogs/business-analytics/diagnostic-analytics-101-why-did-it-happen/.
17. Dumbill, E. (2013). Making sense of big data. *Big Data, 1*, 1–2.
18. Boyd, D., & Crawford, K. (2012). Critical questions for big data. *Information, Communication & Society, 15*, 662–679.
19. Reinsel, D., & Gantz, J. *Extracting value from chaos*. http://www.emc.com/collateral/analyst-reports/idc-extracting-value-from-chaos-ar.pdf.
20. Oracle. *Big data analytics, advanced analytics in Oracle Database*. https://www.oracle.com/assets/bigdataanalyticswpoaa-1930891.pdf.
21. Tsai, C.-W., Lai, C.-F., Chao, H.-C., & Vasilakos, A. V. (2016). Big data analytics. *Big Data Technologies and Applications*, 13–52.
22. Patil, P., & Bhosale, A. *A big data analytics*. https://medcraveonline.com/OAJS/OAJS-02-00095.pdf.
23. Watson, H. J. (2014). Tutorial: Big data analytics: Concepts, technologies, and applications. *Communications of the Association for Information Systems, 34*.
24. Ajah, I. A., & Nweke, H. F. (2019). Big data and business analytics: Trends, platforms, success factors and applications. *Big Data and Cognitive Computing, 3*, 32.
25. Mujawar, S., & Joshi, A. (2015). Data analytics types, tools and their comparison. *International Journal of Advanced Research in Computer and Communication Engineering, 4*, 488–491.

26. Wang, G., et al. (2016). Big data analytics in logistics and supply chain management: Certain investigations for research and applications. *International Journal of Production Economics, 176*, 98–110.
27. Abbasi, A., Sarker, S., & Chiang, R. (2016). Big data research in information systems: Toward an inclusive research agenda. *Journal of the Association for Information Systems, 17*.
28. Casaca, J. *Marketing in the era of big data*. https://www.academia.edu/34354137/Marketing_in_the_Era_of_Big_Data.
29. McKinsey. *Marketing & sales big data, analytics, and the future of marketing & sales*. https://www.mckinsey.com/~/media/McKinsey/Business%20Functions/Marketing%20and%20Sales/Our%20Insights/EBook%20Big%20data%20analytics%20and%20the%20future%20of%20marketing%20sales/Big-Data-eBook.ashx.
30. Vishnoi, S., & Bagga, T. (2020). Marketing intelligence: Antecedents and consequences. *SSRN Electronic Journal*.
31. Hedin, H., Hirvensalo, I., & Vaarnas, M. (2014). *The handbook of market intelligence: Understand, compete and grow in global markets*. Wiley.
32. Verhoef, P., & Lemon, K. (2013). Successful customer value management: Key lessons and emerging trends. *European Management Journal, 3*, 1–15.
33. Neslin, S., et al. (2006). Challenges and opportunities in multichannel customer management. *Journal of Service Research, 9*, 95–112.
34. Baquero, A., Palacios, R., & Molly, O. (2013). Business process analytics using a big data approach. *IT Journal, 5*(11), 7–15.
35. Holtom, B. C., Mitchell, T. R., Lee, T. W., & Eberly, M. B. (2008). 5 turnover and retention research: A glance at the past, a closer review of the present, and a venture into the future. *Academy of Management Annals, 2*, 231–274.
36. Waller, D. *10 steps to creating a data-driven culture*. https://hbr.org/2020/02/10-steps-to-creating-a-data-driven-culture.
37. Verhoef, P., & Leeflang, P. (2011). Accountability as a main ingredient of getting marketing back in the board room. *Marketing Review St. Gallen, 28*, 26–31.
38. McGovern, G., et al. (2004). Bringing customers into the boardroom. *Harvard Business Review, 82*, 70–80.
39. Stubbs, E. (2014). *The value of business analytics identifying the path to profitability*. Wiley.
40. Maguire, J. *Using data science in the real world: Expert tips*. https://www.datamation.com/big-data/using-data-science-in-the-real-world-expert-tips.html.
41. Gunasekaran, A., Papadopoulos, T., Dubey, R., Wamba, S. F., Childe, S. J., Hazen, B., & Akter, S. (2017). Big data and predictive analytics for supply chain and organizational performance. *Journal of Business Research, 70*, 308–317.
42. Kirca, H., Jayachandran, S., & Bearden, W. (2005). Market orientation: A meta analytic review and assessment of its antecedents and impact on performance. *Journal of Marketing, 6*, 24–41.
43. Sammut, G., & Sartawi, M. (2012). Perspective-taking and the attribution of ignorance. *Journal for the Theory of Social Behaviour, 42*, 181–200.
44. Lycett, M. (2013). 'Datafication': Making sense of (big) data in a complex world. *European Journal of Information Systems, 22*, 381–386.
45. Datameer. *Big data use cases to understand your customer journey*. https://www.datameer.com/pdf/Datameer-Customer-Analytics-ebook.pdf.
46. Tan, K. H., & Zhan, Y. (2016). Improving new product development using big data: A case study of an electronics company. *R&D Management, 47*, 570–582.
47. Xu, Z., Frankwick, G. L., & Ramirez, E. (2016). Effects of big data analytics and traditional marketing analytics on new product success: A knowledge fusion perspective. *Journal of Business Research, 69*, 1562–1566.
48. Cui, A. S., & Wu, F. (2016). Utilizing customer knowledge in innovation: Antecedents and impact of customer involvement on new product performance. *Journal of the Academy of Marketing Science, 44*, 516–538.

49. Rejeb, A., Rejeb, K., & Keogh, J. *Potential of big data for marketing: A literature review.* https://www.researchgate.net/profile/Abderahman-Rejeb/publication/339630258_Potential_of_Big_Data_for_Marketing_A_Literature_Review/links/5f3a43d9a6fdcccc43cfc9a0/Potential-of-Big-Data-for-Marketing-A-Literature-Review.pdf.
50. Erevelles, S., Fukawa, N., & Swayne, L. (2016). Big data consumer analytics and the transformation of marketing. *Journal of Business Research, 69*, 897–904.
51. Gnizy, I. (2019). Big data and its strategic path to value in international firms. *International Marketing Review, 36*, 318–341.
52. Javornik, A., & Mandelli, A. (2012). Behavioral perspectives of customer engagement: An exploratory study of customer engagement with three Swiss FMCG brands. *Journal of Database Marketing & Customer Strategy Management, 19*, 300–310.
53. Coulter, K. S., Gummerus, J., Liljander, V., Weman, E., & Pihlström, M. (2012). Customer engagement in a Facebook brand community. *Management Research Review, 35*, 857–877.
54. Ferguson, R. B. (2012). Risky business: How data analytics and behavioral science can help. *MIT Sloan Management Review, 54*, 1–5.
55. Kidwell, B. (2011). Emotional intelligence in marketing exchanges. *Journal of Marketing, 1*, 5–12.
56. Richey, R. G., Morgan, T. R., Lindsey-Hall, K., & Adams, F. G. (2016). A global exploration of big data in the supply chain. *International Journal of Physical Distribution & Logistics Management, 46*, 710–739.
57. Kopp, M. *Seizing the big data opportunity.* Ecommerce Times. http://www.ecommercetimes.com/story/78390.html.
58. Mehra, G. *6 uses of big data for online retailers.* Practical Ecommerce. http://www.practicalecommerce.com/articles/3960-6-Uses-of-Big-Data-for-Online-Retailers.
59. Miller, G. *6 ways to use "big data" to increase operating margins by 60%.* http://upstreamcommerce.com/blog/2012/04/11/6-ways-big-data-increase-operating-margins-60-part-2.
60. Huang, M.-H., & Rust, R. T. (2013). IT-related service: A multidisciplinary perspective. *Journal of Service Research, 16*, 251–258.
61. Kunz, W., Aksoy, L., Bart, Y., Heinonen, K., Kabadayi, S., Villaroel Ordenes, F., Sigala, M., Diaz, D., & Theodoulidis, B. (2017). Customer engagement in a big data world. *Journal of Services Marketing, 31*, 161–171.
62. Vasconcelos, M., Ricci, S., Almeida, J., & Benevenuto, F. *Tips, dones and todos: Uncovering user profiles in Foursquare.* https://www.researchgate.net/publication/221520106_Tips_dones_and_todos_Uncovering_user_profiles_in_Foursquare.
63. Bradlow, E. T., Gangwar, M., Kopalle, P., & Voleti, S. (2017). The role of big data and predictive analytics in retailing. *Journal of Retailing, 93*, 79–95.
64. Rampton, J. *Why small business owners should trust their numbers over their instincts.* https://www.inc.com/john-rampton/why-small-business-owners-should-trust-their-numbers-over-their-instincts.html.
65. Zhan, Y., Tan, K. H., Ji, G., Chung, L., & Tseng, M. (2017). A big data framework for facilitating product innovation processes. *Business Process Management Journal, 23*, 518–536.
66. Zeng, J., & Glaister, K. W. (2018). Value creation from big data: Looking inside the black box. *Strategic Organization, 16*, 105–140.
67. Fernando, Y., Chidambaram, R. R. M., & Wahyuni-TD, I. S. (2018). The impact of big data analytics and data security practices on service supply chain performance. *Benchmarking: An International Journal, 25*, 4009–4034.
68. Wamba, S. F., Akter, S., & Bourmont, M. de. (2019). Quality dominant logic in big data analytics and firm performance. *Business Process Management Journal, 25*, 512–532. https://doi.org/10.1108/BPMJ-08-2017-0218.
69. Baker, B., et al. *Using big data to make better pricing strategies.* https://www.mckinsey.com/business-functions/marketing-andsales/our-insights/using-big-data-to-make-better-pricing-strategies.
70. Sanders, N. R. (2016). How to use big data to drive your supply chain. *California Management Review, 58*, 26–48.

71. De Vries, L., Gensler, S., & Leeflang, P. S. H. (2012). Popularity of brand posts on brand fan pages: An investigation of the effects of social media marketing. *Journal of Interactive Marketing, 26*, 83–91.
72. Hong, W., & Thong, J. (2013). Internet privacy concerns: An integrated conceptualization and four empirical studies. *MIS Quarterly, 37*, 275–298.
73. De Bruin, B. (2015). *Ethics and the global financial crisis: Why incompetence is worse than greed*. Cambridge University Press.
74. Gallo, A. *The value of keeping the right customers.* https://hbr.org/2014/10/the-value-of-keeping-the-right-customers.

Driving Cities as Sustainable Urban Communities

Mauro Romanelli

Abstract Cities of tomorrow evolve as sustainable communities and contribute to shaping and promoting social and economic urban growth and value. The study aims to elucidate that cities can follow certain pathways to develop the city as sustainable urban community. Cities develop a smart city view as a strategic vision to drive a human-centred and community-oriented city able to design citizen-centred services and to strengthen urban organisational networks, thus driving cooperation to develop innovation systems. The main contribution of this study is to provide a framework of analysis about how cities identify some trajectories to make the city a sustainable community looking at the future urban development by supporting urban growth and engaging urban stakeholders in collaborative innovation processes.

Keywords Smart city · Smart community · Sustainable urban development

1 Introduction

Cities of the future will be smart communities, sustainable, healthy and inclusive places for work and life playing a key role to improve and drive urban competitiveness [1, 2]. Rethinking cities of tomorrow relies on rediscovering sustainable and smart-oriented urban communities that promote social and economic growth. As engines and drivers of social and economic change, cities contribute to social and public value becoming inclusive, smart and urban communities that support sustainable growth and develop innovation to drive urban competitiveness and improve the quality of life [1–3]. The future and sustainability of cities rely on smart urban communities that drive knowledge-based and innovation processes [4, 5]. The way how information technology is used in the urban environment can help cities to rethink a smart urban future. A smart city is a city seeking to address public issues via ICT-based solutions on the premise of a multi-stakeholder, municipally based

M. Romanelli (✉)
Department of Business and Economics, University of Naples Parthenope, Via G. Parisi, 13, 80132 Napoli, Italy
e-mail: mauro.romanelli@uniparthenope.it

C. Dal Zotto et al. (eds.), *Smart Technologies for Organizations*, Lecture Notes in Information Systems and Organisation 60, https://doi.org/10.1007/978-3-031-24775-0_11

partnership. It is one that capitalises on the opportunities presented by information and communication technology in promoting its prosperity and influence [6].

Accordingly, a smart city is able to bring together technology, people and community dimensions in order to drive urban innovation and growth [7]. The effort to create a twenty-first-century city is about organising one's community to reinvent itself for living within a knowledge economy and society, and enabling citizens to take ownership of their community [5]. Cities have to follow a smart vision to promote urban innovation by investing in human capital to build collaborative processes between people, business and government within the community [4]. 'A successful smart city can be built from top down or bottom up approaches, but active involvement from every sector of the community is essential' [7] (p. 287). As smart and sustainable communities, cities can identify the smart approach as a vision that makes the city a knowledge-based and technology-enabled engine to design and implement urban development and sustainability [4, 5].

The aim of this study is to elucidate how cities identify a pathway to evolve as sustainable and urban communities. Cities have to select a pathway for sustainability promoting smartness as a human-centred vision to ensure citizen-centred services and drive urban organisational networks and cooperation in order to develop adequate knowledge management and innovation systems and sources. Within a knowledge-based, technology-enabled and learning-oriented society, cities develop as sustainable and smart communities that promote continuous innovation within services and ecosystems. The chapter is structured into six sections. Following the introduction and methodological section, in the third section, cities are considered as sustainable, inclusive and smart communities that evolve by developing technology-enabled and knowledge-driven innovation, constructing the community as an organisational framework for innovation in order to drive growth in the urban landscape. The fourth section elucidates how to drive cities as sustainable and smart communities that follow a smart vision using technology in order to promote innovation. In the fifth section, a framework of analysis is presented in order to elucidate how cities can identify certain pathways that are oriented to leading cities to become sustainable and urban communities. Finally, conclusions are outlined.

2 Methodological Section

The study is theoretical in nature and relies on a review of the literature in order to understand the concepts of sustainable city, smart city and smart community as drivers that enable urban stakeholders to design services and develop innovation, pushing the cities to proceed and evolve as sustainable and urban communities. Smart city and smart community are the keywords and values that help cities to rethink strategy and action to design and shape future sustainable urban development. The articles considered focus on the smart city topic and the smart community trend as drivers of cities' evolution towards sustainability. In other words, the concept of smart city is evolving from a techno-centric to human-centred vision that enhances collaborative and cooperative aspects focusing on social dimensions that involve the urban design to make the city smarter. The smart community concept is moving from using technology to improve the infrastructure and quality of life of the city to developing innovation systems and processes to help make the city more inclusive in proceeding towards sustainable urban development. The selected contributions are interpreted in a narrative synthesis in order to elucidate new perspectives and advance theoretical frameworks on emerging issues [8, 9].

3 Cities as Sustainable Smart-Driven Communities Driving Urban Development

Cities are economic and social systems in space [10] meeting places, as well as service providers and social incubators that drive change and innovation [11]. Sustainable urban development relies on making the city a system of innovation [12]. Cities have to select a knowledge-based vision in order to evolve as smart-driven communities that contribute to building sustainable urban development and shaping social, inclusive and economic growth [13]. Cities have to select a smart-driven and innovation-oriented vision to contribute to the sustainable development of urban communities [13, 14]. Smart growth implies developing an economy based on knowledge and innovation. Sustaining smart growth relies on revitalising economic growth in urban areas by developing smart cities and communities that contribute to promoting innovation processes [15].

Cities are communities made up of people who through their lives, actions, thoughts and beliefs influence and drive history, values and beliefs of their cities [16]. Developing cities as sustainable communities relies on strengthening social capital [17], making the smart cities socially inclusive communities [18] and empowering citizens to actively contribute to urban value creation and sustainability [19]. In *The 2030 Agenda for Sustainable Development*, the focus is on making cities and human settlements inclusive, safe, resilient and sustainable. Smart cities use technology to develop smart industries and smart economies, promoting smart mobility, sustaining smart government, relying on smart people and enhancing smart living and smart environment. Technology helps cities to design a path for sustainability, relying on adopting and following a smart approach for development. Smart cities strategically employ technology to improve quality of life, drive innovation and support the management and delivery of public services based on public–private partnerships [20]. 'A "smart community" is a community in which members of local government, business, education, healthcare institutions and the general public understand the potential of information technology, and form successful alliances to work together to use technology to transform their community in significant and positive ways' (p. 2) [21]. Sustaining smart city and community development relies on using and developing the potential of information technology to empower individuals and groups in promoting change within communities, thus opening up new opportunities for urban innovation and development for improving the wealth of urban communities [5, 21, 22]. Building sustainable urban development relies on improving the quality of life for people in a city without creating a burden for future generations. The use of information technology helps cities to identify a pathway for urban sustainability so as to achieve the long-term goals of preserving the ecosystem and improving the quality of life within a community [23, 24]. Cities are using the potential of information technology in order to develop a pathway for urban sustainability, promoting social interaction and participation within the community [25] without threatening the viability of the natural, built and social system [26].

4 Driving Cities as Sustainable Communities

As evolving and social organisms, cities aliment dynamic processes that involve all the urban stakeholders contributing to making the city a sustainable community. Cities have to follow a smart approach as a vision to drive urban growth and service innovation by using information technology. Cities as smart communities are engines of social and economic innovation that involve all the urban stakeholders. Cities as sustainable urban communities [27] contribute to creating social and economic value. Cities achieve social and policy goals, building innovation-oriented and knowledge-driven inter-organisational relationships [5, 21]. The pathway leading to cities as sustainable communities relies both on developing a smart vision for driving social and economic urban growth and promoting collaborative innovation within smart communities.

4.1 Developing a Smart Vision for Driving Urban Development and Growth

Smart cities perform better and contribute to advancing urban sustainability. Smart cities contribute to open innovation as co-production and co-delivery of services and policies as well [4]. Cities of tomorrow will be smart, sustainable and wealthy communities that support innovation and achieve a better quality of life. Cities are rethinking a smart-oriented approach in order to drive social and economic growth and achieve sustainable development in urban spaces coherently with a knowledge and innovation-led economy. Smart cities are considered to be a socio-technical phenomenon. 'A smart city should be seen as a continuum in which local government officials, citizens, and other stakeholders could think about the initiatives that attempt to make the city a better place to live' [28] (p. 5). Smart cities use information technology to develop smart industries and smart economies, promoting smart mobility, sustaining smart government, relying on smart people and enhancing smart living and smart environment [20]. Technology helps cities to design a pathway for sustainability, relying on following a smart approach for development [18]. Smart cities promote and extend high quality of life in urban environments strengthening the connections between productivity, economic growth and human capital, and in turn enhancing service systems and capabilities in an urban network and ecosystem [29, 30]. The introduction and use of smart urban technologies can significantly contribute to sustainable urban development leading to cities as urban communities that promote innovation processes [15].

Smarter cities develop innovation to promote governance and community development, thus sustaining a collaborative and organisational framework within urban services and ecosystems [30]. Promoting collaboration helps drive urban innovation as a source for value creation within urban communities. Cities promote open innovation as a mindful, strategically driving collaboration between the stakeholders, leading to a change in the way cities grow smart [4]. Driving cities as smart and sustainable places is a long-term change process which must be planned, and which involves policy-makers and key stakeholders [31]. Cities adopt smart city strategies to advance towards urban sustainability [32], and define implementation stages of smart sustainable city design [33]. Cities are rethinking urban planning following a sustainability-oriented and smart strategy, involving stakeholders and citizens strengthening participation and multi-actor decision-making [34, 35]. With the aim of adopting a smart vision of urban development, cities become communities that 'have strong economic growth, are socially inclusive in their growth, and are environmentally responsible' [18], promoting sustainable urban development and improving the wealth of people within a community [14]. Cities promote smartness as a human-centred vision of urban development in order to drive urban transformation of the city [36]. Cities develop a change-oriented vision within a smart community, promoting sustainable inclusive growth and participatory governance in order to make citizens as co-producers of urban policies too [37, 38] by strengthening urban values, cohesion and innovation, and enabling businesses, public bodies and citizens to develop collaborative processes. Smart cities support intelligent, inclusive and sustainable urban communities, enabling citizen participation and intellectual capital [38, 39]. Cities are organisational spaces that enable collaboration and cooperation that involve public and private stakeholders who are involved in designing new ways of urban management and governance. Cities adopt a smart strategy 'for improving the operational and managerial efficiency and the quality of life by building on advances in ICTs and infrastructures' (p. 186) [40], while smarter cities tend to develop innovation for rediscovering governance and community [41].

4.2 *From Services to Collaborative Innovation Within Smart Communities*

Digital technology enables the development of urban communities and effectively supports their sustainable growth. In particular, technology helps empower citizens, thus creating more informed and intelligent behaviour. The use of information technology helps to support collaborative innovation as a framework that strengthens innovation with a variety of actors involved in the process. Cities enable urban organisational spaces as engines of collaborative innovation and rediscover innovative capacity by involving both public actors and companies, social entrepreneurs, nongovernmental organisations, knowledge institutions such as universities and public research institutes, and citizens [42]. Rediscovering cities as sustainable urban communities means that cities have to receive feed forward and feedback information from all stakeholders on whom they depend in order to drive sustainable urban development [27]. Rethinking cities as smart communities helps both to shape the city as a better place for sustainable wellbeing of people living in urban spaces and to empower citizens as co-designers and co-producers of public services [4, 36]. Cities evolve as smart communities understanding the potential of information technology to support collaboration between public and private organisations in order to improve urban competitiveness, support innovation processes and ensure quality of life. Cities as communities drive urban social and economic growth employing technology as a catalyst to transform the economy of an urban community [4]. Cities are smart communities and better places to live and work, while they also proactively promote sustainable wellbeing for people and shape urban community development [34]. Smart communities are cities that interact with citizens sharing data, information and knowledge for problem solving [43]. Smart communities help all the members of a community (local government, business, education, healthcare institutions and the public) to learn to work together in rediscovering the potential of information technology to transform the community in a positive way [21]. Smart communities enable citizen participation in services co-production [44] and support urban innovation, citizen empowerment, knowledge creation and exploitation [45]. In particular, participatory citizen engagement helps a citizen's centricity vision, which enables democratic city decision-making processes and social innovation issues [46]. Government, industry, academy and the citizenry drive urban future development sharing the governing process to help the community face the challenges of the digital era [5]. Cities are *loci* of innovation and innovative milieus [47] where government, business and communities build partnership and collaboration [48]. Cities promote urban growth, innovation and development by embracing a triple helix model and rediscovering the active cooperation among industries, government, university, citizens and civil society to support innovation processes [30, 49]. Cities are the main drivers of economic growth and open innovation, encouraging multi-level private–public partnership and cooperation to co-create, co-design and co-implement innovative solutions and action, thus promoting innovation as a means for public governance and stakeholder involvement [15]. Knowledge-oriented cities use technology to drive urban development and support a learning environment [13, 50]. Cities contribute to strengthening open innovation linking technology, people and territory to develop knowledge and reinforce the cohesion of urban communities [5]. Cities as smart communities acquire intellectual capital attracting and retaining skilled and talented people and thus sustaining innovation systems by reinforcing knowledge and intellectual infrastructures involving public and private actors in driving social and urban development [13, 49]. Cities as smart communities become collaborative spaces that enable dialogue and cooperation between all the stakeholders or *helices* of urban environments (government, industry, university and civil society) [6, 21]. Cities are smart communities promoting sustainable wellbeing for people, building a shared governance relying on participation, dialogue and open debate among all the stakeholders for urban policy options [5]. Cities as smart communities adopt a human-centred vision towards developing democratic and cooperative urban innovation, promoting collaborative decision-making processes [51]. Smart, mature and innovative technologies, and platforms contribute to developing innovation in services and processes that benefit citizens and improve their quality of life. Information technology helps to support urban innovation for the sustainability of cities and communities [52].

5 Discussion

Rethinking cities as sustainable urban communities that shape urban development relies on cities as smart communities which improve the quality of life and competitiveness in urban, local and global environments. Increasingly, cities become incubators and spaces that enhance social and economic sustainability and innovation. As sustainable communities, cities develop the meaning of a community that enables all the stakeholders to actively play a key role to support innovation and knowledge processes and contribute to value creation. As shown in Fig. 1, the main contribution of this study is to identify a framework of analysis to understand how cities tend to identify some trajectories of sustainable development within urban communities. The analysed framework is presented in order to elucidate how cities identify a pathway to drive urban communities to proceed towards sustainability as a source of action and vision for change within urban spaces.

Cities as communities look at the future developing smartness as a strategic view and relying on technology-enabled services by managing knowledge for driving urban development and innovation that rely on collaboration and social networks. Cities develop innovation to help urban communities, drive social and economic

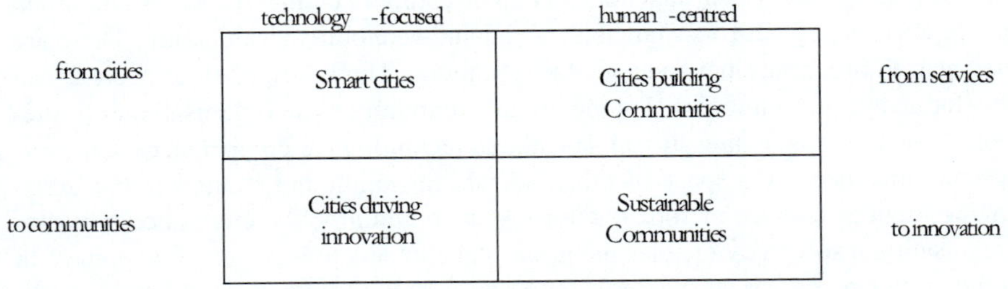

Fig. 1 Driving cities as sustainable urban communities

growth and promote development by involving all the urban stakeholders, and mobilising organisational and managerial competencies that enhance public value creation. As future-oriented communities, cities plan initiatives in urban planning in order to develop innovation-driven and community-oriented pathways. Smart and sustainable cities contribute to building the community as a source for development.

Technology-enabled and services-oriented urban communities develop knowledge helping organisational and individual learning in order to enable value creation within urban service and innovation ecosystems. As sustainable communities, cities promote smart growth to develop the urban community and support a sustainability-oriented pathway moving from ensuring services to involving all the stakeholders in actively partaking in the urban value co-creation process, thus rediscovering knowledge as a source to promote innovation to construct better policies for urban governance and entrepreneurship.

6 Conclusion and Future Research

Cities of tomorrow evolve as communities living in urban spaces in order to build and shape services and innovation ecosystems. Cities as sustainable and smart communities promote a pathway for driving continuous innovation within a knowledge-based, technology-enabled and learning-oriented society. As sustainable ecosystems and engines of economic growth, cities select a strategic agenda by involving different stakeholders. Technology is a necessary but insufficient source which enables cities to acquire, use and manage knowledge for breeding innovation and sustaining urban learning within ecosystems. The smart city and smart community concepts are key drivers of sustainable urban development within cities looking at the future. In particular, the evolution of smart cities from a techno-centric to the human-centred vision stresses the role of the community as a key actor in driving processes of urban development. Strengthening the attributes of smart community within urban spaces helps cities focus on innovation systems and processes as a source through which to make the city both sustainable and inclusive in fostering participatory approaches to urban development and social and economic growth. Cities use technology in order to plan a human-centred and community-oriented urban design for growth and development engaging with urban stakeholders and investing in human resources and urban intelligence as engines of sustainable, social and economic development. There are managerial, organisational and social implications. The future of sustainable urban development relies on mobilising the urban community as an organisational framework that develops organisational, social and technological competences. Cities as communities become a space of urban social innovation and strengthen the urban environment as a space of social action. As a community, the city is becoming an organisational space that enables the urban stakeholders to be involved to engage in collaborative processes. Managing cities as sustainable communities requires mechanisms of governance that involve public and private actors living and acting within urban spaces. Cities develop technological and knowledge infrastructures to enable innovation processes that help drive sustainable growth in urban spaces. This study is conceptual and provides only a theoretical framework of analysis that elucidates the pathway cities can entrepreneur by evolving as smart-driven and innovation-enabled sustainable communities. There are some limitations in this study. It does not provide any empirical analysis. No case studies are presented. Future research may consider how cities use information technology to develop innovation in services, knowledge creation and governance as drivers of value creation.

References

1. Begg, I. (1999). Cities and competitiveness. *Urban Studies, 36*(5), 795–809.
2. Eger, J. M. (2009). Smart growth, smart cities, and the crisis at the pump. A worldwide phenomenon. *I-Ways Journal of E-Government Policy and Regulation, 32*(1), 47–53.
3. European Commission: Report from the Commission to the Council on the Urban Agenda for the E, COM. (2017). 657 final, 20.11.2017.
4. Paskaleva, K. A. (2011). The smart city: A nexus for open innovation? *Intelligent Buildings International, 3*(3), 153–171.
5. Eger, J. M. (2005). Smart communities, universities, and globalization: Educating the workforce for tomorrow's economy. *Metropolitan Universities, 16*(4), 28–38.
6. Odendaal, N. (2003). Information and communication technology and local governance: Understanding the difference between cities in developed and emerging economies. *Computers, Environment and Urban Systems, 27*, 585–607.
7. Nam, T., & Pardo, T. A. (2011). Conceptualizing smart city with dimensions of technology, people and institutions. In *Proceedings of the 12th Annual International Digital Government Research Conference: Digital Government Innovation in Challenging Times* (pp. 282–291). ACM.
8. Denyer, D., & Tranfield, D. (2006). Using qualitative research synthesis to build an actionable knowledge base. *Management Decision, 24*, 213–227.
9. Dixon-Woods, M., Agarwall, S., Young, B., Jones, D., & Sutton, A. Integrative approaches to qualitative and quantitative evidence. Health Development Agency. www.hda.nhs.uk.
10. Clark, D. (2004). *Urban world/global city*. Routledge.
11. Evans, B., Joas, S., Sundback, S., & Theobald, K. (2005). *Governing sustainable cities*. Earthscan.
12. Johnson, B. (2008). Cities, systems of innovation and economic development. *Innovation: Management Policy and Practice, 10*(2–3), 146–155.
13. Knight, R. V. (1995). Knowldege-based development: policy and planning implications for cities. *Urban Studies, 32*(2), 225–260.
14. Trindade, E. P., Hinnig, M. P. F., da Costa, E. M., Marques, J. S., Bastos, R. C., & Yigitcanlar, T. (2017). Sustainable development of smart cities: A systematic review of the literature. *Journal of Open Innovation: Technology, Market, and Complexity, 13*(1), 1–14.
15. European Commission. Smart Cities and Communities—European Innovation Partnership, C. (2012). 4701 final, Brussels, 10.7.2012.
16. Carrillo, F. J. (2004). Capital cities: A taxonomy of capital accounts for knowledge cities. *Journal of Knowledge Management, 8*(5), 28–46.
17. Bridger, J. C., & Luloff, A. E. (2001). Building the sustainable community: Is social capital the answer? *Sociology Inquiry, 71*(4), 458–472.
18. Deakin, M., Hirst, P., Hummerstone, E., Webb, S., Karlsson, A.-K., Blin, A.-S., Duff, M., & Jordanou, M. (2012). JESSICA for smart and sustainable cities. European Investment Bank, Mazars LLP.

19. Haapio, A. (2012). Towards sustainable urban communities. *Environmental Impact Assessment Review, 322*, 165–169.
20. Giffinger, R., Fertner, C., Kramar, H., Kalasek, R., Pichler-Milanović, N., & Meijers, E. (2007). Smart cities: Ranking of European medium-sized cities. Centre of Regional Science (SRF), Vienna University of Technology. http://www.smart-cities.eu/download/smart_cities_final_report.pdf.
21. Lindskog, H. (2004). Smart communities initiatives. In *Proceedings of the 3rd is One World Conference* (Vol. 16).
22. Stratigeia, A. (2012). The concept of 'Smart Cities'. *Towards Community Development? Networks and Communication Studies, 36*(3–4), 375–388.
23. Bibri, S. E., & Krogstie, J. (2017). Smart sustainable cities of the future: An extensive interdisciplinary literature review. *Sustainable Cities and Society, 31*, 183–212.
24. Yigitcanlar, T., Dur, F., & Dizdaroglu, D. (2015). Towards prosperous sustainable cities: A multiscalar urban sustainability approach. *Habitat International, 45*, 36–46.
25. Dempsey, N., Bramley, G., Power, S., & Brown, C. (2011). The social dimension of sustainable development: Defining urban social sustainability. *Sustainable Development, 19*(5), 289–300.
26. Murphy, P. (2000). Urban governance for more sustainable cities. *European Environment, 10*(5), 239–246.
27. Turnbull, S. (2007). A framework for designing sustainable urban communities. *Kybernetes, 36*(9–10), 1543–1557.
28. Gil-Garcia, J. R, Pardo, T. A., & Nam, T. (2016). A comprehensive view of the 21 century city: Smartness as technologies and innovation in urban contexts. In J. R. Gil-Garcia et al. (Eds.), *Smarter as the new urban agenda. A comprehensive view of the 21st century, city* (pp. 1–21). Public Administration and Information Technology, Springer.
29. Albino, V., Berardi, U., & Dangelico, R. M. (2015). Smart cities: Definitions, dimensions, performance, and initiatives. *Journal of Urban Technology, 22*(1), 3–21.
30. Lombardi, P., Giordano, S., Farouh, H., & Wael, Y. (2011). Modeling the smart city performance. *Innovation: The European Journal of Social Science Research, 25*(2), 137–149.
31. Ibrahim, M., El-Zaart, A., & Adams, C. (2018). Smart sustainable cities roadmap: Readiness for transformation towards urban sustainability. *Sustainable Cities and Society, 37*, 530–540.
32. Yigitcanlar, T., Kamruzzaman, M. D., Foth, M., Sabatini-Marques, J., da Costa, E., & Ioppolo, G. (2019). Can cities become smart without being sustainable? A systematic review of the literature. *Sustainable Cities and Society, 45*, 348–365.
33. Huovila, A., Bosch, P., & Airaksinen, M. (2019). Comparative analysis of standardized indicators for smart sustainable cities: What indicators and standards to use and when? *Cities, 89*, 141–153.
34. Komninos, N., Kakderi, C., Panori, A., & Tsarchopoulos, P. (2019). Smart city planning from an evolutionary perspective. *Journal of Urban Technology, 26*(2), 3–20.
35. Komninos, N. (2013). Intelligent cities: Innovation, knowledge systems and digital spaces. Routledge.
36. Lara, A. P., Da Costa, E. M., Furlan, T. Z., & Yigitcanlar, T. (2016). Smartness that matters: Towards a comprehensive and human-centred characterisation of smart cities. *Journal of Open Innovation: Technology, Market and Complexity, 2*(2), 1–13.
37. Caragliu, A., Del Bo, C., & Nijkamp, P. (2011). Smart cities in Europe. *Journal of Urban Technology, 8*(2), 65–82.
38. Deakin, M. (2011). The embedded intelligence of smart cities. *Intelligent Buildings International, 3*(3), 189–197.
39. Dameri, R. (2013). Searching for smart city definition: A comprehensive proposal. *International Journal of Computers & Technology, 11*(5), 2544–2551.
40. Nam, T., & Pardo, T. A. (2011). Smart city as urban innovation with dimensions of technology, people and institutions. In *Proceedings of the 5th international conference on theory and practice of electronic governance* (pp. 185–194). ACM.
41. Allwinckle, S., & Cruickshank, P. (2011). Creating smart-er cities: An overview. *Journal of Urban Technology, 18*(2), 1–16.

42. Meijer, A., & Thaens, M. (2018). Urban technological innovation: Developing and testing a sociotechnical framework for studying smart city projects. *Urban Affairs Review, 54*(2), 363–387.
43. Chourabi, F. J., Gil-Garcia, J. R., Pardo, T. A., Nam, T., Mellouli, S., Scholl, H. J., Walker, S., & Nahon, K. (2012). Understanding smart cities: An integrative framework. In *45th Hawaii International Conference on System Sciences* (pp. 2289–2297). IEEE.
44. Granier, B., & Kudo, H. (2016). How are citizens involved in smart cities? Analysing citizen participation in Japanese 'smart communities.' *Information Polity, 21*(1), 61–76.
45. Concilio, G., Cullen, J., & Tosoni, I. (2019). Design enabled innovation in urban environments. In G. Concilio & I. Tosoni (Eds.), *Innovation capacity and the city* (pp. 85–101). Springer.
46. Castelnovo, W., Misuraca, G., & Savoldelli, A. (2016). Smart cities governance: The need for a holistic approach to assessing urban participatory policy making. *Social Science Computer Review, 34*(6), 724–739.
47. Shearmur, R. (2012). Are cities the font of innovation? A critical review of the literature on cities and innovation. *Cities, 29*, S9–S18.
48. Marceau, J. (2008) Introduction. Innovation in the city and innovative cities. *Innovation, 10*(2–3), 136–145.
49. Leydesdorff, L., & Deakin, M. (2011). The triple-helix model of smart cities: A neo-evolutionary perspective. *Journal of Urban Technology, 18*(2), 53–63.
50. Kunzmann, K. R. (2014). Smart cities: A new paradigm of urban development. *Crios, 4*(1), 9–20.
51. Andreani, S., Kalchschmidt, M., Pinto, R., & Sayegh, A. (2019). Reframing technologically enhanced urban scenarios: A design research model towards human centered smart cities. *Technological Forecasting & Social Change, 142*, 15–25.
52. European Economic and Social Committee. (2015). Opinion of the European Economic and Social Committee on 'Smart cities as drivers for development of a new European industrial policy' C383/24.

FabLab and Digital Manufacturing: Innovative Tools for Social Innovation and Value Co-creation

Marco Savastano, Francesco Bellini, and Fabrizio D'Ascenzo

Abstract Digital manufacturing machines are contributing to the paradigm shift that is determining a new way to design, produce and consume goods. These now happen in a cooperative and/or shared manner allowing the consumer to actively be a part of the process or being him/herself the producer. The aim of this paper is to explore how open production labs such as FabLabs can attract creators and users, boost entrepreneurship and sustainability. We explored the concepts of open production, bottom-up economics and socially sustainable production, identifying which are the key factors for the diffusion and success of FabLabs. Through a number of semi-structured interviews run in the Lazio region FabLabs network, we were then able to identify strengths and weaknesses of the current FabLab models and to set up a framework for future research.

Keywords FabLab · 3D printing · Bottom-up economics · Open production · Social innovation · Makers

1 Introduction

The Third Industrial Revolution, which is currently in progress, is triggered by Internet communication, renewable energies and information [1]. Over the last decade, the disruptive evolution and diffusion of the Internet democratised publishing, broadcasting and communications, increasing the level of participation in everything digital, and creating the world of bits. Nowadays, the same trend is occurring in manufacturing, defining the world of things [2]. The revolution in fabrication is the ability of non-professional users to quickly and cheaply turn data into things

M. Savastano (✉) · F. Bellini · F. D'Ascenzo
Management Department, Sapienza University, Rome, Italy
e-mail: marco.savastano@uniroma1.it

F. Bellini
e-mail: francesco.bellini@uniroma1.it

F. D'Ascenzo
e-mail: fabrizio.dascenzo@uniroma1.it

C. Dal Zotto et al. (eds.), *Smart Technologies for Organizations*, Lecture Notes in Information Systems and Organisation 60, https://doi.org/10.1007/978-3-031-24775-0_12

and things into data through digital technologies [3]. A new economic paradigm is emerging, shifting the focus from hierarchical to consumer power, by fundamentally changing the way business is managed and products are created. The end user, the consumer, becomes part of this ecosystem through a renewed profile: its role changes from passive consumer to active co-creator, more and more originator of ideas and promoter of the customisation process, until becoming the inventor and designer of final products, the "prosumer" [4, 5].

Digital manufacturing machines, such as 3D printers and scanners connected to the Internet, represent good examples of consumer's empowering tools. These technologies reconnect design and manufacturing and can be considered as enablers for the comeback of the designer–producer, together with the concepts of distributed production and entrepreneurship [6, 7]. Moreover, Open source hardware and software, crowdsourcing, big data and analytics are tools that disrupt the way a product is developed, designed and monitored. Likewise, it happens with rapid prototyping and 3D printing. Shortly, culture and technology are two sides of the same revolution [8].

The shift from an industrial to a knowledgeable society is based on the increasing value of the factors "knowledge" and "information" in proportion to the traditional production factors (capital, machines, labour, etc.). Part of the creation process takes place in the virtual sphere, so labour becomes more location-independent and the value creation results cooperative, decentralised and self-organised [9]. In order to keep pace with these innovations and the global competition, companies are pushed to open up to these changes. In fact, the interactive value creation, defined as a partnership between customer and producer, requires strong abilities of knowledge exchange and knowledge management for creating a higher added value [10–12].

More than ten years ago, prototyping equipment such as laser cutters and 3D printers, dramatically dropped in price due to the Open Source Hardware, which further popularised these technologies. This trend allowed machines to become safer and more affordable, and processes simpler and easier to learn, giving the possibility to the owner-maker to re-emerge in a modern shape [13]. Due to these progresses, "digital fabrication" allows people to design and produce tangible objects on demand, wherever and whenever they need them. The widespread availability of these technologies will challenge traditional models of business, foreign aid and education [14].

From 2001, these machines started to become available to the general public in so-called FabLabs (fabrication laboratories), an educational outreach programme promoted by the MIT's Centre for Bits and Atoms in Boston, which in a few years has developed a global network including an increasingly broad range of users at educational institutions and local community centres around the world. A typical FabLab consists of a common space equipped with laser cutters, CNC (Computer Numeric Control) routers, 3D scanners and printers and 3D milling machines, but also electronics workbenches and microcontroller programming tools. It provides stimulus for local entrepreneurship and offers the possibility to learn about versatile digital design and manufacturing technologies, for creating things in individual or collaborative projects [15]. This interdisciplinary set-up and a solidly integrated peer-learning and DIY (do-it-yourself) approach empowered FabLab users to "make almost anything". In this way, fabrication and prototyping is no longer limited to experts or large-scale manufacturing processes: the FabLab becomes an interface to the city, a creative hub opens to different activities and forms of collaboration [6]. In fact, these structures offer services to everybody, which have until recently been reserved to a few professionals only. Accordingly, in parallel to open innovation the concepts of open production and social innovation can be introduced, which have the potential to democratise the production and economic development processes [10]. In addition, each country, and in particular territorial area, presents distinctive and unique characteristics which stimulate a different use and development of technologies and approaches related to these concepts [4].

Therefore, the emerging research question is: can small manufacturing labs, by providing the tools and computing power to make almost anything, infuse new ideas and possibilities into global communities, which share projects and innovative value-added solutions, and give a boost to local entrepreneurship and job creation?

The purpose of this paper is to shed light on the opportunities disclosed by the new models of open production and FabLab glocal network collaboration, in a socially sustainable value creation perspective, addressing the literature gap existing on this topic. The work proceeds as follows: we first briefly review the existing literature on the FabLab history and define its dimensions; then we analyse its implications in economic and social terms and investigate its development in different areas with specific focus on the Italian network. We further present some local initiatives and case studies, in order to evaluate practical benefits of this emerging model.

Conclusively we outline guidelines for the design of technology-enhanced learning and productive environments. We end up with some conclusions and future research avenues.

2 Theoretical Background

2.1 From Craft Manufacturing to Open Production

The meaning of the term "manufacture" is historically and closely linked to the "craft maker" figure: companies and large manufacturing industries often have a common past of creativity, inspiration, dedication and manual labour. The distinguishing features between industry and handicrafts are usually expressed on the one hand by the size and production capacity, and on the other by the uniqueness and specialisation [4].

The manufacturing paradigms have evolved with respect to the origin of design requirements, the number of designers involved, the manufacturing alternatives and the categories of end users.

Craft manufacturing is carried out by artisans in their workshops for specific product consumers. Unlikely, in mass manufacturing, product development is a separated task carried out by a designer, while the production process takes place in a factory with specialised workers. In this configuration, the product's consumers are groups of passive customers which can only select from standardised products, having very little or no influence on their design.

Mass customisation presents many similarities to mass manufacturing, but consumers are provided with a larger selection of goods and may influence the way products are designed through the offered possibilities of personalisation. In fact, products are modularised in order to create a predetermined number of variants for specific segments of end users.

Recently, with the advent of digital manufacturing technologies (e.g. 3D printers, etc.) and additive manufacturing (AM) tools, the design requirements are created by a larger number of subjects. Through the new paradigm of direct digital manufacturing (DDM) parts are no longer produced in a factory, assembled into final products and shipped to customers. These products are manufactured right at or close to the consumption point by using additive manufacturing and directly derived from a digital model [16].

Small-scale enterprises can thrive in this new scenario of distributed manufacturing. We can observe that in some aspects this looks like a return to the earliest days of the First Industrial Revolution, when the spinning jenny changed the world not by creating the manufacturing plant, but by originating the "cottage industry". Cottage industries were powered by wooden-framed machines with foot pedals, able to work with many threads at the same time. They were relatively easy to build or cheap to buy and could be used in a table-sized space: they were the "desktop manufacturing" of the day. Those tools were used at home, creating work for all the family members (both men and women, and lots of children), and making for the first time indoor work more lucrative than outdoor, for most/a large amount of the population. These small-sized enterprises were a distributed form of production, which complemented the big centralised factories by being more flexible and making things in smaller batches, while emphasising and preserving important artisanal skills difficult to replicate by machines at that time. Considering their characteristics, cottage industries—focused on niche markets rather than mass markets—were closer to a Maker-driven New Industrial Revolution than big factories which we now associate with manufacturing [2].

Nowadays, AM is evolving into DDM as an interconnection of additive manufacturing equipment, computers connected to a network and open source designs and tools. DDM offers the possibility of a quicker adaptation of products according to various design values (e.g. usefulness, performance, material selection, high degree of personalisation and aesthetics), stimulating and leading the way to the digital handicraft [5]. Technology and innovation that in the past seemed to have led to the conclusion of the traditional handicraft dimension, strengthening the manufacturing industry, have now become critical success factors for its digital revival. Digital technologies led to an increase in the opportunities for collaboration and design sharing through a direct and continuous communication, in contrast to traditional methods. These networks of individuals consist of the persons involved in producing and consuming the artefacts, the prosumers, which are the main actors of the social innovation and value co-creation processes. Knowledge and information become core production factors of a cooperative and decentralised process, giving the possibility for individuals to act on the same level as companies taking advantage of the opportunities afforded by open production.

Hence, with the progressive adoption of digital manufacturing technologies there is a gradual convergence and rapprochement between the artisan model of creation, design and production, and the large-scale manufacturing dimension, even if it should be noted that DDM is not expected to substitute the established manufacturing paradigms, but rather complement them.

Open Source Hardware (OSH)—which constitutes part of the open production concept—is characterised by source files publicly available for anyone to use, remanufacture, redesign and resell. It is based on the principle that necessary information for the manufacturing of parts to produce goods is published and easily available, potentially allowing everybody to have access to it. This principle represents the foundation of digital making [6, 10].

Open production and DDM paradigms are representatives of a complex and disruptive process of innovation which include, among others, the concepts of dematerialisation, demand driven and collaborative design, social sustainability, on-demand manufacturing and democratisation of production. In fact, these innovations can lead to different outcomes, not only for what concerns the productivity of the manufacturing system, but also for the environment and the entire society [5].

2.2 *Bottom-Up Economics in a Collaborative-Oriented Environment*

The principle of bottom-up economics is based on the collaborative relationships between different stakeholders such as customers and producers. Collaboration is the most intense form of cooperation and relies on reciprocal relationships, emphasising the bidirectional increase of the participants' potential. This mechanism allows to increase the total value creation, leading every participant to reach a win–win relationship [11].

A world famous example of a successful business model, in which the collaborative process in terms of bottom-up economics based on the new ICTs and manufacturing technologies is mastered, is represented by the American company Local Motors, the first open source car company to reach the production [10]. Within this company, the development of new car models has been realised through the cooperation of customers and producers: the design phase was crowdsourced, as was the selection of mostly off-the-shelf components, while the final assembly was done by the customers themselves in local assembly centres as part of the "build experience". The company declared they are able to take a new model from sketch to market in

18 months, about the time it takes Detroit to change the specs of a door trim. Designs are released under a share-friendly Creative Commons license, and end users are encouraged to enhance the designs and produce their own components that they can sell to their peers [17].

These principles constitute the foundations of a socially sustainable economy. First of all, through ICTs and digital technologies private subjects are able to become producers, enabling the creation of small production systems such as micro-factories. Secondly, these producers can interconnect in order to share skills and ideas to develop and produce complex products. Thirdly, based on the development of new business models, their production becomes profitable and a competitive advantage can be gained. Consequently, the paradigm shift based on the new ICTs and manufacturing technologies promotes social sustainability and leads to empowered customers and enabled prosumers. Furthermore, open production—for instance on the basis of micro-factories—has the potential to democratise the production process through a knowledge exchange and a collaboration-oriented industrialisation [10].

Therefore, the idea of Ronald Coase that companies exist to minimise "transaction costs"—since people share a purpose and have established roles, responsibilities and modes of communication it is easier to make things happen—in the era of Internet and new digital technologies has been overcome.

In fact nowadays, working within a company often presents higher transaction costs than running a project online. The web offers an easy access to online community members from a global marketplace of talents, as oppose to turning to the person who happens to work in the desk next to you. Traditional companies are full of bureaucracy, procedures and approval processes, with a structure designed to defend the integrity of the organisation. Communities share interests, ideas and needs, and have no more processes than what they require. The community exists for the project, not to support the company in which the project resides. Hence the new industrial organisational model is built around small players, loosely coupled. Companies are small, virtual and informal. Most participants are not employees. They form and re-form on the fly, driven by ability and need rather than affiliation and obligation. "It doesn't matter who the best people work for; if the project is interesting enough, the best people will find it" [17].

So today we are living a return to a new sort of cottage industry: once again, new technologies are providing individuals with the power over the means of production, allowing for bottom-up entrepreneurship and distributed innovation [2].

2.3 *Sharing Economy and Social Sustainability*

The concept and practice of "sharing economy", which includes a form of "collaborative consumption", suggests making use of market intelligence to foster a more collaborative and sustainable society. Major examples are bike and car-sharing schemes, as well as web-based peer-to-peer platforms covering a broad range of activities, from renting rooms to sharing projects and designs (e.g. Airbnb, Uber, Car2Go, etc.).

This model involves individuals exchanging, sharing, redistributing, renting and donating information, goods, ideas and talent, either organised independently or through commercial organisation on social media platforms. In other words, a socio-economic organisation, no longer based on possession but on sharing and accessing goods and services, that supplies value to users and makes their lives easier by enabling people within and across communities to link together and exchange fundamental skills and information [18, 19].

According to [20], this collaborative environment will deeply change mainstream economies and consumerism, improve social cohesion and contribute to the efficiency of resource use [20]. In this perspective, sharing common spaces and assets for the collaborative creation of value arises as a critical condition for social processes of innovation.

However, the concept of sharing economy should not be limited to collaborative peer-to-peer practices. New developments, especially for sustainability, seem to appear especially at the interface between product service systems, redistribution markets and collaborative consumption. Furthermore, these forms of alternative ownership and usage should not be limited to end-consumer or peer-to-peer sharing but should include business-to-business relationships and the activities of civil society actors and government entities. Following this perspective, the sharing economy might serve as an umbrella concept and encompassing vision, helping to understand and guide new inventions and the institutionalisation of innovative economic practices, roles and interactions of societal actors, enabled by the disruptive diffusion and development of information and communication technologies [21].

The engagement of sustainability science on these topics would contribute to a more detailed understanding of the extent to which the sharing economy could meet expectations regarding effective resource use, strengthening social capital and fostering decentralised value production.

Since social sustainability is not commonly operationalised, different indicators have been developed. The most common key indicators through which the concept is shown are: basic needs, social resources, equality in chances, participation, security of existence, cultural diversity. It can also assume the meaning of global social justice in the sense of good working conditions, fair wage and income equality. Concepts based on open production, instead, take as central criteria of social sustainability the "collaborative participation" (taking part in processes) and "empowerment" (enabling the access, e.g. world market) of the people. These criteria, which will be considered in the next sections, are highly relevant regarding micro-factories and constitute prerequisites of the FabLab model. In fact, open production offers more access points to the value creation chain and therefore increases the amount of intersections and possibilities to participate in international value chains and to compete within the global market [10].

3 FabLab: Origins and Evolution

The Fab Charter provided by MIT's Centre for Bits and Atoms defines FabLabs as "a global network of local labs, enabling invention by providing access to tools for digital fabrication".

A FabLab is a place where ideas become projects, which then become objects or businesses in a knowledge-sharing environment. This process happens in a shared location, through the meeting of creative minds, using common tools and a new generation technology of machines. In these spaces, people come together to learn about digital design and fabrication and create things individually or by collaborating in projects.

FabLab members and visitors are involved in a variety of practices that go beyond digital fabrication and computation processes [15]. In fact, on the one hand it can be considered a technical prototyping platform for innovation and invention, which provides stimulus for local entrepreneurship; but on the other hand it also works as a platform for learning and innovation: a place designed for creating, learning, mentoring, inventing and collaborating.[1]

The first lab was established in 2001 as part of the MIT'S Interdisciplinary Centre for Bits and Atoms from an intuition of Professor Neil Gershenfeld. The FabLab became representative of a principle which until then had never been heard of or implemented in practice: "How to Make (almost) Anything", which expresses how, with new technologies and the diffusion of the relevant knowledge, any individual can create almost anything by himself, or better, by working in a global network with other people who share the same principles. Nowadays, FabLabs constitute a growing global connected network, including over 200 labs around the world, which incorporates communities of makers, students, educators, local inventors, designers, tinkerers, researchers, small businesses and entrepreneurs: a knowledge-sharing network that spans 30 countries and 24 time zones [6]. They share projects and knowledge through ICTs or social media and meet up at international events [15].

Individual FabLabs are open to a wider array of users than had traditionally been the case in industrial design or even DIY communities, allowing a mix of widely diversified activities. By sharing a core set of capabilities and instruments, projects started in one FabLab can be continued and modified in others. In this way, FabLabs can combine entrepreneurial innovation, research and education under a single roof [22]. Moreover, FabLabs can give entrepreneurs a low-cost space for designing and building prototypes as well as acquiring new skills. They can be spaces where students are engaged in design and technology education. They are centres of community-driven innovation, where problems that governments and corporations are not able to address can be solved, and solutions can later be shared with similar communities around the world. At the heart of this idea is the belief that the most sustainable way to bring the deepest results of the digital revolution to developing communities is to

[1] http://www.fabfoundation.org.

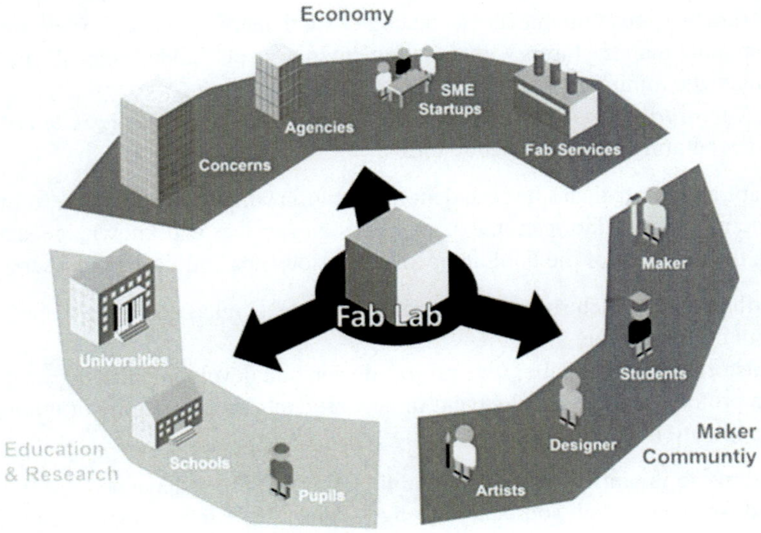

Fig. 1 FabLab's environment and stakeholders. *Source* FabLab Berlin

enable them to participate in creating their own technological tools in order for them to find the solutions to their own problems [23].

Some labs are run voluntarily, while others receive institutional support (e.g. from universities, libraries or local authorities), but all share common values towards providing labs that can be freely accessed by the wider public [15].

Figure 1 shows a representation of the environment where a FabLab operates and the network of stakeholders around it.

Even though it is a technology-driven platform network, the governance of FabLabs is still extremely loose. In order to achieve its full potential it could require facilitating collaboration across different labs and types of users [22].

Although there is no real formal process for setting up a FabLab, which is a movement based on the "open culture", a common set of requirements exists.

In fact, over the years some defining characteristics of a FabLab have been set, resulting in four main conditions to be observed:

1. *Access to the laboratory must be public*, at least in a part of the week. There may be different business models, but the access must be public and preferably free at certain times, in order to provide access to all.
2. *The laboratory must join and show (physically and on the website) the FabLab Charter*, the document that outlines the values of the FabLab community.
3. *The laboratory must have a set of tools and processes shared with the entire network of FabLab*: the idea is that a project carried out in Rome can be easily reproduced in all the other laboratories, wherever they are in the world. There is a list that defines the types of equipment (i.e. machines such as computer controlled laser cutter, numerically controlled milling machine and 3D printers),

instruments and components, but machinery and instruments can be also adopted from other manufacturers. Other equipment may be also added since the list only defines the minimum instrumentation.

4. *The laboratory should be active and involved in the global network of FabLab*, by cooperating with other laboratories.

A FabLab is more than a brick and mortar location where objects can be produced; it is a community of fabrication expertise, design services and knowledge sharing. Indeed, the potential of the FabLab opens up various possibilities for its users:

- distributed manufacturing: the assembly of products released on the web;
- in situ customisation;
- brainstorming and collaborative project design and development;
- open prototyping: the development of new products by community of innovators in cooperation with business.

Similarly to the infrastructure of a public library, FabLabs have become the new standard for public open source knowledge share. These resources for knowledge sharing are decentralised and networked. The development of a decentralised network of FabLabs is just the next step of the trend of user-centred innovation in services and products, who argued that democratic innovation is possible through the rethinking of manufacturing processes that begins with the open sharing of knowledge.

As we progress along that research path, knowledge and best practices are disseminated throughout the FabLab network, making it a distributed cutting edge laboratory for R&D [24].

3.1 Key Factors

Some key factors are critical for the diffusion and success of the FabLab phenomenon [24]:

- *Physical location*: the proximity to infrastructure and industrial clusters is critical. Fab labs should be located within a reasonable distance of key artisanal industrial clusters. In addition, locating a FabLab within an urban area is much better than in the countryside. In this case, students, researchers and the general public would be able to access the facility. The best choice would be to locate it close to a school, business incubator or research institute (e.g. university, etc.). This location would give a better visibility to the FabLab. In addition, in an urban area there would be more closer infrastructures like trains or metro. This would also simplify reintegration activities of resources, such as machines and materials.
- *Facility design*: Building a FabLab is more than just purchasing a machine and plugging it in.
- Machining techniques involved in digital fabrication include to consider critical issues such as ventilation, structural stability and fire safety become critical. Moreover, in addition to standard equipment, additional machines and tools have to be

chosen according to the specific uses and activities (taking into account not only artisans but also research institutions in the proximity of the FabLab), meeting the changing demands of the community it engages.

- *Marketing initiatives:* as with any community, visibility is important. Many different strategies can be deployed for the success of a FabLab. External marketing strategies should be varied depending on different community members' target. Digital media will reach youths, while in order to attract the attention of artisans and community leaders many more traditional marketing strategies must be deployed. Therefore, it is crucial for the FabLab to properly manage such diverse communication channels (e.g. online and offline).
- *Research initiatives:* The types of research conducted within the sphere of digital fabrication, and more in general inside the FabLab, are numerous. FabLabs must be connected to universities and take on these research initiatives in order to succeed. Indeed students and researchers will be attracted by the prospect of gaining professional experience through the FabLab, and of being connected to industry leaders.
- *The community model:* clarifying the organisational structure of such an entity is critical for its success. The knowledge produced within the FabLab should be "open" in nature and create a community of innovators, inventors and entrepreneurs. The culture of the lab and its role in community engagement is fundamental to its success.

3.2 FabLab Global Network

Although the maker and FabLab movements find their origins mainly in the U.S., after a few years these concepts arrived in Europe and a fast diffusion was then triggered in our continent. FabLabs and hackerspaces initially found fertile ground especially where technological culture is deeply rooted even among young people, that is in Germany and the Netherlands; but recently there has been a proliferation of initiatives in many European countries (especially in Italy, as we will further analyse). For instance, one of the first and most famous FabLab in Europe was founded in Barcelona. It is very active and, thanks to its promoters, participants and leaders, it became a global benchmark [25].

FabLabs are individually useful, but what makes them a promising global solution network is that their capabilities and impact on individuals, communities and companies are multiplied by collaboration and shared innovation. Projects initiated at one FabLab can be adopted, modified and improved for local conditions by other nodes in the network [22].

There are a number of uses for digital fabrication spanning many market sectors. Each discipline within the arts, design and sciences approaches digital fabrication machines in a new way. Different disciplines and subjects with specific capabilities will see the same machine differently, while in other (more traditional) markets specialised applications require entirely different types of tools [24].

The most interesting initiatives within the FabLab community can be grouped into five categories, connected to the same number of sectors [26]:

- medicine (e.g. development of low-cost biomedical technologies);
- archaeology (e.g. restoration projects);
- design complements (furnishing printable in 3D);
- robotics (e.g. development of new models of 3D printers);
- education and research (e.g. innovation in materials, collaboration with universities).

The FabLab Network is an open, creative community of fabricators, artists, scientists, engineers, educators, students, amateurs, professionals, located in more than 40 countries in approximately 200 FabLabs with participants aged from 5 to 75 and more. From community-based labs to advanced research centres, the FabLabs connected to the network share the mission of democratising access to the tools for technical invention. This community is simultaneously a manufacturing network, a distributed technical education campus, and a distributed research laboratory working to digitise fabrication, inventing the next generation of manufacturing and personal fabrication and providing everyone with access to knowledge.

Among the main activities carried out by the global networked FabLabs there is the FabAcademy,[2] an advanced technical training course held by professor Gershenfeld in streaming. Practical exercises are performed physically at the various local workshops, implementing the principle of collaboration and simultaneity made possible by the digitisation of manufacturing processes.

3.3 The Italian Market

Following Gershenfeld's intuition of 2001, two essential facts subsequently determined the future of the makers movement. They both took place in Europe, one of which in Italy.

In 2005, Adrian Bowyer—mechanical engineering professor at the University of Bath—launched the "RepRap" project (Replicating Rapid Prototyper), which created a 3D printer capable of producing its own components. Since the project was open source, anyone can now build his/her own 3D printer on a low budget. The rise of cheap and increasingly advanced desktop printers was born in Bath.

The second event took place in Italy, during the same period: the School of Interaction Design of Ivrea created Arduino, an Open Source Hardware and Software platform (produced in Olivetti factories) from which a series of cheap and easy microcontrollers have been developed. It allows anyone, with little money and no technical expertise, to realise electronic projects making innovation accessible to everyone. Indeed, it is not a coincidence that Italy is one of the most important countries for the makers movement. Italy has always been recognised worldwide as a

[2] http://fabacademy.org/.

country of inventors, artisans, small businesses and entrepreneurs able to combine precision engineering, high quality and design. This has always been behind the "Made in Italy", that the digital revolution can further strengthen.

Moreover, since 2013 the most important maker exhibition in Europe, the Maker Faire, is organised in Rome. According to the official data from the Fab Foundation, Italy is also the second country in the world for number of FabLabs, after the U.S.A.: in the last three years it grew up from zero to one hundred [8].

During 2013 some researchers from MIT, together with the project Prospera, engaged a number of Italian artisans, students and small business owners in Tuscany for a collaborative digital fabrication workshop, to test what a FabLab and digital maker community might do for this area. Results of the research conducted confirmed the initial hypothesis: "the small-scale production offered within FabLabs would afford artisans the potential to expand their skill sets and prototype new and innovative products. These products could then be produced at a larger scale within their own workshops". Moreover "FabLabs could connect artisans with advanced contemporary digital production techniques that better inform their processes and inspire new innovations in manufacturing. Additionally, FabLabs could be excellent places for artisans to produce tools that they would traditionally purchase at a high cost or with long lead times from outside vendors. This would allow artisans to innovate their production processes and become more competitive with their fabrication methods" [24].

3.3.1 FabLab Trend

As mentioned before, the FabLab phenomenon in Italy shows strong vitality and a fast growth rate. This positive trend finds evidence in the investigation "Censimento Laboratori Fabbricazione Digitale" [25].

This study shows that the number of FabLabs in Italy had a steady growth, exceeding a hundred units in 2015 and presenting a widespread distribution over the country. Reasons of the growth are both the explosion of the Makers' phenomenon and the activation of supportive public policies in some regions of the country. Furthermore, the study reveals/points out that the potential of FabLabs are yet to be explored and developed in their relationship with the manufacturing companies [8].

The chart (Fig. 2) shows the number of FabLabs in Italy divided by region.

From a further qualitative analysis, two different models of FabLab emerge:

- a part of the FabLabs is set to become a stable partner for businesses and entrepreneurs;
- another part focuses on the task of promoting the digital manufacturing culture for businesses, artisans and citizens.

In fact, each initiative presents a unique and different story: there are groups of enthusiasts who share expenditures and give life to a local community, public agencies that invest in training, companies that believe in new methods of research and

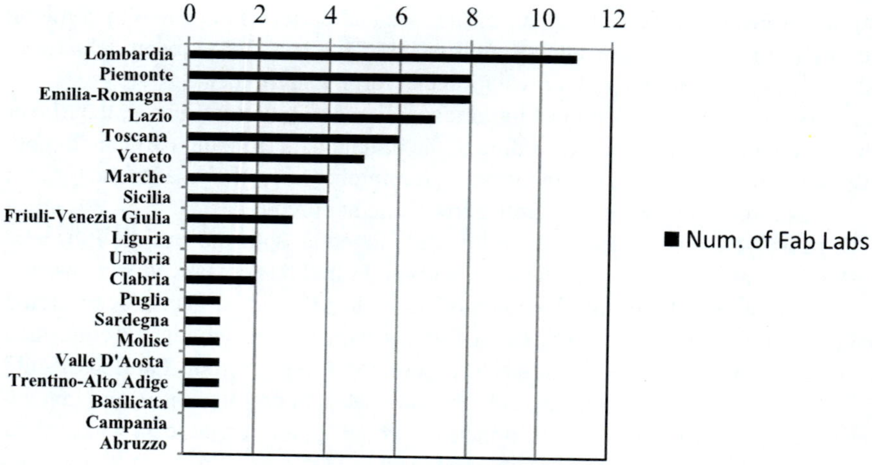

Fig. 2 Number of FabLabs per Region in Italy

design, schools leading practical experience in teaching activities, FabLabs within universities and more. Some initiatives are born with a clear hobby intention, whereas others as start-ups or economic development points. There are spaces developed to support other existing realities as co-working, companies and universities, providing them with a pool of people and complementary activities, and others born to offer technical training to a wide audience [25].

4 Case Histories

To complete this study, we introduced some case histories of FabLabs located in Lazio area (in central Italy), which in the last couple of years have been able to interpret the opportunities offered by this innovative model and present themselves as benchmarks within the national scene and beyond. For this reason, we organised in-depth interviews with FabLabs' responsible/founders, in order to understand the drivers of success and the vision behind this, by obtaining a direct testimony from those who work daily in this environment. Interviews (which last about an hour each) are based on a semi-structured track with open questions, in order to define the contents to be treated without, in any way, limiting the respondents to express freely their opinions and experiences [27]. The respondents, in fact, were solicited to describe the values and objectives of the FabLab in great detail and in-depth explanations, with a specific focus on the benefits and the potential that these initiatives can offer to the local entrepreneurship, artisans and maker communities, from an economic development and sustainability viewpoint. A data triangulation has been implemented combining data from the interviews with data gathered from secondary sources (websites, reports, articles, etc.), in order to guarantee greater validity to the investigation.

We first interviewed Riccardo Panunzio, Director-general (General Director) for the BIC Lazio FabLab of Rome; Rosita Esposito, Lab Manager in the same office/structure and Stefano Capezzone, Co-Founder of Roma Makers, one of the first and most important independent FabLab in Rome, and President of Blu Sistemi Spa, which managed the launch and the start-up period of the Lazio "diffused FabLab". We then met Marco De Carolis, Lab Manager in the FabLab of Viterbo, hosted inside the ICult incubator of BIC Lazio.

FabLab Lazio is a project financed by the Regional Operational Plan (POR FESR) that the Department of Culture and Youth Policy of the Lazio Region has decided to develop in order to pursue the goal of rethinking the territory as a place for social innovation and experimentation of new production models. Following this mission, the "diffused FabLab" was created, which operates on the territory simultaneously in several places, closer to its users. In fact, it presents four different locations connected to many themes:

- FabLab Roma, specialised in Creative Industry, Fashion, Design and Interactive Arts;
- FabLab Bracciano, specialised in Agrifood;
- FabLab Viterbo, specialised in the Cultural Industry and Protection of Artistic and Cultural Heritage;
- FabLab Latina, specialised in Life Sciences and the Economy of the sea.

The four FabLabs have been equipped with laboratories and advanced machines in compliance with the Fab Foundation guidelines but declined according to the specialisation allocated to the different locations. Each FabLab is strongly linked to the culture of the area where it is located and is embedded inside the business incubator of BIC (Business Innovation Centre) Lazio, whose mission is to support the development of the region through the creation of new businesses (e.g. start-ups) and the expansion of existing ones, by providing effective managerial and technical consulting tools and a direct connection to the business world through local chambers of commerce. As part of the "Spazio Attivo" project, a new regional network of innovation hubs is emerging; large meeting spaces, open to the territories and cooperation with local authorities, in which there will be access to all services for businesses, start-ups, training and work. In these places, entrepreneurial training, business incubation, co-working and FabLab activities will co-exist through a cross-fertilisation, divided into three areas: training, digital and interactive.

Another element of uniqueness that characterises them is that all the services offered (courses, machines and the majority of materials) are entirely free for interested citizen-users, since they are financed by the regional administration. This characteristic can be found only in a few FabLabs based in universities, mainly in the U.S. or Israel (but it is usually limited to students and researchers).

Riccardo Panunzio and Stefano Capezzone observed that these four FabLabs have been created and developed as a "living machine", based on a perfect combination of mechanical instruments and human capital, i.e. the community, emphasising all its values and cultural traits through contamination and interdisciplinary approach. In fact, the performance and importance of a FabLab can be defined and assessed based on some specific key indicators as follows:

- Number of members (i.e. people who use machines and tools);
- Community (number of users and visitors);
- Number of activities and education hours (i.e. programming and design, 3D printing, etc.);
- Number of projects managed within the FabLab (partially shared with the network).

Projects are driven by the open source model and shared through the community, but intellectual property of what has been invented and developed belongs to the people who did it.

With regard to the relevant set of laws, in Italy FabLabs are regulated by the specific standards for the type of company/association formed (e.g. S.r.l., S.p.a., no-profit or public activity, etc.). Evidently, all the instruments and machines installed in the labs have to respect European safety standards. According to the format, the organisation chart of the four FabLabs foresees a fab manager and a fab technician.

To use the machines and tools provided by the structure, registered users must book in advance, checking for their availability on the FabLab calendar: during the educational activities, machines are reserved for teaching, whereas normally they can be booked for personal/experimental activities or to realise specific projects developed within that FabLab. The only limitation is that users are not allowed to use machines for personal commercial activities, producing parts in more than one piece. Indeed, machines are made available for rapid prototyping only, not for a real low-scale production (which is provided in some private FabLabs by paying the rent of machines and the material used). Therefore, it is important to declare initially the purpose of their use and the project specifications.

Projects developed at the FabLab are supported by the structure facilities and mentored by staff from a technical and commercial point of view. The aim of the FabLab is to demonstrate the substantial feasibility of a project or product, leading it up to the stage of pre-industrialisation. Through the promotion of the digital culture and the practical training provided to the users, the FabLab makes them aware of the great potential offered by these new instruments and the growth path to be taken in order to fulfil their needs.

On the contrary, alternative models of FabLab, mainly private or born spontaneously, have a different life cycle and distinctive features: in a first stage, the beta version of the FabLab, they consist of an open space with basic machinery (usually a cutting plotter and a 3D printer) made available by some members and by an emerging community. In this early stage, the main purpose is the divulgation of digital knowledge. In a second phase structured training activities start and a contribution to the expenses is asked to members (for instance through a membership fee and a minimum charge for the use of machines). In this way, the FabLab can obtain better machines (designed also for the production) and move to a larger space and becomes a reference point on the territory. For example, this happened when craftsmen were in crisis due to an absent market for their activities and they joined together to form the Roma Makers. They, consequently, discovered the power of digital manufacturing tools, in which they could never have previously directly invested. Therefore, the FabLab reaches the next phase: it becomes a shared laboratory, where local artisans can use laser machines all day long sharing costs and ideas in a sharing economy perspective. This will give an impulse to contamination and support the local economy. Recently, for instance, from the cooperation of a dressmaker artisan—with an innovative idea—and an entrepreneur in crisis—owner of laser cutting equipment—who met at Roma Makers FabLab, a successful business was set up. In addition, BIC FabLabs host periodically hackathons, which is an extreme form of brainstorming/challenge that often leads to unexpected results by speeding up the process of creating a project: in 24 h participants pass through stages of team building, brainstorming and challenges in order to meet a need or solve a problem. Sponsors can find interesting resources by observing the way participants work and collaborate.

5 Results and Implications

Evidence resulting from our interviews points out that knowledge sharing and diffusion are the key principles for a FabLab and in particular for the model under consideration. Education and training are strictly focused on the digital culture and oriented to the use of digital manufacturing machines (i.e. 3D modelling/CAD software, rapid prototyping, 3D printing and fabrication, digital programming—e.g. Arduino—etc.), in order to fulfil users' needs and support them through the different phases of project development.

The BIC FabLab model has a specific focus on self-employment, in synergy with the opportunities offered by the surrounding space (business planning, market analysis, incubation, etc.). It provides users with all the necessary instruments and knowledge essential to pass from a simple idea (i.e. "talent working" spaces) to a product (through the prototyping process), to a real business based on that product. As a result, the incubator is evolved in "social innovation space", in which the community is provided with different types of tools:

- "hard" tools: laboratory machines, the FabLab;
- "soft" tools: mentoring and economic support to start the business.

This is not just a simple evolution of the classic incubator but a real transformation of the entire economic system, which increasingly relies on the ability of the individual and the power of the crowd (e.g. crowdsourcing and crowdfunding models) since more and more often these are the sources of the best solutions, and provide costs hugely lower than traditional processes of R&D.

If the community receives the right tools and the necessary support, there is then a spontaneous and collaborative development of innovative projects. In this context, we can consider that the growing co-design within the international community of

FabLabs, bases its strong potential mainly on human bonds supported by sophisticated ICT tools. Nevertheless, the moments of greatest creativity are reached during the physical meeting between community members belonging to different FabLabs. The starting point is often an international challenge launched within the world network, to trigger a shared process of "open problem solving". Projects can be launched from one part of the world and developed in a different part.

Another important connection can be created with universities and research centres. The FabLab is in fact a structure open to everyone involved in the research process, which the academic world could use to give a practical validation to theoretical researches, in a cross-fertilisation model. In this process, FabLabs work as a social hub which links the research and the go-to-market phases. In fact, BIC opened connections with all the universities, and framework agreements are still ongoing; but the real contamination depends on the single behaviours and strategies of each institution and professor. Similarly, important players such as companies and entrepreneurs are invited to meet start-ups and makers who are developing products to sell on the market, during different weekly activities designed ad hoc (i.e. "small talk", "share or die", etc.). This collaboration can become a win–win relationship in which companies invest in new ideas that can be translated into innovative products or services to delight customers.

The traditional top-down pattern of development is thus reversed in a bottom-up process which leads to a huge reduction in time-to-market through a rapid sharing of expertise.

From a sustainability point of view, the concept of FabLab is characterised by embedded values of environmental, economic and social sustainability:

- Environmental:
  - processes use recycled, recyclable and compostable materials (i.e. PLA filament for 3D printers);
  - reduction of pollution linked to transport;
- Economic:
  - sharing economy principals applied to the use of machines and tools, allowing the access to innovative technologies to people who otherwise could not use them;
  - through the open manufacturing, production process becomes locally distributed, reducing logistics costs;
  - optimisation of the use of a common space to carry out different activities
- Social:
  - sharing and dissemination of digital culture through training courses and activities;
  - support for start-ups and artisans or SMEs in crisis;
  - creation of added value for local entrepreneurs.

Table 1 FabLab's KPIs

KPIs/FABLAB	Roma	Viterbo	Bracciano	Latina	TOT
Courses participants	169	193	62	11	435
Workshops participants	155	151	52	11	369
Visitors	1.160	1.525	543	110	3.338
Machine hours	804	205	27	8	1.044
Ongoing projects	27	27	12	–	66
Published projects	16	5	3	–	24
Number of courses	20	18	17	1	56
Number of workshops	18	17	12	1	48

Results achieved by BIC Lazio FabLabs after less than one year of activity are very positive and consistent with the FabLab principles:

- the two projects that have followed the first "open lab" training in February 2015 (the course designed to create a start-up), changed from a business idea to a commercial company in 8 months.
- data related to the abovementioned FabLab key performance indicators are shown in Table 1 (to take into account that the FabLab of Latina opened on the 9th of November 2015).

We can observe that, in less than one year of activity, the community of visitors and participant members grew rapidly (almost 3500 visitors until November 2015) as well as the number of courses and projects carried out at the Lazio FabLabs.

6 Conclusions and Future Research Directions

This study examined the state of art of the FabLab model, by analysing existing studies and successful case histories, with a specific focus on the Italian market. We tried to deeply understand which values and enabling factors drove this disruptive movement and its rapid global expansion.

We verified that through the knowledge sharing and the rapid diffusion of the digital culture, based on the principles of open source and cooperation, a social innovation process is possible and can lead to a new phase of economic development.

In fact, the real strength of a FabLab is not only technical, it is social. The innovative people that drive a knowledge economy share some common beliefs. FabLab is a common space, characterised by few rules and a standard equipment, where different ideas and skills can be contaminated to build up innovative proposals in a creative bottom-up approach.

The theme of the importance of organisational culture as a core part of the competitive advantage has been also much debated in scholarly literature.

Users and consumers become producers and makers, giving a tangible meaning to the term prosumer.

Advanced tools such as ICTs and digital manufacturing technologies (3D printing, etc.) dramatically reduce the distance between an idea and its realisation, by connecting people in communities and giving them the opportunity of sharing projects all around the world and developing them on a collaborative basis. It is not necessary to invest in a massively expensive plant or acquire a vast workforce to turn ideas into reality. Manufacturing new products is no longer the domain of the few, but the opportunity of the many [2].

The aim of the FabLab, from its first conceptualisation made by Prof. Gershenfeld at MIT, and through its evolution, is to provide everyone with the same instruments and free the human intellect, through its differences, to exploit them in different ways. It can be considered as the starter of an engine: it is the spark of self-employment and the boost for a digital renaissance of the rapidly declining traditional craft activities.

As the present research pointed out, in the FabLab concept all the three dimensions of education, business incubation and research are present and mixed in the most balanced manner. In fact, the FabLab is configured as a hub where the diffusion of digital culture, the mentoring of innovative ideas and the connection with the world of research and business can lead to the emergence of successful projects. Global proposals can be developed locally, through the open network.

The power of crowd-based innovation enables a rapid and steady growth, and overcomes the limitations and the high costs associated with a top-down process of R&D.

Our data from the Italian market, and in particular concerning the region of Lazio, gave us a picture of a constantly growing movement, supported by the number of FabLabs on the territory, members of the community, major events (i.e. the Rome Maker Faire), courses provided and projects developed. The Italian economy, consisting mainly of small businesses and a strong artisan culture, appears as a fertile ground for these initiatives.

Despite these important insights, some limitations emerged. In fact, more work is needed to integrate and develop the FabLab approach in school education, and particularly the collaboration with universities and research centres needs to be improved. There is still the lack of an organic state support, not limited to a regional level. Moreover, although it is a global phenomenon, the FabLab model will disclose its full potential when it will become a stable reality for the neighbourhood, through a broader distribution of local proximity structures. This is the only way to fully achieve the goal of social innovation. Finally, since the present work is focused mainly on the Italian dimension of this phenomenon, further comparative studies are needed in order to highlight differences with other countries and identify common drivers of growth to invest in for future developments.

Acknowledgements We would like to thank Stefano Capezzone (Roma Makers and Blu Sistemi Spa), and Marco De Carolis (FabLab of Viterbo), for their kind availability and professional contribution.

A special thanks goes to Riccardo Panunzio and Laura Tassinari, at the General Directorate of BIC Lazio, for their fundamental collaboration and support to our research.

References

1. Rifkin, J. (2011). *The third industrial revolution: How lateral power is transforming energy, the economy, and the world*. Macmillan.
2. Anderson, C. (2012). *Makers: The new industrial revolution*. Random House New York.
3. Reifschneider, L., Kaufman, P., Langrehr, F. W., & Kaufman, K. (2015). Marketers understanding engineers and engineers understanding marketers: The opportunities and constraints of a cross-discipline course using 3D printing to develop marketable innovations. *Marketing Education Review, 25*(3), 259–273.
4. PwC. (2015). Digital manufacturing. Cogliere l'opportunità del Rinascimento Digitale.
5. Chen, D., Heyer, S., Ibbotson, S., Salonitis, K., Steingrímsson, J. G., & Thiede, S. (2015). Direct digital manufacturing: definition, evolution, and sustainability implications. *Journal of Cleaner Production, 107*, 615–625.
6. Mostert-Van Der Sar, M., Mulder, I., Remijn, L., & Troxler, P.: FabLabs in design education. In *Internaional Conference on Engineering and Product Design Education* (pp. 629–634).
7. Rifkin, J. (2012). The third industrial revolution: how the internet, green electricity and 3-D printing are ushering in a sustainable era of distributed capitalism. The World Financial Review.
8. Make In Italy CDB: Make in Italy. Il Made in Italy al tempo dei Makers (2015).
9. Wulfsberg, J. P., Redlich, T., & Bruhns, F.-L. (2011). Open production: Scientific foundation for co-creative product realization. *Production Engineering, 5*(2), 127–139.
10. Basmer, S., Buxbaum-Conradi, S., Krenz, P., Redlich, T., Wulfsberg, J. P., & Bruhns, F.-L. (2015). Open production: Chances for social sustainability in manufacturing. *Procedia CIRP, 26*, 46–51.
11. Ichijo, K., & Nonaka, I. (2007). *Knowledge creation and management. New challenges for managers*. University Press.
12. Krishna, A., Lazarus, D., & Sunishtha, D. (2013). Co-creation-channel: A concept for paradigm shift in value creation. *Journal of Management Science and Practice,* (1/1), 14–21.
13. Blikstein, P. (2013). Digital fabrication and " Making " in education: The democratization of invention. In: J. Walter-Herrmann & C. Büching (Eds.), FabLabs: Of machines, makers and inventors. Transcript Publishers.
14. Gershenfeld, N. (2012). How to make almost anything: The digital fabrication revolution. *Foreign Affairs, 91*(6), 58.
15. Hielscher, S., Smith, A., & Fressoli, M.: WP4 case study report: FabLabs. Report for the TRANSIT FP7 Project (2015).
16. Gibson, I., Rosen, D. W., & Stucker, B. (2010). Additive *Manufacturing technologies—Rapid prototyping to direct digital manufacturing*. Springer
17. Anderson, C. (2010). The next industrial revolution, Atoms are the new bits. Wired.
18. Felson, M., & Spaeth, J. L. (1978). Community structure and collaborative consumption: A routine activity approach. *American Behavioral Scientist, 21*(4), 614–624.
19. Nica, E., & Potcovaru, A.-M. (2015). The social sustainability of the sharing economy. *Economics, Management, and Financial Markets, 10*(4), 69–75.
20. Botsman, R., & Rogers, R. (2010). *What's mine is yours*. Co.
21. Heinrichs, H. (2013). Sharing economy: A potential new pathway to sustainability. *GAIA, 22*(4), 228–231.
22. Global Solution Networks: The FAB LAB Network (2014). A global platform for digital invention, education and entrepreneurship.
23. Mikhak, B., Lyon, C., Gorton, T., Gershenfeld, N., Mcennis, C., & Taylor, J. (2002). FabLab: An alternate model of ICT for development. Development by Design (DYD02) (pp. 1–7).

FabLab and Digital Manufacturing: Innovative Tools for Social Innovation and Value Co-creation

Marco Savastano, Francesco Bellini, and Fabrizio D'Ascenzo

Abstract Digital manufacturing machines are contributing to the paradigm shift that is determining a new way to design, produce and consume goods. These now happen in a cooperative and/or shared manner allowing the consumer to actively be a part of the process or being him/herself the producer. The aim of this paper is to explore how open production labs such as FabLabs can attract creators and users, boost entrepreneurship and sustainability. We explored the concepts of open production, bottom-up economics and socially sustainable production, identifying which are the key factors for the diffusion and success of FabLabs. Through a number of semi-structured interviews run in the Lazio region FabLabs network, we were then able to identify strengths and weaknesses of the current FabLab models and to set up a framework for future research.

Keywords FabLab · 3D printing · Bottom-up economics · Open production · Social innovation · Makers

1 Introduction

The Third Industrial Revolution, which is currently in progress, is triggered by Internet communication, renewable energies and information [1]. Over the last decade, the disruptive evolution and diffusion of the Internet democratised publishing, broadcasting and communications, increasing the level of participation in everything digital, and creating the world of bits. Nowadays, the same trend is occurring in manufacturing, defining the world of things [2]. The revolution in fabrication is the ability of non-professional users to quickly and cheaply turn data into things

M. Savastano (✉) · F. Bellini · F. D'Ascenzo
Management Department, Sapienza University, Rome, Italy
e-mail: marco.savastano@uniroma1.it

F. Bellini
e-mail: francesco.bellini@uniroma1.it

F. D'Ascenzo
e-mail: fabrizio.dascenzo@uniroma1.it

C. Dal Zotto et al. (eds.), *Smart Technologies for Organizations*, Lecture Notes in Information Systems and Organisation 60, https://doi.org/10.1007/978-3-031-24775-0_12

Exploration of Transformational Leadership in Innovation—The Case of the Lebanese Banking Sector

Dina Sidani and Bissane Harb

Abstract Perceived as the driving force of organizational performance, innovation has become today the major challenge of all institutions. Based on leadership theories, the objective of this study is to analyze the contribution of the components of transformational leadership to the development of different types of innovation within the banking sector in Lebanon. Exploiting a qualitative approach based on semi-structured interviews conducted with seven senior executives occupying key positions in five major commercial banks in Lebanon, our research confirms the essential role of intellectual stimulation and inspirational motivation of transformational leadership in the banking innovation, while the other two dimensions, the idealized influence and individualized consideration, play a very limited role in the banking sector. This study allows us to better understand the role of each of the transformational leadership dimensions in order to promote and escort innovation in the banking sector.

Keyword Transformational leadership · Innovation · Banking sector in Lebanon

1 Introduction

At the dawn of the twenty-first century, with the abolition of borders and globalization, innovation is no longer a choice but an absolute necessity for the growth and survival of any institution that evolves in an increasingly unstable and fluctuating environment. According to Damanpour et al. [1], innovation is now unavoidable because it allows the organizations to adapt to their environment and the resulting technological changes and encourages them to do so. Having become a major issue as a key factor of success and more competitive, innovation is perceived as the driving force of institutional development [2].

D. Sidani (✉) · B. Harb
Faculty of Business and Management, Saint-Joseph University of Beirut, Beirut, Lebanon
e-mail: dina.sidani@usj.edu.lb

B. Harb
e-mail: bissan.harbbaghdadi@usj.edu.lb

C. Dal Zotto et al. (eds.), *Smart Technologies for Organizations*, Lecture Notes in Information Systems and Organisation 60, https://doi.org/10.1007/978-3-031-24775-0_13

Leadership style has been identified as the most important criterion for influencing innovation. Thus, many studies have shown that leaders are a key factor in organizational creativity and innovation [3, 4]. In fact, leadership plays a key role in the good governance of corporate businesses, and consequently in the implementation and concretization of the adopted strategies, particularly that of innovation. According to Sarrors et al. [5], the leader is the one who can influence the emergence, acceptance, and adoption of new ideas, and this through the setting of specific objectives and the creation of a work environment promoting a culture of innovation. Under the conditions of increased global competition, fast-changing market situations, and continuous customer demand for quality services [6], information and communications technologies innovation is seen today as a determining factor in the success and competitiveness of organizations [7].

Shown in the official speeches as a source of growth and progress, innovation is the preferred means for both sustainable development in the so-called developed countries and an improvement in the standard of living in developing countries. Thus, innovation can be a lever to put a developing country like Lebanon in development and prosperity trajectories. Furthermore, banks around the world face multiple challenges such as finding new sources of growth, controlling their operating costs, and meeting the constantly changing expectations of their customers. In order to meet these multiple challenges, banks are concentrating their investments more on information and communications technologies and are increasing their investment in this type of innovation.[1] ICTs have profoundly transformed the banking industry and have enabled the banks to increase their profits by reducing transaction, research, and marketing costs, reduce the risks associated with financial intermediation, cope with competition and circumvent the regulations [8]. In addition, one of the most significant developments in the banking industry in the recent years has been the development of new distribution channels made possible by innovation in information and communications technologies. Electronic banking services have been rapidly established by major international banking institutions both as autonomous operations and as part of an established banking service [9, 10].

According to Sobreira [11], the development of information and communications technologies has also contributed to the rise of financial innovation, which has profoundly transformed the banking industry. Innovation has thus enabled the banks to increase their profits by reducing transaction, research, and marketing costs, reduce the risks associated with financial intermediation, cope with competition, and circumvent regulations [8].

The Lebanese banking sector, which plays a key role in the development and modernization of the Lebanese economy,[2] is an interesting field of study because any innovation in this sector is reflected in the growth and economic development of the entire country. In this context, our communication aims to suggest reflection trails and provide answers to the following questions: How does transformational leadership promote the implementation of innovations within the Lebanese banking

[1] Innovation in retail banking 2016, EFMA innovation report.

[2] Commercial bank assets account for 400% of GDP, Banque du Liban report, 2017.

sector? How do transformational leadership dimensions influence innovation within this sector?

Thus, by referring to leadership theories, particularly the one developed by Bass [12], we want to examine how the theory of transformational leadership makes a significant contribution to the process of ICT implementation within the framework of the Lebanese banking sector. The objective is to verify the extent to which theories of style and transformational leadership are valid in the context of ICT implementation. As a first step, we will elucidate the two key concepts for our research, namely transformational leadership and innovation, and conclude the theoretical part with the results of previous research, which have examined the relationship between transformational leadership and innovation.

As a second step, we will expose the empirical study carried out with the senior leaders of the banking sector in Lebanon. Finally, we will present the main contributions of this article by gathering the elements likely to answer to our research questions. This research fills an important gap in the literature and adds new insights to our understanding of how to achieve more performance in the process of integrating the ICTs in the banking sector.

2 The Concept of Innovation

The concept of innovation has been addressed by several researchers. It was firstly introduced by Shumpeter [13] and was applied as the process of creating new business brands. Since then, several researchers have offered several explanations for this concept during the life of the organizations. Roger [14] defined innovation as an idea, practice, or project that is perceived as new by an individual or other unit of adoption. In this definition, the notion of new idea introduces the design process, which is a mental process that allows generating new ideas or concepts. In addition to this design process, the implementation process, modeled in 1986 by Kline and Rosenberg [15], is "a succession of steps that go from the identification of the need to the writing of a specific notebook for manufacturing". The activities encompass both processes (design and implementation) and go beynd through manufacturing and placing on the market. Hatchuel et al. [16] distinguish between design and implementation (development) and emphasize on design as a characteristic feature of innovation. Durieux [17] considers innovation as an important ongoing process for the survival and development of the enterprises, which should now be able to innovate in all aspects of their organization.

The OECD (https://www.economie.gouv.qc.ca) defines innovation as the implementation of a product (good or service) or a new or significantly improved process, a new marketing method, or a new organizational method in business practices, workplace organization, or external relations.

According to De jong and Hartog [18], innovation is the creation of new ideas, products, and processes that have concomitant effects on organizational performance. According to Baregheh [19], innovation is a process that manifests itself at different levels through which organizations transform ideas into new improved products, services, or processes in order to be competitive. Nusair et al. [20] support this definition by viewing innovation as the development and implementation of new ideas, methods, and procedures in order to achieve organizational goals effectively.

In its broadest sense, innovation is defined as "the generation and adoption of new ideas or new behaviors pertaining to a device, system, process, policy, program, product or service which is new to the adopting organization [1]". For the purposes of our study, we will define innovation as all activities leading to the launch of new products and/or new processes, as well as the implementation of new management methods and practices in the business due to this innovation.

Innovation takes many forms. Thus, it can be considered from several angles that vary according to the researchers' point of view and according to their field of research. Saenz et al. [21] distinguish between radical innovation (rupture), and incremental innovation (progressive). Radical innovation profoundly changes the usual references and involves the use of new skills and technologies. Technological choices are then guided by the values of the founder-innovator. As for incremental innovation, it leads to a progressive improvement of references and does not require new know-how. It is an adaptation and is perceived as an acceptance/openness to new ideas. Furthermore, Lin [22] distinguishes between innovation in terms of process and the product innovation. Process innovation provides tools to protect and improve quality and cut costs [23] and includes the creation of new methods of production, distribution, or delivery. Product innovation is the creation of a new product or a major improvement of an already existing product. Damanpour and Avellaneda [1] agrees more or less to this distinction by retaining two dimensions of innovation: technological innovation and administrative innovation. Technological innovation refers to the technical sphere of the business and is usually related to technology. It can be a new product, service, or process. In addition, administrative innovation refers to the social sphere of the business. It occurs in the social system of an organization. This concerns the recruitment, authority, rewards, and structuring of tasks or the allocation of resources. Edquist et al. [24] integrated the two typologies of product/process and technological/administrative innovation into a mixed typology (Fig. 1).

In this study, we will base ourselves on the integral approach to the concept of innovation proposed by Edquist et al. [24] because it seems wider to us and reflects the definition of innovation that we retain within the framework of our study.

Researches emphasize the importance of innovation for the success and survival of our institutions. For Liao et al. [25], for example, innovation in terms of products and processes can determine the success or failure of the organization. Dannels [26] goes further and proves that innovation allows translating competencies into new products and processes enabling organizations to realize a competitive advantage. Product and process innovation, that tend to be present in organizations where creativity and problem-solving thrive [27], would be a means to enhance the adaptability of organizations to environmental and technological changes.

In this research, we are interested in technological innovation in banks. One thing is clear: the banking sector, especially in Lebanon, is a high-volume industry, with

INNOVATIONS		Abernathy & Utterback typology (1978)	
		Technological	**Organizational**
Evan typology (1996)	**Product**	PRODUCT	**ORGANIZATIONAL INNOVATION AS OUTPUT**
	Process	TECHNOLOGICAL PROCESS	**ORGANIZATIONAL PROCESS INNOVATION**

Fig. 1 Organizational Innovation in the main existing typologies

an increasingly wide range of products, a high rate of innovation, and complex relationships. For several decades, the French banking world has undergone profound changes involving considerable cultural, strategic, and organizational repercussions. These developments have several origins: the globalization of the banking market, the intensification of the competitive pressure coming from foreign banks and insurance companies, the multiplication of distribution and information channels, the commoditization of banking products, the volatility of customers with better-informed customers that are more mature in terms of banking needs, more demanding and not hesitating to call on several banks and put them in competition. Conscious of these changes and challenges, the banks have been led to set up operational strategies. As a result, the information system, and hence its improvement and efficiency, are an imperative for processes, computer applications, and their operation. At this level, ICTs and especially technological innovations are part of this strategy in order to rethink the relationship with customers, develop the potential of existing customers, conquer new market shares, and improve performance even more.

3 Transformational Leadership Theory

According to Bass [28], "Leadership is the ability to influence the attitudes and behaviors of individuals within a group and between the members of a group to achieve the desired goals". For Chemers [29], leadership would be a process of influence whereby one person, the leader, is able to mobilize the support and assistance of others for the accomplishment of a common goal. We will retain these two definitions of leadership that have been widely recognized and validated by researchers, for the purposes of our article. Leadership is in fact the way by which a person influences others in order to achieve a common goal by giving them a good vision of the future for effective action.

Transformational leadership has been central to leadership literature for almost three decades [30]. According to Bass [12], whose works are a continuation of that of Burns' [31], transformational leadership is defined as the leader's ability to transform the energies deployed by his collaborators so as to bring them to be more motivated in achieving the desired results and objectives, both individually and organizationally [32].

The transformational approach is based on the transformation exchanges between the leader and the follower. Transformational leaders motivate their employees to go beyond their personal interest in working for the common good of the organization through four dimensions: **inspirational motivation, idealized influence, intellectual stimulation, and individualized consideration** [12, 30, 33]. The inspirational motivation consists of developing and articulating a vision while demonstrating optimism and confidence in the ability to realize that vision. It represents the faculty of the leader to give challenges and a meaning to the work of his followers, which arouses team spirit but mainly motivates followers to surpass themselves for the sake of the organizational mission [34]. The idealized influence also called charisma can be defined as the ability of the transformational leader to act with the articulation of the vision in an admirable and coherent way, in order to trigger imitation and identification behaviors among his collaborators [12]. Such leaders are recognized for their exceptional will, which earns them the admiration, respect, and trust of their followers [35]. Intellectual stimulation consists of inciting the followers to give suggestions and ideas, encouraging them to question some ideas, and to take risks. Finally, individualized consideration refers to the leader's ability to develop an orientation toward his team, to pay particular attention to his followers, to be aware of their personal concerns, and to foster a two-way communication with them [12].

Transformational leadership helps to increase employee satisfaction and motivation, and leads them to achieve higher performance, as they not only meet work expectations, but rather exceed them [36]. This type of leadership also promotes innovative thinking and creativity.

4 Transformational Leadership and Innovation

Many works have demonstrated and proven that transformational leadership is the most important and most valued leadership style for the development of innovation [37, 38]. Previous research link TL with innovation. Thus, Redmond and Mumford [39] show that by defining objectives and controlling critical resources to achieve these objectives, TLs can create the context that allows motivating employees to engage in the innovation of products and processes. According to Amabile [40], transformational leadership is a precursor to organizational creativity and innovation. Recent research also shows that TL acts as a lever that facilitates innovation [41].

Such leaders have an interactive vision and can foster an appropriate environment for innovation [21]. Gumusluoğlu and Ilsev [42] also argue that, unlike charismatic leaders, transformational leaders not only promote innovative activities within the organization, but also ensure their success in the marketplace.

De Jong and Hartog [18] establish that leadership as well as the leaders' behaviors of consultancy and have a positive influence on employees' efforts in terms of innovation, more specifically in terms of the generation of new ideas and their implementation. Chang [43] proves that leaders who advise, train, and coach their followers can strengthen and hone their skills by encouraging them to try new ideas and methods. However, Jamaludin and Rahman [44] are much more skeptical. In a recent study, they conclude that transformational leadership seems to be more appropriate for stimulating creativity and generating ideas than for implementing innovations.

Finally, we could conclude that TL would be a kind of catalyst for innovation, by promoting motivation, inspiration, self-esteem, intellectual stimulation, and self-confidence among members of the organization [35].

5 Research Methodology

Previous studies have associated leadership style with performance or effectiveness [45]. Yet, the role of the leader in the implementation of ICTs has received little attention in the literature. Thus, to understand the associations between transformational leadership and the implementation of ICTs, we will adopt an exploratory approach. The concept of innovation is a broad phenomenon that is defined in different ways in different contexts. Thus, to understand the relationship between transformational leadership and innovation, we will take an exploratory approach. Our methodology is based on the collection and analysis of qualitative data. These data were collected through a series of focused semi-structured interviews with 11 senior managers (G1, G2, G3, G4, G5, G6, G7, G8, G9, G10, G11), heads of departments, working in the five largest banks in Lebanon. These banks are always under pressure in order to find new sources of revenue and control their operating costs. In order to cope with these multiple challenges, they focus more on innovation and regularly increase their investments in innovation.

To the data collected from the interviews, we will add secondary data from several written (reports published by financial institutions) or verbal (discussions with banking experts in charge of the research and development programs) sources. People were met and interviewed face-to-face at their workplaces. The face-to-face interviews were conducted in a largely unstructured format consisting of two parts. We first asked each respondent to describe his own leadership style, the role that ICTs play in their business, and to determine if and how employees participate in innovation. After these general questions, we encouraged participants to talk freely about their attitudes and behavior as leaders. We have always tried to ask for more details about the potential impact of the leadership style an interviewee has himself described on the generation of ideas and employees' behavior. In the second part of the interview, we revealed to the interviewees that the purpose of the interview was to explore and understand how leaders stimulate the innovative generation of ideas and behavior among their employees. We then asked more direct questions about how a person could stimulate (or discourage) these behaviors.

The duration of the interviews varies between 45 min and one hour. The data collection was spread over 3 months. The information collected during the interviews and combined with the data related to both concepts of transformational leadership and innovation have shown saturation from the eleventh interview, which indicates that the collected information accurately reflects a good comprehension of the phenomenon under study. In the analysis of our interviewees' discourse, we have adopted the content analysis approach by counting the number of times each transformational characteristic was related to innovation. The interview guide (see Box 1) includes 8 questions that focus on two main themes: the challenges related to the implementation of different types of innovation within the public sector in Lebanon, the relationship between the components of transformational leadership and the promotion of these innovations within the said sector.

Box 1: Guide to the interview

(1) Tell me about your career path.
(2) Has your bank experienced innovations lately? What types of innovation (of product, process, and service)?
(3) What are the employed means in order to promote the implementation of innovation?
(4) What is your superiors' attitude toward innovation?
(5) What can you say about the work environment in terms of innovation? Do your leaders encourage you to find new ideas? How?
(6) Does your team include creative people? How do you work with them?
(7) Do you encourage people to approach new ways of thinking?
(8) What are, in your opinion, the challenges related to the application of an innovation strategy within your organization?

6 The Context of the Study: The Banking Sector in Lebanon

The banking sector is the backbone of the Lebanese economy. It is reputed to be one of the most resilient sectors in Lebanon. It survived numerous economic and financial crises, maintained a solid reputation, and played a major role in cushioning the country. The main source of strength of the Lebanese banking sector has been and continues to be its large base of depositors.[3] In 2018, bank deposits reached more than 3 times the volume of GDP.[4] The transfer of migrants' financial flows to Lebanon accounts for a major share of these deposits. Indeed, transfers of funds to Lebanon as a percentage of GDP are among the highest in the world reflecting the importance of the Lebanese diaspora.[5] This has helped it survive during the multiple political and economic crises it has faced in the recent years. The banking sector, agent of the progressive economy, was well placed to manage the money of the diaspora as well as the regional and foreign money long before the neighboring countries could embark on the banking sector.

The banking sector currently consists of 42 commercial banks, 11 foreign banks, and 17 investment banks. It remains the main growth engine of the economy with a contribution of 6–7% to Lebanon's GDP estimated at 56.7 billion USD in 2018.[6] Furthermore, the sector is currently facing multiple challenges. Lebanese banks will face the dilemma of achieving sufficient growth to meet both the growing financing needs of the state and the private sector. Indeed, the banking sector assumes most of the public debt that weighs heavily on the economy (150% of GDP). In addition, it is exposed to the challenge of finding new markets to improve its performance and grow further. Besides, banks are expected to consolidate their capital base in accordance with international capitalization requirements and to build additional general provisions according to international financial standards.

Currently, with the massive explosion of information and communication technologies, Lebanese banks are more and more concerned about the integration of these technologies into their work. They heavily invest in innovation in order to remain competitive, achieve higher returns, expand their customer base, and meet the increasingly sophisticated needs of this customer base. These transformations prompt us to focus on the behavior and role of the leaders in the adoption and integration of innovations in the banking sector in Lebanon.

7 Obtained Results

The results of our exploratory study highlight the following three points: the nature of innovations operated in the banking sector in Lebanon, the role of transformational behaviors of senior leaders in the implementation of innovations, and the challenges pertaining to the adopted processes of innovation.

[3] Mantach, Maya, (2015), "The banking sector in Lebanon: Rising up to the challenges of a conflict zone", Global Banking and Financial Policy Review.

[4] www.expert-comptable-international.info/fr/pays/lebanon/economie.

[5] Rapport/ pays FMI n'9/131.

[6] www.expert-comptable-international.info/fr/pays/lebanon/economie.

7.1 *The Type of Innovations Applied in the Lebanese Banking Sector*

Internet has significantly transformed the banking service process and disrupted innovative marketing approaches opening up new opportunities for Lebanese banking institutions [46]. Bank distribution channels are currently the main vectors of innovations, especially with the development of smartphones. Technological diffusion also encourages banks to implement product innovations. Unlike industrial innovations, financial innovations allow the diversification of the range of existing products without causing them to become obsolete. Furthermore, another type of innovation was mentioned by our interviewees, it is the business model innovation. "*Innovation is not necessarily about new products, but about adding new activities, linking activities in new ways, or changing the part that does the activity*" (G5).

These financial innovations are most often incremental, in response to environmental changes [47]. These incremental innovations are useful but should not radically change the banking industry and are easily imitated.

Thus, according to the theory of the diffusion of innovation in services [48, 49], technological progress generates progressive innovations (step change) allowing to change the ways of interacting with customers and improving the functioning of the bank. "*With the development of e-banking and mobile banking, innovations that include a new device have emerged. These innovations have allowed customers to interact with the bank in a very different way, and in the future, they could radically change the functional possibilities of the banking sector*" G1).

However, banks are able to radically innovate [50], but it turns out to be rare. "*We have not yet seen the use of technology in a disruptive way, in a way that could radically change the banking industry, as is the case, for example, in the music sector, where changes have affected the large musical companies*" (G4).

7.2 *The Role of Transformational Components in Innovation*

Matrix 1 presents all leadership behaviors identified during our exploratory study and associated with innovation.

Bass and Avolio [51] identified 4 main components of transformational leadership: inspirational motivation, idealized influence, intellectual stimulation, and individualized consideration.

7.2.1 The Role of Intellectual Stimulation

According to Bass [12] and Bass and Avolio [52], intellectual stimulation refers to the development of the capacity of followers to solve problems by adopting new methods. It is defined as the ability of the leader to encourage followers to innovate.

Matrix 1 Leadership behaviors associated with innovation

Leadership behaviors	Verbatims	Theoretical references
• Intellectual stimulation	*"Workshops have gathered employees from all departments in order to identify the causes of the problems that affect the banking experience. We were thus led to learn how to suggest innovative solutions in order to improve the banking processes and reinvent the customer experience"* *"Tops managers support us a lot and often encourage us to think out of the box, they keep pushing the limits of our knowledge and mobilizing our skills in order to rethink the current relations with the customers in different fields as the loans, deposits and other transactions"*	Bass [12], Bass and Avolio [52]
• Inspirational motivation – Providing vision	*"… there were at the top of our bank visionary managers who understood very early the importance of digital revolution and the important role it could play in the banking industry"* *"…he made things easier for us as he had a clear and well-defined vision of the process of integrating the elements of advanced technologies into the bank's work"*	Bass [12, 28], Bass and Bass [34]
• Inspirational motivation – Support for innovation	*"…my followers know that I love creativity, the search for new ideas, and that is why they never hesitate to make new propositions because they are sure that it interests and excite me at the same time"*	

(continued)

Matrix 1 (continued)

Leadership behaviors	Verbatims	Theoretical references
• Idealized influence – Innovative role modeling	*"i am always searching for new solutions in order to make the customer experience more digital and improve the results, and this encourages some of my employees to act in the same way"*	Bass [28], Bass and Riggio [35]
• Individualized consideration	*"In order to motivate them to work and encourage them to adhere to change policies, I listen to them and pay attention to their problems and complaints"*	Bass [12]
• Rewards	*"The management has sent e-mails to all the employees of the bank asking them to propose new ideas in order to face many dysfunctions, Some employees have suggested new ideas which the management adopted, and of course the winners have been rewarded"*	
• Delegation	*"I had to go through several departments of the bank, and I noticed that employees are more likely to engage in various innovations when they are led by leaders who make a lot of delegations, who encourage them to assume their responsibilities and allow them to have some leeway"*	According to Krause, the fact of granting employees more autonomy and some flexibility is positively associated with different types of innovation

Managers do everything they can to make the most of their followers' potential. This component is the most quoted by our interviewees.

"Tops managers support us a lot and often encourage us to think out of the box, they keep pushing the limits of our knowledge and mobilizing our skills in order to rethink the current relations with the customers in different fields as the loans, deposits and other transactions" (G1).

"Our superior entrusts each of us with the same mission in order to achieve a better result, so each one of us tries to work to find the most innovative solution. The best suggested solution will then be adopted by the management" (G3).

Managers go far in introducing a culture of open innovation that takes into consideration individual knowledge and ideas and promotes the exchange of ideas. They offer their employees all the opportunities that allow them to develop their innovative ideas.

"Workshops have gathered employees from all departments in order to identify the causes of the problems that affect the banking experience. We were thus led to learn how to suggest innovative solutions in order to improve the banking processes and reinvent the customer experience" (G11).

"The management has sent e-mails to all the employees of the bank asking them to propose new ideas in order to face many dysfunctions, Some employees have suggested new ideas which the management adopted, and of course the winners have been rewarded" (G4).

7.2.2 The Role of Inspirational Motivation

According to Bass [12, 28], inspirational motivation is to involve the followers in a vision while showing optimism and the ability to realize that vision. This is the second component that was quoted by our interviewees.

"Our bank engaged very early in an innovative approach because there were some visionary leaders among the top managers. Those leaders were capable to see very early, what the others could not see. They had a strong belief in the necessity to go into innovative projects in order to attain a large customer base and stay competitive in the market" (G10).

"The leader of innovation is not just a profit-oriented person, but he rather has a holistic vision of the work and the future developments of the business; he is thus able to create an environment that is appropriate for creativity and innovation" (G9).

7.2.3 The Role of Idealized Influence

According to Bass [28], the idealized influence is to provide vision, pride of duty, and to gain the respect and trust of one's followers. Thus, the leader plays a model role and shows ethical behavior. This component appears timidly in our interviewees' speeches.

"We trust our superior because he always tries to eliminate the barriers that can prevent the implementation of innovations" (G2).

7.2.4 The Role of Individualized Consideration

According to Bass [12], individual consideration emerges when the leader listens to his followers, makes delegations, develops their potential, and follows their development and progress.

"I had to go through several departments of the bank, and I noticed that employees are more likely to engage in various innovations when they are led by leaders who make a lot of delegations, who encourage them to assume their responsibilities and allow them to have some leeway" (G3).

7.3 Challenges Related to Innovation in the Banking Sector in Lebanon

According to our interviewees, the main factor that limits the implementation of innovation is the computer system of the bank.

"*The rigidity of the computer system delays the integration of new digital programs in the digital system of the bank, whereas banks need to access the markets with their innovations in a faster way*" (G2).

In addition, organizational silos (vertical organizational structure) also pose a problem.

"In *a large bank like ours, the organizational structure is very complicated, which makes it more difficult to integrate innovations into the organization*" (G3).

"Organizational *silos, especially for large banks, have a significant impact on the time to market new distribution channels), the costs of innovation and their functionality*" (G5).

Regulations are also a barrier, but it is of lesser importance for large banks.

"*Our sector operates in a particularly regulated universe subject to several constraints imposed by the financial authorities. This regulatory abundant and strict framework has not hindered the financial innovations, especially in large banks, but it made it more difficult*" (G1).

Furthermore, fintechs play an important role in the financing of the economy and cause a decline in the traditional activity of the bank in the developed countries. These virtual companies are an important issue for the Lebanese banking sector in the coming years.

Senior bank executives try to develop entrepreneurial skills among their employees and esteem their innovative capabilities in order to remain a key financial player. The two components of transformational leadership, inspirational motivation, and intellectual stimulation play a vital role in the creativity and implementation of innovations within the Lebanese banking sector. The invested means and technology, the working atmosphere, the role and the personality of the leader encourage employees to innovate so as to establish operational strategies in order to rethink the relationship with customers, develop the potential of existing customers, conquer new market shares, and further improve performance.

We also note among the interviewees the importance of the customer orientation factor in the innovation processes; most of the innovations aim at rethinking the customer relationship not only for better performance, but also greater competitiveness in the Lebanese market, where the banking sector is dominant and accounts for more than two-thirds of the Lebanese economy. Based on that, transformational leadership has a key role in terms of motivation for the implementation of ways to drive and foster the creativity of its collaborators.

8 Conclusion

Based on style-oriented leadership theories, specifically the one developed by Bass [12], the purpose of this study is to examine how the components of transformational leadership could make a significant contribution to innovation processes in the context of the Lebanese banking sector. In order to achieve this goal, we adopt an exploratory approach based on a series of focused semi-structured interviews conducted with eleven senior managers working in 5 large commercial banks in Lebanon.

Our exploratory study partially confirmed the role of transformational leadership theory in implementing change since only the two components related to intellectual stimulation and inspirational motivation were well identified. As for the other two components of transformational leadership, namely idealized influence, and individualized consideration [12], we have noticed that they play a limited role in the development of banking innovations. At this stage, it would be appropriate to set up all the material and human resources in terms of intellectual stimulation and motivation, aiming at stimulating, promoting, facilitating, and implementing innovations within the organization.

However, several factors intervene to limit the role of transformational leadership in conducting and implementing different types of innovation. In fact, the top managers showing transformational leadership are confronted in their transformational projects with multiple structural and regulatory constraints. Moreover, they are actually threatened by the fintechs that are not subject to the same constraints and are thus driven to develop a culture of innovation and continuously motivate their employees in order to engage in different types of financial innovation. The results of the study show that top managers and those in charge of the human resources should encourage and work for the development of all the dimensions of transformational behaviors among the leaders of the banking sector.

We are aware of the limitations of our results in terms of generalization to the extent that the validity of our conclusions remains contingent within the studied organizational environment. Despite these limitations and other limitations pertaining to the external validity of the qualitative studies, this study could be a starting point for future research that would be conducted in other financial institutions or in companies working in other sectors. Finally, we suggest that the future research be more oriented toward the way in which the dimensions of transformational leadership would separately affect innovation in the banking sector.

References

1. Damanpour, F., & Avellaneda, C. N. (2009). Combinative effects of innovation types and organizational performance: A longitudinal study of service organizations. *Journal of Management studies*.
2. Mounir, A. (2014). La gestion de l'innovation dans les entreprises algériennes: enjeu majeur pour obtention d'un avantage concurrentiel durable. Dissertation of Masters in management sciences. Tizi Ouzou University.
3. Montes, F., Moreno, A., & García-Morales, V. (2005). Influence of support leadership and teamwork cohesion on organizational learning, innovation and performance: An empirical examination. *Technovation, 25*(10), 1159–1172.
4. Mumford, M., Scott, G., Gaddis, B., & Strange, J. (2002). Leading creative people: Orchestrating expertise and relationships. *The Leadership Quarterly., 13*, 705–750.
5. Sarrors, J., Cooper, K., & Santora, J. (2008). Building a climate for innovation through transformational leadership and organizational culture. *Journal of Leadership and Organizational Studies, 15*(2), 145–158.
6. Tushman, M., & O'Reilly, C. (1996). The ambidextrous organization: Managing evolutionary and revolutionary change. *California Management Review, 38*, 1–23.
7. Kombe, S., & Wafula, M. (2015). Effects of internet banking on the financial performance of commercial banks in Kenya a case of Kenya Commercial Bank. *International Journal of Scientific and Research Publications, 5*(5). ISSN 2250-3153. www.ijsrp.org.
8. Asli, M. H. (2012). Open innovation: Quels enjeux pour le secteur bancaire. *Innovations, 3*(39), 27–48.
9. Hughes, T. (2003). Marketing challenges in E-banking: Standalone or integrated? *Journal of Marketing Management, 19*(9–10), 1067–1085.
10. Hilal, M. (2015). Technological transition of banks for development: New information and communication technology and its impact on the banking sector in Lebanon. *International Journal of Economics and Finance, 7*(5).
11. Sobreira, R. (2004). Innovation financière et investissement: Le cas de titrisation. *Innovations, 19*, 15–29.
12. Bass, B. (1985). *Leadership and performance beyond expectations*. Free Press.
13. Schumpeter, J. A. (1934). The theory of economic development: an inquiry into profits, capital, credit, interest and the business cycle. In *Harvard economic studies* (Vol. 46). Harvard College.
14. Rogers, E. M. (1983). *Diffusion of innovation* (3rd ed.). The Free Press.
15. Kline, S., & Rosenberg, G. (1986). An overview of innovation. In R. Landau & N. Rosenberg (Eds.), *The positive sum strategy: Harnessing technology for economic growth* (pp. 275–305). National Academy Press.
16. Hatchuel, A., Le Masson, P., & Weil, B. (2006). Building innovation capabilities. The development of design-oriented organizations. In J. Hage & M. Meeus (Eds.), *Innovation, science and industrial change, the handbook of research* (pp. 294–312). Oxford University Press.
17. Durieux, F. (2000). In Vuibert (Ed.), *Management de l'innovation*. Coll. Fnege.
18. De Jomg, P. J., & Hartog, N. D. (2007). How leaders influence employees' innovative behaviour. *European Journal of Innovation Management, 10*(1), 41–64.
19. Baregheh, A., Rowley, J., & Sambrook, S. (2009). Towards a multidisciplinary definition of innovation. *Management Decision, 47*(8), 1323–1339.
20. Nusair, N., Abaneh, R., & Bae, Y. K. (2012). The impact of transformational leadership style on innovation as perceived by public employees in Jordan. *International Journal of Commerce and Management, 22*(3), 182–201.
21. Saenz, J., Aramburu, N., & Rivera, O. (2009). Knowledge sharing and innovation performance: A comparison between high-tech and low-tech companies. *Journal of Intellectual Capital, 10*(1), 22–36.
22. Lin, H.-F. (2007). Knowledge sharing and firm innovation capability: An empirical study. *International Journal of Manpower, 28*(3/4), 315–332.

23. Jiménez-Jimenez, D., Sanz Valle, R., & Hernandez-Espallardo, M. (2008). Fostering innovation: The role of market orientation and organizational learning. *European Journal of Innovation Management, 11*(3), 389–412.
24. Edquist, C., Hommen, L., & McKelvey, M. (2001). *Innovation and employment: Process versus product innovation*. Edward Elgar.
25. Liao, S. H., Fei,W. C., & Liu, C. T. (2008). Relationships between knowledge inertia, organizational learning and organization innovation. *Technovation, 28*(4) 183–195 (Elsevier).
26. Dannels, E. (2002). The dynamics of product innovation and firm competences. *Strategic Management Journal, 23*(12), 1095–1122.
27. Tsai, H. C., Huang, C. K., & Kao, C. (2001). The relationships among organizational factors, creativity of organizational members and innovation capability. *Journal of Management, 18*, 527–566.
28. Bass, B. M. (1990). From transactional to transformational leadership: Learning to share the vision. *Organizational Dynamic, 18*, 19–32.
29. Chemers, M. (1997). *Leadership, change, and organizational effectiveness*. University of California.
30. Avolio, B., Walumbwa, F., & Weber, T. (2009). Leadership: Current theories, research and future directions. *Annual Review of Psychology, 60*, 421–449.
31. Burns, J. M. (1978). *Leadership*. Harper & Row.
32. Molines, M. (2010). *Leadership transformationnel, stress et performance: Une étude multi niveaux*. https://www.agrh.fr/assets/actes/2010molines.pdf.
33. Judge, T., & Piccolo, R. (2004). Transformational and transactional leadership: A meta-analytic test of their relative validity. *Journal of Applied Psychology, 5*, 755–768.
34. Bass, B. M., & Bass, R. (2008). *The bass handbook of leadership: Theory, research, and managerial applications* (4th ed.). Free Press.
35. Bass, B., & Riggio, R. (2006). *Transformational leadership* (2nd ed.). Lawrence Erlbaum Associates.
36. Wang, G., Oh, I. S., Courtright, S. H., & Colbert, A. E. (2011). Transformational leadership and performance across criteria and levels: A meta-analytic review of 25 years of research. *Group & organization management, 36*(2), 223–270.
37. Hussain, H., Abu Talib, N., & Shah, I., (2014). Exploring the impact of transformational leadership on process innovation and product innovation: A case of Iraqi Public Universities. *Asian Social Science, 10*(21) (Published by Canadian Center of Science and Education).
38. Kesting, P., Ulhøi, J., Song, L., & Niu, H (2015). Becoming a leader who foster innovation. The impact of leadership styles on innovation, a review. *The International Journal on Multidisciplinary Approaches on Innovation, 3*(4). https://www.researchgate.net/publication/299533980_Becoming_a_Leader_Who_Foster_Innovation.
39. Redmond, M. R., Mumford, M. D., & Teach, R. (1993). Putting creativity to work: Effects of Leader behavior on subordinate creativity. *Organizational Behavior and Human Decision Processes., 55*, 120–151.
40. Amabile, T. M., Conti, R., Coon, H., Lazenby, J., & Herron, M. (1996). Assessing the work environment for creativity. *Academy of Management Journal, 39*(5).
41. Eisenbeib, S., & Boerner, S. (2013). A double-edged sword: Transformational leadership and individual creativity. *British Journal of Management., 24*, 54–68.
42. Gumusluoglu, L., & Ilsev, A. (2009). Transformational leadership, creativity, and organizational innovation. *Journal of Business Research, 62*, 461–473.
43. Chang, I.-H. (2012). The effect of principals 'technological leadership on teachers' technological literacy and teaching effectiveness in taiwanese elementary schools. *Educational Technology & Society., 15*(2), 328–340.
44. Jamaludin, Z., & Rahman, N. (2011). Do transactional, transformational and spiritual leadership styles distinct? A conceptual insight. *Journal of Global Business and Economics, 2*(1), 73–85.
45. Shalley, C., & Gilson, L. (2004). What leaders need to know: A review of social and contextual factors that can foster or hinder creativity. *Leadership Quarterly, 15*, 33–53. https://doi.org/10.1016/j.leaqua.2003.12.004

46. Azzam, E. (2010). *L'impact des nouvelles technologies de l'information et de la communication sur le marketing bancaire.* Proche Orient Studies in Management, Faculty of Business and Management, Saint-Joseph University.
47. Fasnacht, D. (2009a). *Open innovation in the financial services: Growing through openness, flexibility and customer integration,* Springer.
48. Barras, R. (1986). Towards a theory of innovation in services. *Research Policy, 15,* 161–173.
49. Barras, R. (1990). Interactive innovation in financial and business services: The vanguard of the service revolution. *Research Policy, 19,* 215–237.
50. Favre-Bonte, V., Gardet, E., Thevenard-Puthod, C. (2009). Parler d'innovation bancaire a-t-il un sens? *Gestion: Revue internationale de Gestion, 34*(1): 10–19.
51. Bass, B. M., & Avolio, B. J. (1994). *Improving organizational effectiveness through transformational leadership.* Sage.
52. Bass, B. M., & Avolio, B. J. (1995). The multifactor leadership questionnaire. Mind Garden.

Enhancing Inclusiveness Through Remote Working: What Happened During Covid-19 Experience in Italy?

Enrico Cori, Daria Sarti, and Teresina Torre

Abstract This study aims at deepening the employees' inclusion perception produced by the intensification of the technologies used in the remote working (RW) experience lived during the Covid-19 lockdown. We believe that ad hoc mechanisms of support have assumed a greater significance by virtue of suddenness and absence of preparation for the switching in RW due to the lockdown. In detail, our research aims to examine how employees' perception of inclusion is fostered by control and coordination mechanisms implemented for supporting the new (and forced) way of working remotely; further, we aim to understand if and how previous experience in RW gathered by workers may affect such relationship. The survey, developed in May 2020 on a sample of 254 employees in Italy, demonstrates the importance of control and coordination mechanisms implemented for smart workers in fostering their perception of inclusiveness. Further, previous experience in RW of employees was investigated as a moderating variable.

Keywords Remote working (RW) · Covid-19 · Inclusiveness · Process control · Output control · Coordination · Workers' previous experience in RW

E. Cori (✉)
Polytechnic University of Marche, Ancona, Italy
e-mail: e.cori@univpm.it

D. Sarti
University of Florence, Florence, Italy
e-mail: daria.sarti@unifi.it

T. Torre
University of Genoa, Genoa, Italy
e-mail: teresina.torre@economia.unige.it

C. Dal Zotto et al. (eds.), *Smart Technologies for Organizations*, Lecture Notes in Information Systems and Organisation 60, https://doi.org/10.1007/978-3-031-24775-0_14

1 Introduction

In the first years of the 2000s, remote working (RW) has emerged in international literature as an innovative approach to work organization [1–3], challenging the conventional work design and re-design, and to control processes [4]. RW is essentially based on a greater flexibility and on a larger discretion in work activities, in place and time they are carried out, and on an increased responsibility toward results. Anyhow, the intensive introduction of such factors is believed to favor better performances by employees and coherently to foster competitiveness [5, 6].

At the same time [7] the acceleration in the development and diffusion of new communication technologies, together with the pervasive dissemination of powerful and easy-to-use mobile devices, helps the growth of portable systems [8, 9]. This offers new opportunities for innovative solutions regarding when, where, and in which way to work.

Enterprises have been practicing for some years solutions that combined in office and remote working, as a feasible response enabling to balance often conflicting needs, such as productivity and work–life balance [10–12].

In Italy, remote workers represented a small percentage of the workforce [13]. One of the first decisions taken by the Italian government to face the pandemic was to foster RW diffusion. Suddenly, it became the new normal way of working for most workers. However, one of the basic characteristics of RW—that is the "wherever"—was discharged and, in the specific contingent condition, it had to be performed at home, with all the inevitable implications it entailed.

This new scenario offers the opportunity to explore a wide range of issues. This work focuses on the role played by coordination and control tools used during the Covid-19 lockdown. We wonder if tools implemented to make RW effective fostered the employees' feeling of inclusion and if and how the previous experience in RW has (or not) influenced the perception of inclusion of workers.

The paper is organized in the following manner. In the second section, the theoretical background is offered. In the third part, we present the analysis and its most relevant results. In the last section, some preliminary suggestions concerning our question are introduced together with the main limitations of our research; also, considerations useful for future development are proposed.

2 Theoretical Background

2.1 Key Dimensions of Workplace Inclusion

Several different inclusion constructs have been presented and discussed in the literature [14]. In the present study, we accept the definition of inclusion "as a collaborative, supportive, and respectful environment that increases the participation and contribution of all employees," proposed by the Track selected for this paper. The meaning is such as to evoke some basic conditions for the development of an inclusive working environment (propensity for collaboration, support from superiors and colleagues, etc.), as well as some possible impacts on the behavior of employees (participation, contribution), which in turn, presumably, will result in a variety of outcomes.

Mor Barak et al. [15, p. 73] represent the inclusion–exclusion construct "as a continuum of the degree to which individuals feel a part of critical organizational processes such as access to information, connectedness to co-workers, and ability to participate in and influence the decision making process."

However, for the purposes of our research, the concept of "perceived organizational inclusion" [14, 16] seems to be more pertinent. This refers to individual-level perceptions of an employee's inclusion in the organization. Shore et alii [16] propose a framework in which some contextual antecedents (among which inclusiveness practices, aimed at satisfying needs of belonging and needs of uniqueness) contribute to employees' perception of inclusion, that in turn results in several potential outcomes. As regards inclusiveness practices—especially those aimed at knowledge sharing, communicating, participating in decision-making, and providing support from the direct supervisor—they are in some way attributable to the choices concerning work coordination and control.

For their part, Jansen et alii [17] state that a sense of belonging and the perception of being a distinct and unique individual may be particularly predictive of affective outcomes, such as mood within the group, work satisfaction, and interpersonal trust.

From this research, it seems evident that the sense of belonging represents a key concept in the evaluation of the inclusiveness of a work context.

2.2 *Coordination and Control of Remote Workers: The Effects on Isolation and Inclusion*

To explore in depth the impact of coordination and control methods on inclusion at work we also took into consideration the concept of (social and professional) isolation; indeed, it can be regarded as predictive of low levels of inclusion. For this purpose, we first identified the few studies in which the constructs of isolation and inclusion are explicitly or implicitly associated.

Golden, Veiga, and Dino [18] outline a significant relationship between the length of the RW experience and feelings of isolation and inclusion, the latter declined in terms of satisfaction of the need to belong.

For their part, Allen et al. [19] highlight a greater sense of inclusion on the part of office workers compared to those who work at home or remotely. Finally, Feldman and Gainey [20] underline the impact of social isolation determined by telecommuting on the ability to satisfy their needs for affiliation.

The concept of isolation is implicitly associated with that of inclusion also by Orhan et al. [21]. Along the lines of Sacco and Ismail [22], they affirm how the rarefaction of face-to-face contacts is positively correlated with the sense of social isolation and negatively with the sense of social belongingness.

Subsequently, we moved to analyze the studies that, even if only indirectly, aim to observe the possible effects of coordination and control modes on the perception of inclusion and isolation. About this topic, a laudable attempt to arrive at a conceptual framework was recently made by Errichiello and Pianese [23], who place the feeling of isolation among the individual outcomes of organizational control choices related to the implementation of RW.

Previously, Kurland and Cooper [24] had pointed out how choices concerning managerial control are unequivocally linked to the perception of professional isolation. According to them, RW implies a more intense use of output control and this, in turn, can negatively affect the opportunities for learning and professional development of distance workers, resulting in their progressive isolation.

However, not all scholars agree that a more intense use of output control in a RW context is inevitable. In its recent analysis, Downes [25] observed how many managers, instead of moving toward a tighter output control, prefer to strengthen relationships and feel closer to their subordinates when face-to-face interactions become less frequent. This would help to prevent the possible consequences associated with an excessive emphasis on output control, due to the impersonal character of the latter.

For their part, Sewell and Taskin [26] observed a shift in coordination and control practices that constrained teleworkers, through the replacement of ad hoc and informal coordination with formalized team meetings, thus inducing feelings of isolation. Hence, it could be assumed that a more intense use of formal control tools has negative effects on the perception of inclusion.

3 Empirical Analysis

3.1 Aim of the Analysis

According to the literature examined, the role of output control, process control, and intensity of coordination (frequency of feedback, frequency of virtual meetings and personal interaction with superiors) on the employees' sense of inclusion (measured through a sense of belonging, interpersonal trust, task-based job performance) was investigated. Further, the role of previous experience with RW was hypothesized as moderator in the relationship between coordination and control tools adopted and the sense of inclusion.

3.2 Method

The empirical research was based on data gathered across a population of "remote workers" during the Covid-19 pandemic lockdown in Italy, by means of an online survey software (i.e., Lime Survey). The questionnaire was developed using both established scales and new ones that are: organizational trust and sense of belonging to the group (see Table 1).

320 questionnaire access codes were distributed; 254 questionnaires were fully filled; so the response rate is 79.4%.

Table 1 The variables of the study

Variable name	Nr. of items	References	Example of item	Alpha Cronbach
Output control[1]	3	Weibel et al. [28]	There is greater attention to monitoring the achievement of the objectives assigned	0.841
Process control[1]	3	Weibel et al. [28]	There is a greater number of written rules to follow concerning the activities performed	0.916
Coordination[1]	3	Foss et al. [29]	Formal (virtual) meetings with managers/bosses/supervisors are more frequent	0.866
Task based job performance[1]	6	Goodman and Svyantek [27]	I was more inclined to perform activities not expressly requested by the organization, but which can improve customer satisfaction	0.877
Organizational trust[1]	1	*Authors' own elaboration*	The experience of SW has increased my sense of trust toward the company	n.a
Sense of belonging to the group[1]	1	*Authors' own elaboration*	The experience of SW has increased my sense of belonging to the group	n.a
Previous experience of smart working[2]	1	*Authors' own elaboration*	Before Covid-19 pandemic, did you already work in SW?	n.a

Note [1] Five-point scale ranging from 1 = "decreased a lot" to 5 = "increased a lot."
[2] Four-point scale ranging from 1 = "no previous experience of SW," 2 = "yes, but less than 1 day per month," 3 = "yes, one day a week," 4 = "yes, 2–3 days a week."
n.a., Not available, for one-item scales the Cronbach's alpha was not computed

3.3 Variables

Three dependent variables are used to measure inclusion perception of employees that are: task-based job performance, which was previously developed in the study by Goodman and Svyantek [27], trust in organization and sense of belonging to the group, one-item scales developed for this study (see Table 1).

Three independent variables were used: output control and process control, previously developed by Weibel et al. [28], and coordination mechanisms, developed in the study by Foss et al. [29].

Scales' reliability was proved and Cronbach's alpha was reported in Table 1.

The moderating variable used is having previous experience of RW and it is measured with the question: "Before Covid-19 pandemic did you already work remotely?".

In addition, some control variables were included, since they might have an impact on dependent variables considered in our analysis. These are: *age* (1 = 25–34, 2 = 35–39, 3 = 40–44, 5 = 50–54, 6 = 55 and more); *gender* (1 = female; 2 = male); *education level* (ranging from 1 = high school education to 7 = tertiary education, advanced level).

3.4 Analysis and Results

Data were empirically analyzed using SPSS and the PROCESS macro developed by Hayes [30].

The sample is composed of 44.8% of females. The 39% of respondents are aged between 25 and 34 years old, 12% between 35 and 39 years old, 8% between 40 and 44 years old, 9% between 45 and 49 years old, 14% between 50 and 54, and 18% are aged 55 or more. Approximately 35% of respondents held a high school diploma, 21% a bachelor's degree, 41% a master's degree, and 3% a Ph.D.

Results show that control and coordination mechanisms implemented for RW during the Covid-19 pandemic positively affect the perception of inclusion of employees, as well, the less control and coordination are implemented the lower the levels of inclusion perceived.

In particular, it was proved that process control ($r = 0.151$, $p < 0.05$), output control ($r = 0.139, p < 0.05$), and coordination ($r = 0.515, p < 0.01$) relate positively and significantly to task-based job performance. Also process control ($r = 0.243$, $p < 0.01$), output control ($r = 0.180$, $p < 0.1$), and coordination ($r = 0.524$, $p < 0.05$) relate positively and significantly to trust in organization. As for sense of belonging to the group, output control ($r = 0.153$, $p = 0.07$) and coordination ($r = 0.288$, $p < 0.001$) are positively and significantly related to the DV while process control does not demonstrate a significant relationship ($r = 0.124$, $p > 0.1$).

Further, it was demonstrated that the experience of RW moderates significantly and negatively the relationship between the three independent variables—process control ($r = -0.099$, $p < 0.1$), output control ($r = -0.128$, $p < 0.05$), and coordination ($r = -0.129$, $p < 0.05$)—and *task-based job performance*. This means that if the experience of RW by employees is low, the benefits of the implementation of coordination and control mechanisms on the DV are positive and relevant and the relationship is stronger compared to conditions in which RW experience is high.

Results demonstrate that the experience of RW moderates the relationship between process ($r = -0.182$, $p < 0.05$), output control ($r = -0.144$, $p < 0.1$), and *trust in organization* in a significant and negative way. Also, it moderates the relationship between process ($r = -0.210$, $p < 0.05$), output control ($r = -0.247$, $p < 0.01$), and *sense of belonging* in a significant and negative way. So that, if the experience of RW by employees is low, the benefits of the implementation of control mechanisms on the DVs are positive and relevant and the relationship is stronger compared to conditions in which RW experience is high. Furthermore, results demonstrate that no moderation effect of RW experience exists in the relationship between coordination mechanisms and both the DVs.

4 Conclusions, Limitations, and Further Research

This study's aim is twofold: first, shedding light on the influence exerted by the use of control and coordination tools on the employees' feeling of inclusion during the experience of RW; second, understanding whether the relationship between control and coordination and perception of inclusion was moderated by any significant RW experiences before the Covid-19 emergency. The results of the analysis show a positive and meaningful relationship between the three independent variables and one of the three dimensions measuring inclusion perception by employees (task-based job performance), and between two independent variables (output control and coordination) and the other two dimensions of inclusion: trust in the organization and sense of belonging to the group.

It is noteworthy that—despite the fact that the RW mode was not chosen by every worker, but imposed by contingencies—virtuous behavioral dynamics were activated, oriented toward the strengthening of the sense of inclusion at work.

In this sense, the results seem to attenuate the evidence that has emerged so far in the literature. In fact, it especially highlights the negative impacts of the coordination and control mechanisms used in RW contexts on the dimensions of isolation and inclusion.

Further, previous experience with RW positively played an interesting role. Despite it could have been considered that previous experience might have contributed to develop digital skills, so that in the new context workers were facilitated, and thus would express higher levels of inclusion perception, rather the higher the experience with RW, the lower the effect of control and coordination mechanisms on inclusion perception.

As regards managerial implications, results underline the importance of an appropriate design to prepare people and to organize and manage both cultural and organizational conditions.

Some limitations to the work presented deserve discussion. First, the study was limited to a sample of Italian employees, so that generalized conclusions regarding the model's predictions cannot be made.

Second, as the variables were measured via a common method and source, there may be some systematic bias in asking the same respondents about both dependent and independent variables.

In addition, the cross-sectional design adopted is a limitation that precludes coming to any causal conclusions. However, in order to guarantee the anonymity of the respondents on such sensitive issues, the chosen method is the only viable one for collecting such data. The use of multiple sources and the longitudinal analysis is thus impossible in this context.

Future paths of research could concern an in-depth qualitative analysis on the forms of support to be assumed—e.g., technology-mediated or face-to-face—and the intensity it has to play as a premise for proactive behaviors.

References

1. Boorsma, B., & Mitchell, S. (2011). Work-life innovation, smart work. A paradigm shift transforming: How, where, and when work gets done. Cisco Internet Business Solutions Group (IBSG).
2. Chartered Institute of Personnel and Development—CIPD & Capgemini: Smart working: The impact of work organization and job design. Research insight. (2008).
3. Lee, J. (2013). Cross-disciplinary knowledge: Desperate call from business enterprises in coming smart working era. *Technological and Economic Development of Economy, 19*(sup1), 285–303.
4. Brewer, A. M. (2000). Work design for flexible work scheduling: Barriers and gender implications. *Gender, Work & Organization, 7*(1), 33–44.
5. Haines, V. Y., III., & St-Onge, S. (2012). Performance management effectiveness: Practices or context? *The International Journal of Human Resource Management, 23*(6), 1158–1175.
6. Wood, S., Van Veldhoven, M., de Croon, M., & Menezes, L. M. (2012). Enriched job design, high involvement management and organizational performance: The mediating roles of job satisfaction and well-being. *Human Relations, 65*(4), 419–445.
7. Morgan, J. (2014). *The future of work: Attract new talent, build better leaders, and create a competitive organization*. Wiley.
8. Ahuja, M. K., Chudoba, K. M., Kacmar, C. J., McKnight, D. H., & George, J. F. (2007). IT road warriors: Balancing work-family conflict, job autonomy, and work overload to mitigate turnover intentions. *Mis Quarterly* 1–17.
9. Barnes, S. J., Pressey, A. D., & Scornavacca, E. (2019). Mobile ubiquity: Understanding the relationship between cognitive absorption, smartphone addiction and social network services. *Computers in Human Behavior, 90*, 246–258.
10. Torre, T., & Sarti, D. (2018). Into smart work practices: Which challenges for HR department? In E. Ales, Y. Curzi, T. Fabbri, O. Rymkevich, I. Senatori, & G. Solinas (Eds.), *Working in digital and smart organizations* (pp. 249–275). Palgrave MacMillan.

11. Torre, T., & Sarti, D. (2019). Themes and trends in smart working research. An systematic analysis of academic contributions. In B. Imperatori & R. Bissola (Eds.), *HRM 4.0 for human-centered organizations. Advanced series in management* (Vol. 23, pp. 177–200). Emerald Publishing Limited.
12. Torre, T., & Sarti, D. (2020). Innovative approaches to work organization and new technologies. First insight from the Italian context. In Y. Bagdadi, A. Harfouche & M. Russo (Eds.), *ICT for an inclusive world, Industry 4.0. LNISO* (pp. 133–145). Springer.
13. Milan Polytechnic: Osservatorio Smart Working, report dei risultati della ricerca. (2019).
14. Shore, L. M., Cleveland, J. N., & Sanchez, D. (2018). Inclusive workplaces: A review and model. *Human Resource Management Review, 28*(2), 176–189.
15. Mor Barak, M. E., Nissly, J. A., & Levin, A. (2001). Antecedents to retention and turnover among child welfare, social work, and other human service employees: What can we learn from past research? A review and metanalysis. *Social Service Review, 75*(4), 625–661.
16. Shore, L. M., Randel, A. E., Chung, B. G., Dean, M. A., Holcombe Ehrhart, K., & Singh, G. (2011). Inclusion and diversity in work groups: A review and model for future research. *Journal of Management, 37*(4), 1262–1289.
17. Jansen, W. S., Otten, S., van der Zee, K. I., & Jans, L. (2014). Inclusion: Conceptualization and measurement. *European Journal of Social Psychology, 44*(4), 370–385.
18. Golden, T. D., Veiga, J. F., & Dino, R. N. (2008). The impact of professional isolation on teleworker job performance and turnover intentions: Does time spent teleworking, interacting face-to-face, or having access to communication-enhancing technology matter? *Journal of Applied Psychology, 93*(6), 1412–1421.
19. Allen, T. D., Golden, T. D., & Shockley, K. M. (2015). How effective is telecommuting? Assessing the status of our scientific findings. *Psychological Science in the Public Interest, 16*(2), 40–68.
20. Feldman, D. C., & Gainey, T. W. (1997). Patterns of telecommuting and their consequences: Framing the research agenda. *Human Resource Management Review, 7*(4), 369–388.
21. Orhan, M. A., Rijsman, J. B., & Van Dijk, G. M. (2016). Invisible, therefore isolated: Comparative effects of team virtuality with task virtuality on workplace isolation and work outcomes. *Revista de psicología del trabajo y de las organizaciones, 32*(2), 109–122.
22. Sacco, D. F., & Ismail, M. M. (2014). Social belongingness satisfaction as a function of interaction medium: Face-to-face interactions facilitate greater social belonging and interaction enjoyment compared to instant messaging. *Computers in Human Behavior, 36*, 359–364.
23. Errichiello, L., & Pianese, T. (2016). Organizational control in the context of remote work arrangements: A conceptual framework. In *Performance measurement and management control: Contemporary issues (Studies in managerial and financial accounting)* (pp. 273–305).
24. Kurland, N. B., & Cooper, C. D. (2002). Manager control and employee isolation in telecommuting environments. *The Journal of High Technology Management Research, 13*(1), 107–126.
25. Downes, R. (2020). The proximity paradox: How distributed work affects relationships and control. In *Proceedings of the 53rd Hawaii International Conference on System Sciences* (pp. 440–450). HICSS.
26. Sewell, G., & Taskin, L. (2015). Out of sight, out of mind in a new world of work? Autonomy, control, and spatiotemporal scaling in telework. *Organization Studies, 36*(11), 1507–1529.
27. Goodman, S. A., & Svyantek, D. J. (1999). Person–organization fit and contextual performance: Do shared values matter. *Journal of Vocational Behavior, 55*(2), 254–275.
28. Weibel, A., Den Hartog, D. N., Gillespie, N., Searle, R., Six, F., & Skinner, D. (2016). How do controls impact employee trust in the employer? *Human Resource Management, 55*(3), 437–462.
29. Foss, N. J., Lyngsie, J., & Zahra, S. A. (2013). The role of external knowledge sources and organizational design in the process of opportunity exploitation. *Strategic Management Journal, 34*(12), 1453–1471.
30. Hayes, A. F. (2013). *Introduction to mediation, moderation, and conditional process analysis: A regression-based approach.* Guilford publications.

Customized Artificial Intelligence for Talent Recruiting: A Bias-Free Tool?

Eleonora Veglianti, Matteo Trombin, Roberta Pinna, and Marco De Marco

Abstract In recent years, technological innovations in e-recruitment systems have seen an explosive expansion, allowing Human Resource professionals to find the talents who are supposed to be the most suitable to their organizations. In particular, the purpose of this paper is to explore the contribution that Artificial Intelligence Technologies can give in order to increase the efficiency of the recruitment process and overcome human errors, by comparing theoretical convergences among the various approaches and platforms addressed to companies. A case study was conducted to explore the research questions of this study. A brief study of limits, risks, as well as managerial and business implications linked to the use of AITs in HRM will be also conducted.

Keywords Talent acquisition · Gamification · AI in recruitment process · Predictability · Bias

1 Introduction

The recent transition from a discrete paradigm of information processing (programming for Turing machines) to a continuous paradigm (learning of artificial intelligence) allows faster and more accurate adapting to environmental requirements. In the modern conditions of business, it becomes more relevant for organizations to use the Artificial Intelligence Technologies (AITs) for decision-making. This also led to the application of AITs to Human Resource Management (HRM), since AITs enable the prompt analysis (cleaning, investigating, and making conclusions) of data by people that do not have special skills in data analysis.

E. Veglianti (✉)
FGES, Université Catholique di Lille, Lille, France
e-mail: eleonora.veglianti@univ-catholille.fr

M. Trombin · M. De Marco
Department of Economics, University Uninettuno, Rome, Italy

R. Pinna
Department of Economics and Management, University of Cagliari, Cagliari, Italy

C. Dal Zotto et al. (eds.), *Smart Technologies for Organizations*, Lecture Notes in Information Systems and Organisation 60, https://doi.org/10.1007/978-3-031-24775-0_15

The HR sector is aware of the progressive change that HR boards are stimulating, by shifting the focus from automation to the impact upon business and engagement, that are supposed to define HR strategies and mold the future of the sector since 2018. Many analyses conducted by HR professionals agree that one of the major trends in HRM will be Artificial Intelligence to serve HRM, and that three key areas of recruitment will be revolutionized by AI:

1. AI will bring more efficient and fairer candidate screening, in order to free up time to focus on the human aspect of recruitment and offer a more personal service to clients and candidates.
2. AI will ensure a better candidate fit through online job boards.
3. AI will help safeguard future talent pipelines in order to improve employee retention and development.

There is a growing body of scientific research pointing toward the use of behavior, rather than self-report, as a better assessment tool. This is because behavior can be objectively measured, and this direct measure can be used to counteract some of the conscious and unconscious bias that is inherent to all human self-report measures. In many big companies the validation of selection process through AITs is now in place, from screening/indexing of the resume (almost obsolete in comparison to social profiles), up to automated interviews by screening facial mimics and eye movements. Common opinion in recent research states that a correct parameterization of AI guarantees better hiring (efficient and durable) compared to hiring made by managers, since it allows to eliminate cognitive biases and judgment subjectivity, and to seek in the profiles just the features that have proved successful by individuals embedded in the organization and acknowledged as effective members of the team. Although these technological innovations have generated a transformation of the recruitment process [1] and their implementation has been widely embraced by large companies, many outstanding issues with regard to AITs have remained unanswered. Despite the pervasive impact of technology on HR in organizations, and the enormous potential to improve talent recruitment process [2], there has been relatively little research examining its effectiveness and its capacity to remove the potential bias in the selection process and to evaluate a wide variety of knowledge, skills, abilities, and other characteristics of candidates. Thus, we believe that further research is needed to assess to what extent AITs enable organizations to achieve their goals. In particular, based on the case study of Pymetrics platform, the aim of this paper is to explore the contribution that AITs can give in order to overcome human errors by comparing theoretical convergences among the various approaches and platforms addressed to companies. A brief study of the limits linked to the use of AITs will be also investigated. Actually, there are still many open questions. If it is true that the possible errors of assessment, inevitable when the human component comes into play, represent a cost and a problem for companies, it is also true that the selection algorithm is based on data collected by humans, with the same possibility of error.

In particular, our goal was to answer the following research questions:

- How accurate is the system in order to remove bias and judgment subjectivity from selection process?
- Do Artificial Intelligence Technologies attract and select the talented who can perform successfully in organizations?

Often then, in writing down the characteristics of an open position one can be tempted to resort to too general descriptions, which artificial intelligence, unlike the selector who knows the company well, may not be able to grasp. For this reason, we will examine the "cognitive traps" in relation to skills. By comparing the methodology and the theoretical background adopted by the above-mentioned actors, especially the case of Pymetrics platform, we suggest however that at this stride these systems can present limits as to parameterization, that there is an urgent need to theoretically define the relation between actual skills and desired skills (the latter being a request from companies), that the pretense of unbiased hiring can be partially criticized and finally, that bias must not be confused with subjectivity, which has to be preserved instead.

The rest of the paper is organized as follows. First, key concepts are defined and situated within the academic literature. Next, the research methodology employed is detailed. Finally, we provide a summary of the key findings and discuss the implications for organizations.

2 Literature Review

2.1 *Talent Management*

In the world economy, human capital is one of the most valuable assets of an organization, and employee recruitment and selection cover the important function of attracting and retaining talented workforce [3, 4]. These goals have become extremely important in recent years because organizations compete on the basis of the skills and talents of their workers [5]. For these reasons, the field of recruitment and selection has traditionally been one of the most important domains of research and practice in the field of Human Resource Management in organizations. Michaels [6] defined talent as a person's ability, which Include a person's abilities, knowledge, and potential for future development. Talent consists of those individuals who can make a difference to organizational performance, either through their immediate contribution or in the longer term by demonstrating the highest levels of potential (chartered institute of personnel and development). Over the years, there has been a dramatic shift in the role of human resources in an organization. Nowadays companies focus on their potential/talented employees rather than having a neutral attitude for all the employees working in the organization as talented people act as a critical resource for achieving success. Furthermore, in many parts of the global economy there is a talent shortage and many companies still struggle to recruit and retain key talent [7]. Demographic trends such as a smaller supply of younger workers and retirements among baby boomers indicate that recruitment and applicant attraction will be even more important in the future [8]. As a result, recruitment has become one of the most critical human resource functions for organizational success and survival [9]. Talent management is the process of identifying, attracting, engaging, and retaining highly qualified and talented individuals. Acquiring and retaining the best talent acts as competitive advantage, but it is equally important and difficult to find the right person for the right job. The increased competitiveness in the recruitment market has led to organizations spending more time, effort, and resources in order to develop tools for the efficient selection of employees with the required skills and aptitude to meet current and future organizational needs [6, 10]. A critical task in the talent management is to identify all the essential skills that are required by employees for their development in all the key areas of business [6, 11] and to ensure that the organization has a continuous supply of highly productive individuals in the right job, at the right time. Rather than a one-time event, talent management is a continuous process that plans talent needs, builds an image to attract the very best, ensures that new hires are immediately productive, helps to retain the very best, and facilitates the continuous movement of talent to where it can have the most impact within the organization [12].

Among strategic priorities, diversity is an important aspect of any organization. It is no longer just a legal compliance issue but a key business strategy [13]. Many studies based on worldwide collected data point out that the top quartile of gender-diverse companies is 21% more likely to experience above average profitability than companies in the bottom quartile [14]. Diversity has also been linked with higher rates of employment retention [15]. This fact couples with the corporate need to prop up the organizational change in a context of lacking appropriate leadership (so-called "leadership drought") and market instability at a global level.

But a diverse workforce does not happen on its own [13]. In general, companies are not always aware of the perspective they are looking at things from and keep on being self-centered. Anyway, companies know they have to reach the above-mentioned goals. Often, though, they run after shortcuts: they tend to use innovative methods and technology, while not changing their organization by assessing themselves in relation to the adoption of such innovations. Since diversity is not just about hiring, a gap has been emerging between HR managers (and middle managers in general) and the adoption/acceptance of the new methods of recruiting/retention. Organizational behavior can cause diversity to dauntingly fail. Tech evolution should be understood, integrated, tailor-designed, gradually implemented—if necessary, underpinned by supporting measures—undoubtedly.

In fact, in the Internet Era the traditional method of recruitment and selection has been revolutionized. The internet helped attract potential candidates to an organization from the recruitment process, which is referred to as e-Recruitment. It is usually defined as "a way of implementing strategies, policies and practices in organizations through a conscious and direct support of and/or with the full use of web-based channels" [16]. Over time, various forms of technology were introduced to attract applicants, ranging from passive, one-way technologies (e.g. web-based job ads, job boards) to more interactive techniques (such as virtual job fairs). Recent developments in digital technologies have enabled new possibilities for improving recruitment and selection processes. In particular, the talent acquisition highlights how leading organizations use social networking, analytics, and cognitive tools to find people in new ways, attract them through a global brand, and determine who will best to the job, team, and company. In particular, Internet facilitates the use of tools suitable for gamification, that is the use of games and game mechanisms in a non-game context [17–20], such as business. The more innovative ideas and solutions are focused on cognitive technologies such as artificial intelligence (AI), neuroscience assessment, predictive algorithms, and data science analytics and algorithms.

2.2 Gamification in Talent Selection

Recently, there is an increasing number of large organizations that used gamification in the selection process in order to assess cognitive, social, and personality traits. Gamification is the use of game attributes to drive game-like behavior in a non-game context [21]. Many organizations are turning to a job simulation software which could improve hiring by giving candidates tasks they would do on the job. Allowing potential candidates to experience various jobs within different departments of the company. To fill the gap between the need to measure soft skills and the inadequacy of traditional interviews, some companies have developed pre-screening online assessments—user-friendly tests that take candidates about 20 min to complete and give employers a sense of their soft-skill strengths and weaknesses. Whether candidates play neuroscience-inspired games or answer straightforward questions, these online tests let companies screen large pools of talent quickly and make smarter hiring decisions. A series of behavior games collect millions of data points, measuring cognitive and personality traits. There is no right or wrong way to play the games because every trait makes a great fit for a certain job. These tools can be utilized for different goals. First and foremost, they guarantee that these models will help build a more diverse workforce by carefully detecting and removing bias from the selection process. Nonetheless, they will raise efficiency by increasing the yield, reducing missed talent, and expanding the reach of recruiting team. Furthermore, they help companies fight attrition and retain employees by modeling different roles within the company and matching current employees who are ready to leave with a new position.

Despite the growing popularity of gamification in recruiting and the apparent success in some cases, little is known about how these processes influence applicant attraction or if the return on investment is positive. Companies can support this new approach to talent acquisition by starting with a degree of centralization to gain the benefit of scale and efficiency and create a strong and competitively differentiated candidate experience.

Some researchers found gamification to be providing a number of generally positive outcomes: use people's behavior to assess cognitive, social, and personality traits. In research contexts, behavior-based assessments have largely replaced self-report instruments wherever possible. There is a growing body of scientific research pointing toward the use of behavior as a better assessment tool. This is because behavior can be objectively measured, and this direct measure can be used to counteract some of the conscious and unconscious bias that is inherent to all human self-report measures. These games provide a snapshot of a person's unique characteristics. For example, some games focus heavily on traits that are the hardest to train—such as flexibility, learning ability, and decisiveness. Some scholars [22, 23] point out that gamifying the employee recruitment process has the potential reduce bias in employment decisions and help employers make better decisions in hiring, performance evaluations, and promotions. For most organizations, the use of games in recruitment and selection process offers massive economic savings to companies, and also the opportunity to assess large number of their applicant pool.

The use of data-driven algorithms does not automatically guarantee fairness—the conclusions drawn from a data set can only be as inclusive as the input data itself. Meaning, the use of sophisticated algorithms alone cannot remove bias. Fairness will only be achieved through active debiasing of the data on which the tools rest. To guarantee the removal of all bias when evaluating a candidate, some tools allow the companies to the removal of any residual bias found in the data. Adding this third step to the bias removal methodology is essential for ensuring a bias-free final outcome. This last step in bias elimination starts after data are acquired from science-based games. Data are acquired from candidates and from high performing incumbents at a specific job. Statistical methods are then used to measure bias in the incumbent dataset. Once bias has been identified, this information is used to guide the model parameter selection process so that the final candidate recommendation model is bias-free.

2.3 *Development of AI Systems for Recruiting: An Overview of Multifarious Approaches*

AI use in recruiting is a quite new phenomenon linked to the creation of candidate databases through modern software, which has made storing and retrieving of resumes easier. Its implementation is considered crucial, in particular by multinationals, which have an overall urgent need for scalability, also in HRM. Indeed, with the advent of AI, there is a vision for HR where successful innovations are combined with existing point solutions to form a suite that can deliver a user-friendly experience for both candidates and administrators [24].

The recruitment software landscape has further expanded with the evolution of many tech companies catering to various activities in the recruitment value chain. The recruitment landscape has thus seen extensive changes over the years and has evolved with the creation of new recruitment tools and processes. For example, Xref and Skill Survey conduct candidate reference checking; Hackerrank, Hackerearth, Aspiring Minds, ELitmus, and Cocubes conduct online tests; Ideal, IBM Watson, X.ai, Glider.AI, belong.co, and param.ai handle Artificial Intelligence and automation in recruitment; Skype, SparkHire, WePOW, Google Duo conduct video interviewing; and BambooHR, KinHR take care of onboarding [25]. Many are also cloud-based SaaS analytics solution providers, such as PredictiveHire and other SaaS architectures [25–27], which reinvent the applicant tracking systems (ATS), and which are augmenting the ATS with other TA technologies, including candidate relationship management, video interviewing, and analytics [25].

Then, implemented systems comprise a wide spectrum of solutions. In many big companies, the validation of selection process through AITs is now in place, from screening/indexing of the resume (almost obsolete in comparison to social profiles), up to automated interviews by screening facial mimics and eye movements. Common opinion states that a correct parameterization of AI guarantees better hiring (efficient and durable) compared to hiring made by managers, since it allows to eliminate cognitive biases and judgment subjectivity, and to seek in the profiles just the features that have proved successful by individuals embedded in the organization and acknowledged as effective members of the team. In order to reach this goal, companies are using some tech platforms trying to:

1. Simplify the operations: ClearFit (automatic candidate search), Wade & Wendy (virtual assistant at the first contact between candidate and company);
2. Improve the effectiveness of the process: Textio (creation of distinctive and customer-specific job ads), Engage Talent (search of candidates not actively looking for a job), Ansaro (predictive model aimed at helping companies hire new employees), Helena (the first AI-powered Headhunter, looking for passive job seekers);
3. Suggest ad-hoc tests: Harver (that generates simulation job-like tests to assess the candidate's skills), Filtered (that automatically generates coding activities to test potential candidates).

On the one hand, chatbots are now clearly established. They are considered as "Partially Intelligent Systems" that exhibit Perception of the data, Heuristic values, Memory, Consciousness in the surroundings [26]. For this reason, they behave as first screening actors in the recruitment process.

On the other hand, more refined forms of predictive analytics aimed to exploiting big data are advancing, in order to meet the urgent demand for a customized recruitment strategy based on AI. This need is covered by companies like Pymetrics, which combine neuroscience games, customized AI, and the use of client-specific algorithms, rejecting candidates who do not match the characteristics required by that company.

For sure from companies' perspective, there is a tendency to automating processes, to saving costs, and to maximizing human capital in a whirling and fast-paced labor market, which especially big companies are willing to dominate and mold according to their own exigences. In general, though, there is a lack of amalgamation throughout the industry related to AI.

In the academic literature, several analyses have been conducted in relation to organizational implications caused by AIT. In Strohmeier [28, 29]. AIT use in HRM is explored in six selected scenarios: turnover prediction with artificial neural networks; candidate search with knowledge-based search engines; staff rostering with genetic algorithms; HR sentiment analysis with text mining; resume data acquisition with information extraction; employee self-service with interactive voice response [28, 29]. In Bissola and Imperatori [30], the main focus is on transitional HRM, so on building trust in an organization through e-HRM. In Furtmueller [10], an interesting research is developed on the requirements and traits of resume contents in relation to e-HRM. Provided that all of the aforementioned approaches and outcomes are taken into account, in the present paper we aim to explore the contribution that Artificial Intelligence Technologies can give in order to overcome human errors by comparing theoretical convergences among the various approaches and platforms addressed to companies. In this case, we will link our study to some research to identify how companies are utilizing pre-hire assessments, and how these assessments are evolving in the technology and talent age [31]. Though, we will provide evidence that the enthusiastic business-driven aim at obtaining aseptic outcomes could hide some perilous and biased consequences.

3 Research Method

Basically, we needed to question whether the human component, which is a critical component in the hiring process, was considered. We wanted also to ask ourselves, how can we assess and evaluate whether individual constraints, subjectivity, and biases have been accounted for while designing the recruitment tool. Finally, in an overall overview, if we contemplate the vision of the elimination of frictional unemployment, albeit an aspirational, we should question the unavoidability of frictional unemployment [27], since economy is empirical and inefficient, as labor market is. Given the scarcity of research works in that field a qualitative methodology can be considered a proper solution to investigate a contemporary phenomenon within its real-life context. The single case study approach has been used as it is well suited to exploratory investigations where phenomena are not well-understood [32, 33] from a qualitative point of view, since our goal was not to explore the quality of the implemented scientific method (which is unquestionably excellent), rather to debate the intimate justification of the whole as a successful and optimistic system. The case under consideration is Pymetric platform, a neuroscience-based assessment developed by Pymetric an innovative company worldwide in the field of neuroscience applied to HRM through AITs. The data mainly consists of primary data collected through qualitative explorative and semi-structure interviews. Secondary data such as reports, patents, and other materials retrieved on the web are also used.

3.1 The Case Study

The case study setting is Pymetric platform, a neuroscience-based assessment developed by Pymetric an innovative company worldwide in the field of neuroscience applied to HRM through AITs, working with the most important multinational corporations, like as Unilever, Randstad, Accenture, LinkedIn, Tesla, Mercer, Arla, Sutherland. Pymetrics platform is based on neuroscience games assessing different qualities of the candidate (cognitive traits and emotional traits) in a framework of a methodology aiming at predictive and debiased staffing services. Pymetrics claims to bring innovation through its platform in terms of overriding the key loopholes of traditional hiring practices, which are considered ineffective (70%), biased (50%), and offering poor experience (50%). Pymetrics platform can be also defined as agame-based recruiting tool that assesses the strengths of candidates and recommends the right careers and companies for them. On the background, Pymetrics assumes that traditional hiring is ineffective, biased, and offers a poor experience. That is the reason why they are proud of four own innovations: Neuroscience games, Customized AI, Bias-free Algorithms, and Common Application. Hiring is based on neuroscience and data science, with the aim of democratizing career search and hiring. In particular, Neuroscience Games measure established building blocks of cognitive and emotional functioning, so that measuring traits through behavior allows Pymetrics to profile people very accurately and in high-dimensional space. Pymetrics defines itself as the Netflix-like recommendation algorithm for jobs, since they build a custom, cross-validated profiles for each role and company, based on top performers. What is more interesting, Pymetrics poses itself the question of bias in recruitment processes, that is why its team contrived Bias-free Algorithms, i.e. algorithms extracted from an iterative algorithmic auditing technique aimed to identify and remove bias. This technique is made of a reference set of tens of thousands of people to check for any potential biases, and of reweighting inputs in the model. When candidates play the games, 77 different traits are assessed, both emotional and cognitive, in order to identify tasks that would most reveal the traits that recruiters and hiring managers to want to know (Fig. 1).

Cognitive Traits	Emotional Traits
Processing speed	Risk profile
Continuous attention	Reward sensitivity
Memory span	Emotional identification
Cognitive control	Tolerance for ambiguity
Planning	Ability to delay gratification
Sequencing	Learning from feedback
Learning	Learning from mistakes

Fig. 1 Some of the candidate's most important traits after Pymetrics' collaboration with the company Mercer. Retrieved from www.mercer.com

When candidates play the games, there is no losing or winning, just the assessment of different qualities. The attributes required are vastly different. The game is designed in such a way that it helps an organization assess candidates' capabilities, like processing speed, attention to details, memory, emotion detection, risk-taking, fairness, and focus. Then there is also an assessment of emotional traits of candidates by analyzing an image of person's eye while viewing the given conception. Together, data coming from trait assessment provide a snapshot of a person's unique characteristics. Pymetrics games are non-directional, meaning that unlike school or standardized tests, there are no right answers or preferred scores. Instead, they measure traits where either end of the spectrum can be beneficial based on the demands of a particular profession. That is why they also provide career recommendations based on these traits. The assessment also includes a top career report, suggesting the best professions for you based on your strengths. In addition, Pymetrics also suggests jobs based on these reports and lets candidates connect with recruiters and companies with suitable job offers for them.

4 Results and Discussion

The analysis of literature and the case study shows that philosophical discussions and decisions about the proper use of AI need to be based on a proper understanding of the way AI-scientists achieve their results—in particular, in their dependence on the initial planning input of human beings. In this regard, evidence outlined by data and by the interview(s) are particularly interesting as long as they allow to shed a brighter light on the data and the theory provided by Pymetrics' white paper, patent, and scattered data to be found on websites and articles related to the platform.

In fact, the most highlighted outcomes by client companies around the world are related to cost shrinking, which is considered the most significant advantage. In the interview to Unilever Italy that is mentioned below, one can see that emphasis is put on data, few are said about the organizational change needed to reach diversity inclusion or HR professional inclusion in the new processes. Data related to the new pre-hiring assessment method are not coupled nor matched with organizational data.

Most emphasized data by Unilever are those related to the hiring process, in which Pymetrics platform was used to assess 280,000 applicants in 68 countries and in 15 languages, replacing the resume as a first-pass filter and meeting the following KPIs:

- 100% increase in hire yield;
- 75% reduction in time to hire;
- 25% decrease in recruiting costs;
- Record diversity across gender, ethnicity, and socioeconomic status.

In a more general framework related to all clients, Pymetrics itself declared in the interview that some of the clients have seen the following outcomes:

- 75% reduction in time to hire: 4 months–4 weeks;
- 75% reduction in recruiter time;
- 6x improvement in applicant to offer yield: went from 150 resumes to fill 1 role to only 25 resumes at a global financial institution;
- 3x improvement in interview to offer yield: baseline of 8.5% improved to 25% for global consulting firm.

These outcomes were confirmed by HR managers of the client companies, like the HR manager of Unilever Italy. The HR manager explained that the resume amount the company receives has notably increased (in the first 90 days, comparing this period with the same in the previous year, applications rose from 15.000 up to 30.000 in the United States only). Nonetheless, save up time and resources are greater than through traditional recruiting, and selected talents can be hired in roles where they can exploit their added value. Unilever, Google, and Uber use gamification in recruiting with the goal of maximizing their operational efficiency, increasing candidates' engagement, cutting time spans, and building brand awareness.

On the side of the candidate, the key area of investigation is actually the definition of talent, linked to the concept of bias. Requested to define talent on the basis of Pymetrics motto "Matching talent to opportunity, bias-free," Pymetrics replied: "For us, talent means a job candidate's inherent cognitive, emotional and social traits. Our role is to identify these traits using our neuroscience-based games and then match the candidate to a job where their traits predispose them for success."

This might seem a quite neutral and consistent assumption. Instead, we should stress a couple of words here, i.e. "inherent" and "predispose." In the framework of the background bias-free assumption, those words should be reassuring. Instead, they help us better investigate the risk of intrinsic bias. Outside of medicine, the use of artificial intelligence in predictive policing, bail decisions, and credit scoring has shown that artificial intelligence can actually exaggerate racial and other bias. As a matter of fact, it is difficult to construct a virtual environment that accurately portrays real-world dynamics and stochasticity without error, leading to simulation bias. What should be stressed is, for example, that games are available in 15 languages (English, Spanish, French, Italian, Mandarin Chinese, etc.), but human biases are baked right into our language and the language-usage data AI systems learn from.

Applying machine learning to ordinary human language results in human-like semantic biases. In addition, predisposition and commonly desired traits by companies (this is meant by 'inherent') leads to doubt that companies do not hire the best candidates in general but the best candidates among a selected group based on a dataset of strict parameters which may seem a kind of Darwinian trap.

Finally, this last point reveals a risk of intimate tendency to hindsight bias.

In order to give a more colorful picture, we are going to focus also on aspects related to the latter. Indeed, when it came to questions regarding the choice of emotional and cognitive traits, the parameterization method, and the determination of useful and not useful data, we also found some interesting aspects.

When asked about the choice of the 77 traits, Pymetrics points out that that two things are taken into consideration:

(1) traits that are most commonly desired by companies evaluating job candidates; and
(2) which traits can reliably be measured by neuroscience-based games?

Furthermore, at the question of whether there are other traits which are initially selected, and later left behind for being considered useless, we were replied that Pymetrics regularly reviews the traits to be measured with the aim to determining whether to add or remove any traits. This determination is made using factors such as client demand, and developments in neuroscience regarding trait measurement. Further we asked, either emotional traits are measured only by using gamification, or this process is also deployed with aptitude tests, for instance. The reply was that Pymetrics platform solely uses neuroscience-based games to assess traits, and no other assessment tool is used. A final aspect we were willing to scrutinize regarded how Pymetrics attributes the weight to every parameter, and how parameters are measured.

The answer was quite expected, i.e. weights are data-driven, determined by commonalities in existing top-performing employees. Pymetrics has developed a proprietary machine learning algorithm which finds the combination of cognitive and emotional traits that distinguish top-performing current employees from the general population. The weighting of the traits is determined by the data: for some roles, memory might be a crucial trait, while other roles might emphasize attention or risk. Just for this reason, it was also crucial to investigate if they are aware to miss some quintessential components in human behavior by using predictability and people analytics, that could be also necessary while working in a company. At this last question, the answer was that the Pymetrics platform is the preliminary screening process for clients looking to hire talent. Employment decisions are not made solely based on the data our platform provides. All candidates that match with a given role via our platform then move on to interview(s) with hiring managers on the client side. During these subsequent interview(s) managers further assess candidates to determine if they should be offered the position. However, this point leads us to a pivotal argument raised by recent literature regarding about human–AI interaction, i.e. machine's accountability.

Still, meaningful was the answer given with regard to the definition of "error," and to the errors Pymetrics aims to avoid in the selection process.

For Pymetrics, errors are any factors that are considered during the matching process that are not predictive of a candidate's success in a given role, such as a candidate's gender, socioeconomic status, and ethnicity. However, this counter-bias effort can mirror "socially legalized" biases [34, 35].

Related to the notion of error, a question was devoted to the future of AI in recruitment, namely if one day these methods could possibly eradicate any kind of error from the selection process. The reassuring answer is that there will likely always be a human element to the hiring process, meaning that platforms like Pymetrics will always be paired with evaluations by hiring personnel during the hiring process.

Because eradicating bias in humans is nearly impossible, introducing bias or errors into the hiring process is always a possibility. Unlike what Russian hiring robot Vera aims to do in the future, i.e. taking the final decision of hiring a person, Pymetrics is cautiously confident that human component is still "likely" to play the decision-making role. Though, that "likely," and the sentence following it, could betray a certain degree of wariness toward human inefficiency. However, some scholars point out that it is imperative not to forget the "human" in HR, since we still desire a personal interaction and this topic is more than just about an efficiency game [36].

5 Conclusions and Managerial Implications

Provided that the advent of AI poses major challenges to the skill set, competence, and capability of humans on the one hand, and that machines have the ability to learn, adapt, make decisions, accumulate experience, and even take actions on the other hand [37], we should focus on organizational change as key to success of AI. Organizational change can be achieved by two factors:

1. collaboration human—AI;
2. accompanying measures to reciprocal understanding.

As to point 1, to take full advantage of such collaboration, companies must understand how humans can most effectively augment machines and, conversely, how machines can enhance what humans do best and how to redesign business processes to support the partnership [38].

As to the point 2, companies (and even governments) should contrive and implement some accompanying measures to foster mutual understanding between humans and AI, and between staff/management and company.

From an organizational point of view, human–AI collaboration involves redesigning business operation and improving five elements of the business process: flexibility, speed, scale, decision-making, and personalization. The hiring process belongs to the third element, since poor scalability is the primary obstacle for improvement [38]. Indeed, e-recruiting practices should succeed in integrating them within the organizational vision, and to do so a fair discussion should be set up in order to decide how to interpret data, with regard to both recruiters, and middle/top managers. Also, another point is intertwined with the latter, i.e. machine's accountability. Machine's authority is considered unquestionable and fair: what kind of consequences can this sort of impunity have upon management and society? What relationship should an, as human, "inefficient" manager (recruiter, top manager, etc.) establish with an authoritative machine?

There is an urgent need to realign, reshape, and reinvent the profession [37], and integrate e-HRM and i-HRIS (Integrated Human Resources Information System) into HRM practice [39]. In order to achieve this, we should ask ourselves how we can complement and empower the human capability and improve the processes that surround it, by means of an honest assessment of the pros, the cons, and the actual output [36]. As a matter of fact, technology in and of itself is of limited utility if companies and government organizations lack people who can use it, effective strategies for how to use it, and training to be good at using it [40].

Actually, the HR industry is still in the trial phase of figuring out how to implement the technology and decide how swiftly to deal with it [41]. This hinges on a couple of key factors:

1. need of career reinvention, bound to generation and expertise gap.
2. company-centered implementation.

As to point 1, we see that career reinvention will be critical, as being much more fast-paced than it was in the past and based on fluidity. As a consequence, companies are centering their recruitment efforts on Millennials and the new selection processes follow this track, as we have seen. Millennials have in-demand STEM skills and fresh perspectives but almost all types of organizations say they are confronting a leadership drought, simply because Millennials lack experience [42]. Conversely, experienced people are required to manage complex chameleonic processes, which they should be able to rapidly grasp and learn. Paradoxically, companies think they are preparing the future by hiring Millennials, but they will not have the in-house experience to manage their creativity and bridge the gap between generations, and between workers. In addition, large segments of the global workforce risk being excluded and discriminated by a too rigid hiring scheme that is hidden behind a customized, goal-targeted AI-based recruitment process.

Point 2 leads us to focus on the concept of company itself to be envisioned for the future. Traditionally many HR practices take the needs of the organization as the starting point, and recruitment is a major example of this maybe outdated (mis-) conception: there is a fixed structure, a defined hierarchy, a top-down approach, and well-defined jobs, so how do we find candidates that can fill this specific, pre-defined vacancy? [43]. For sure, more efforts should be done toward the following goals:

- more interaction with candidates in the pre-screening and hiring phase but also in the retention phase;
- machines cannot be allowed to determine how good people are, corporate ethics should be structurally embodied in evaluations in order to generate real value: this means changing the approach toward what someone has lately defined "analethics" (see, [43]), as well as toward flipped hiring practices (see [44]);
- gauge the expectations of workforce analytics and technology outcomes to a well-rounded, aware, and people-centered corporate leadership, which is lacking nowadays.

Strategizing is key to both recruitment and talent acquisition but far less strategy is considered to be necessary in the recruitment process. For this reason, during talent acquisition companies should formulate preventive strategies to find and maintain a group of potential candidates. But there is also an urgent need to kick off a fair discussion about practices in managing reward systems, job performance evaluation systems, and diversity systems [45].

Last but not least, it should be noticed that the interview question about a possible relation between the word "quality" and word "skill" a/o "performance" was skipped by Pymetrics. Indeed, the word "skill" does not occur too many times in Pymetrics' reports and papers, and this pushes us back to the urgent need of theoretically defining the relation between actual skills and desired skills. This point is crucial to shape the collaborative intelligence we would like to see in the future of our society.

6 Future Research Directions

From previous literature, it is observed that some pivotal topics need to be further investigated. Some scholars have pointed out the following points to be scrutinized in the future: transition push model to pull model in the assessment market; efficiency vs. fairness; evolving legal compliance; companies' tendency to mirror software the way work is structured; unconscious biases perpetuation; risks related to society uberization, consistency of language related to interpretation; recalibration and upkeep; risks linked to scalability and customization; awareness of candidates/recruiters about the hiring process; machine learning vs. human learning; actual skills vs. required skills; strategic competition; holistic approach (i-HRIS); companies awareness in the fast changing context.

Central issue will be to evaluate the impact of these trends, which are sometimes a critically acknowledged, and to better research the implications of Human–AI interaction. This has to be achieved both on organizational and on societal level, with the broadest possible multidisciplinary approach.

If it is true that Huzinga's [46] concept of *homo ludens* can lead us to gamification, nonetheless we should try to avoid that the game we are playing would eventually turn into a Russian roulette.

References

1. Sylva, H., & Mol, T. S. (2009). e-Recruitment: A study into applicant perceptions of an online application system. *International Journal of Selection and Assessment, 17*(3), 311–323.
2. Stone, D. L., Deadrick, D. L., Lukaszewski, K. M, & Johnson, R. (2015). The influence of technology on the future of human resource management. *Human Resource Management Review.*
3. Katz, D., & Kahn, R. L. (1978). *The social psychology of organizations* (2nd ed.). Wiley.
4. Chapman, D. S., Uggerslev, K. L., Carroll, S. A., & Jones, D. A. (2005). Applicant attraction to organizations and job choice: A meta-analytic review of the correlates of recruiting outcomes. *In Journal of Applied Psychology, 90*(5), 928–944.
5. Huselid, M. A. (1995). The impact of human resource management practices on turnover, productivity, and corporate financial performance. *Academy of Management Journal, 38*, 635–672.
6. Michaels, E., Handfield-Jones, H., & Axelrod, B. (2001). *The war for talent*. Harvard Business School Press.

7. Hiltrop, J. M. (1999). The quest for the best human resource practices to attract and retain talent. *European Management Journal, 17*, 422–430.
8. Breaugh, J. A. (2008). Employee recruitment: Current knowledge and important areas for future research. *Human Resource Management Review, 18*, 103–118.
9. Saks, A. M. (2005). The impracticality of recruitment research. In A. Evers, N. Anderson, & O. Voskuijl (Eds.), *Handbook of personnel selection* (pp. 419–439). Blackwell.
10. Furtmueller-Ettinger, E., Wilderom, C. P. M., & Tate, M. (2011). Managing recruitment and selection in the digital age: E-HRM and resumes. *Human systems management, 30*(4), 243–259.
11. Axelrod, B., Handfield-Jones, H., & Michaels, E. (2002). A new game plan for C players. *Harvard Business Review*, 81–88.
12. Sullivan, J. (2004). Eight elements of a successful employment brand. *ER Daily, 23*(2), 501–517.
13. Berkowitz, J (2018). Is diversity in recruiting a business strategy or empty talk? https://www.ibm.com/blogs/insights-on-business/gbs-strategy/recruiting-diversity/.
14. Hunt, V., Prince, S., Dixon-Fyle, S., & Yee, L. (2018). *Delivering through diversity*. McKinsey and Company.
15. Nishii, L. H., & Mayer, D. L. (2010). Paving the path to performance: Inclusive leadership reduces turnover in diverse work groups (CAHRS ResearchLink No. 3). Cornell University, ILR School.
16. Girard, A., & Fallery, B. (2010). Human resource management on internet: New perspectives. *The Journal Contemporary Management Research, 4*(2), 1–14.
17. Deterding, S., Dixon, D., Khaled, R., & Nacke, L. (2011). From game design elements to gamefulness: defining gamification. In *Proceedings of the 15th International Academic MindTrek Conference: Envisioning Future Media Environments* (pp. 9–15).
18. Paharia, R. (2013). *Loyalty 3.0: How to revolutionize customer and employee engagement with big data and gamification*. McGraw Hill Professional.
19. Zichermann, G., & Linder, J. (2013). *Gamification revolution*. McGraw-Hill Books.
20. Raczkowski, F. (2013). It's all fun and games... A history of ideas concerning gamification. In *DiGRA Conference*.
21. Wu, M. (2012). The gamification backlash + two long term business strategies. Retrieved from https://lithosphere.lithium.com/t5/Science-of-Social-Blog/The-Gamification-Backlash-Two-Long-Term-Business-Strategies/ba-p/30891.
22. Chow, S., & Chapman, D. S. (2013). Gamifying the employee recruitment process. In *Proceedings of the First International Conference on Gameful Design, Research, and Applications* (pp. 91–94).
23. Collumus, A. B., Armstrong, M. B., & Landers, R. N. (2016). Game-thinking within social mediat o recruit and select job candidates. In R. N. Landers & G. B. Schmidt (Eds.), *Social media in employee selection and recruitment* (pp. 103–124). Springer International.
24. Mathis, J. (2018). Innovations are coming to the human resource. *The Journal of Private Equity, 21*(4), 14–17.
25. Bersin, J., McDowell, T., Rahnema, A., van Durme, Y. (2017). The organization of the future: Arriving now. In *Global human capital trends 2017: Rewriting the rules for the digital age* (pp. 19–28). Deloitte University Press.
26. Mishra, D., & Shekhar, S. (2018). Artificial intelligence candidate recruitment system using software as a service (SaaS) architecture. *International Research Journal of Engineering and Technology (IRJET), 5*(5), 3804–3808.
27. Roy, R. (2017). Corporate recruiting opening its doors to AI: The performance opportunity? *International Society for Performance Improvement, 56*(10), 43–44.
28. Strohmeier, S., & Piazza, F. (2015). Artificial intelligence techniques in human resource management—A conceptual exploration. In C. Kahraman & Ç. Onar (Eds.), *Intelligent techniques in engineering management* (pp. 149–172). Springer International Publishing.
29. Strohmeier, S., & Piazza, F. (2015b). Human resource intelligence und analytics. Grundlagen, Anbieter, Erfahrungen und Trends. Gabler Verlag.

30. Bissola, R., & Imperatori, B. (2014). The unexpected side of relational e-HRM: Developing trust in the HR department. *Employee Relations, 36*(4), 376–397.
31. Chiang, J., & Berkoff, R. (2017). How are pre-hire assessments contributing to unbiased and more targeted, successful hires? Cornell University, DigitalCommons@ILR.
32. Eisenhardt, K. M. (1989). Building theories from case study research. *Academy of management review, 14*(4), 532–550.
33. Yin, R. K. (2014). *Case study research: Design and methods.* Sage Publishing.
34. Masuch, M. (1985). Vicious circles in organizations. *Administrative Science Quarterly, 30*(1), 14–33.
35. Fridell, L., & Lim, H. (2016). Assessing the racial aspects of police force using the implicit- and counter-bias perspectives. *Journal of Criminal Justice, 44*, 36–48.
36. Lee, H., Lee, S., & Tarpey, M. (2018). *CAHRS partners' implementation of artificial intelligence.* Cornell University, DigitalCommons@ILR.
37. Munoz, J. M., & Naqvi, A. (Eds.). (2018). *Business strategy in the artificial intelligence economy.* Business Expert Press.
38. Wilson. H. J., & Daugherty, P. R. (2018). Collaborative intelligence: Humans and AI are joining forces. *Harvard Business Review,* 1–11.
39. Latif, A., Ullah, R., Din, M. S., & Anjum, Z. A. (2014). The future of human resource professionals. *City University Research Journal, 4*(2), 208–222.
40. Horowitz, M. C., Allen, G. C., Kania, E. B., & Scharre, P. (2018). Strategic competition in an era of artificial intelligence. Report for the Center for New American Security.
41. Benemann, K. (2017). Enhancing talent acquisition and retention with AI: 3 tips for HR executives. https://www.hr.com/en/magazines/all_articles/enhancing-talent-acquisition-and-retention-with-ai_j5724fh0.html.
42. Bagai, S. (2018). Prejudices in the hiring process extend beyond traditional notions of diversity. https://www.crowdstaffing.com/blog/workforce-diversity-the-benefits-of-hiring-experienced-talent.
43. Haak, T. (2019). 10 inspiring HR trends for 2019. Retrieved from https://hrtrendinstitute.com/2018/11/26/10-inspiring-hr-trends-for-2019/.
44. Triant, B., Craig, R. (2018). The flipped hiring revolution. Retrieved from https://www.edsurge.com/news/2018-10-10-the-flipped-hiring-revolution.
45. Manuti, A., & De Palma, D. (2018). *Digital HR: A critical management approach to the digitilization of organisations.* Palgrave Macmillan.
46. Huizinga, J. (2008). Homo ludens: Proeve eener bepaling van het spel-element der cultuur. Athenaeum Boekhandel Canon.

Employee Perceptions of Electronic Performance Monitoring: A Multi-Level Analysis

Karma Sherif and Mazen El-Masri

Abstract Research on Electronic Performance Monitoring (EPM) has taken an integrative perspective toward employee perception of the technology without consideration of the effect of organizational culture and specifically subcultures on the acceptance of an EPM system. There is a dearth of literature on the interplay between subcultures and how organizational mechanisms enforce a dominant culture to facilitate the acceptance of IT systems. In this paper, we address the gap through a differential perspective of organizational culture and its effect on the acceptance of EPM systems. We conduct a case study of two multi-cultural organizations: one in Qatar and another in the US, and examine the influence of the dominant culture on the design of the EPM system and how organizational mechanisms facilitate acceptance by the different subcultures. Our study demonstrates the importance of incorporating a subunit level of analysis when examining EPM systems and the different mechanisms that can help reshape employee perceptions.

Keywords Electronic performance management system · Employee perception of monitoring · Organizational culture · Subculture

1 Introduction

Organizations consider Electronic Performance Monitoring (EPM) as a critical control mechanism that aims at focusing employees' attention toward achieving organizational goals [1, 2]. EPM systems capture data on employee performance on the job through monitoring productivity [3], electronic communication, internet activity [4, 5], and employee location and speech [6]. As employees become aware of the scope of monitoring activities, they are likely to conform to the organizational values of accountability, efficiency, and effectiveness [7, 8]. While several studies

K. Sherif · M. El-Masri (✉)
Department of Accounting and Information Systems, Qatar University, Al Tarfa, Doha, Qatar
e-mail: mazen.elmasri@qu.edu.qa

K. Sherif
e-mail: ksherif@qu.edu.qa

C. Dal Zotto et al. (eds.), *Smart Technologies for Organizations*, Lecture Notes in Information Systems and Organisation 60, https://doi.org/10.1007/978-3-031-24775-0_16

have drawn attention to employee reaction to EPM systems [6, 9–12], very few examine the role of culture in shaping the perception of the technology [13–15]. In addition, those who examined culture presented an integrative perception where all members of the organization manifest a shared and consistent view. Overlooking the different and conflicting cultural views that may exist within an organization fails to depict the true dynamics behind the acceptance or resistance, resulting in a forfeit of the opportunity to converge views and generate value from the technology [16].

Over the years, researchers have examined the tension EPM creates concerning several social issues like privacy, trust, and fairness [6, 12, 17, 18]. These tensions have different meanings to the various groups. While one group may perceive the EPM as beneficial, giving it credit for the work completed and the value created, another may perceive it as a threat to its power relations and a violation of its privacy. It is thus important to identify ways to establish a cultural fit in a cross-cultural context where there are different views of the relevance, applicability, and value of the EPM system. A set of organizational mechanisms needs to be established to channel the different perceptions toward convergence and create a dominant culture aligned with the design features of the EPM [14].

This study adopts a differential perspective on the role of subcultures in the acceptance of EPM systems and examines the role of dominant groups in aligning the different and conflicting employee values and beliefs. The differential perspective beholds that different groups within the organization are likely to have different frames of reference based on their motivation and needs [19]. These frames affect how employees perceive information systems, especially ones that evaluate performance and control rewards [20]. We use the social construction of technology (SCOT) [21, 22] as our theoretical lens to understand the position of relevant social groups (RSG) for EPM. SCOT is particularly helpful in understanding why RSGs differ in their perspectives and how the dominant group has "a decisive influence when constructing a dominant 'official' narrative which congeals quickly for the whole organization." [23, p. 115]. This is crucial for multinational organizations where RSGs exercise different levels of power [24]. Successful technology adoption requires organizations to establish mechanisms that help create a dominant culture compatible with the technology [23, 25].

In this study, we shift the focus from the normative question of what needs to be done to an interpretative inquiry of how the different subcultures perceive EPM within organizations and how organizational mechanisms instituted by the dominant group help converge these perceptions to align with the design of the technology. In particular, this study attempts to answer the following research questions:

(1) How do RSGs differ in their perception of an EPM system?
(2) How does the dominant group help create a dominant culture aligned with the design features of the EPM system?

To answer the research questions, we conduct a case study of two multinational organizations in the US and Qatar and examine the difference in the interest and perspective of RSGs in the two countries toward EPM systems. We take an in-depth view of the interplay between the different subcultures to understand how a dominant culture emerges in each organization studied. We also examine the different organizational mechanisms that help converge employee perceptions toward the technology. Our interpretive study provides a rich understanding of the multifaceted and dynamic nature of culture as values collide and shift over time.

The rest of the paper is organized as follows: first, we review the literature on the acceptance of EPM systems in different national and organizational cultural settings drawing on two conceptual frameworks that examine how the dimensions of EPM systems can be aligned with different national [13] and organizational cultures [14]. Second, we develop our theoretical model based on Social Construction of Technology (SCOT) as a background to shed light on frame inconsistencies that employees of multinational organizations may face when interacting with IT-enabled change. Third, the study methodology and design are outlined, followed by a discussion of the empirical results. Finally, we conclude by providing recommendations for academics and practitioners.

2 Literature Review

2.1 Electronic Performance Monitoring Systems

Electronic Performance Monitoring (EPM) are devices that monitor employee behavior and performance on the job [26]. Monitoring data includes video camera observation of employees, computer work monitoring, recording of phone calls, keystroke monitoring, location monitoring, and screen sharing capabilities on computer networks [26]. The data is analyzed to optimize resource utilization, enhance situational awareness, and improve business processes [27]. While EPM systems significantly reduce the organizational cost of monitoring, their effect on employee perception of fairness is questionable. While some studies report a negative impact of EPM creating a depersonalized atmosphere and leading to the erosion of trust and perceptions of unfairness, others observed a positive impact that leads to higher efficiency and improved employee performance [27].

Several studies have contended that the acceptance of an EPM will depend on how well the dimensions of the monitoring systems have been designed to align with the dominant culture. Furthermore, the effective organizational mechanisms in place will help reshape negative perceptions of EPM systems and engender employee self-monitoring [28]. Five dimensions of EPM impact employee acceptance of the technology: (1) the target of an EPM either the individual or the group [6]; (2) the level of pervasiveness of an EPM depending on whether the monitoring is intermittent or ongoing [29]; (3) the recipient of the results of the EPM be it the employee, the supervisor, and/or the group [13]; (4) The purpose of the EPM to assess the performance of an employee or provide informative feedback [18]; and (5) Concomitant practices that support the EPM system like the reward system.

Apart from the design of the EPM, organizational procedures surrounding electronic monitoring will affect employee trust in the organization and their acceptance of EPM systems [28]. Employee prior knowledge of the monitoring processes, their scope, and how the data collected will be used [6] affect how employees perceive EPM. In addition, their active engagement in the design of the monitoring systems and their ability to voice concerns regarding the processes involved [30] or how the output of the system is used tend to create favorable perceptions of EPM.

2.2 The Effect of Culture on Employee Acceptance of EPM

Research on organizational behavior has long established the premise that variations in cultural beliefs have a significant effect on employee behavior and attitude [31, 32]. Schein [33] defined culture as "the sum total of all the shared, taken-for-granted assumptions that a group has learned throughout its history" [34, p. 29]. It is a shared set of values that are acquired over time and may be subject to change reflecting changes in culture [34]. Employees within a multinational organization contribute to cultural diversity that makes it hard to assume that organizational culture will have a unified presence. While some aspects of culture may be collectively shared, we expect different subgroups within the same organization to uphold different patterns of basic assumptions that shape members' beliefs and behaviors with respect to IT systems.

Cross-cultural variance in the perception of fairness has been widely documented [35–38]. Employees with a different cultural frames of reference differ in their perception of fairness because of changes in the meaning of social relations, personal goals, and behavioral norms [38]. Thus, acceptance of an EPM is affected by the corporate beliefs and values that most organizational members hold [13, 39], and response to the pervasiveness of PM systems will differ. EPM will be accepted in organizations whose culture is aligned with the design features of the EPM. Using Hofstede's cultural taxonomy, Panina and Aiello [14] propose a conceptual model of the effect of the dimensions of national culture on employee acceptance of EPM. The authors maintain that the dimensions of the EPM need to be aligned with the employee's cultural values. An individualistic culture with a large power distance, a masculine orientation, and a short-term focus is more likely to accept an EPM that monitors the individual and ensures procedural justice [40]. In an individualistic culture, it is better to monitor the individual, who over the years seems to be less concerned with the pervasiveness of monitoring [41]. A collectivist culture, on the other hand, with small power distance, a feminine orientation, and a long-term focus is less likely to accept EPM unless it is formative focusing on interpersonal justice [42]. In a collectivist culture, it is better for the EPM to monitor group activity and group performance, provide feedback to the group, focus on mentoring and developing the employee, and reward the process rather than the outcome. In their cross-cultural study of organizational controls, Rocha Flores et al. [43] found that while Sweden implements organizational controls that are aligned with their employees' needs and are concerned with how these controls are perceived, the US prefers a more centralized control with less emphasis on how employees perceive the control. In multinational organizations, the unambiguous, individualistic, and bureaucratic controls are likely to be perceived positively by both employees of individualistic and collectivist cultures [40].

Focusing on the organization level, Alder [13] differentiates between bureaucratic and supportive organizations and proposes that employees in bureaucratic organizations perceive pervasive electronic monitoring as fair without their participation in its design. This is mainly because participation appears counter-cultural [44] and leads to suspicion [45]. In a bureaucratic culture, individual performance is valued over group cohesion and teamwork. Individuals also welcome continuous monitoring because it produces a comprehensive view of their effort on the job and a fair evaluation of their performance [46]. Employees also do not worry about the relevance of the data collected by the monitoring system to their task performance. Supportive cultures, on the other hand, do not generally accept electronic monitoring, unless employees participate in its design and the monitoring is intermittent. The target of the monitoring is the group performance (Table 1).

In addition to the major differentiation between bureaucratic and supportive cultures, Jeng-Chung and Ross [39] looked at the value of objectivity within corporate cultures and proposed that managers who appreciate the importance of basing decisions on objective data will favor electronic monitoring of employees. A positive perception of the technology will have a substantial impact on adoption. The belief that technology can prevent theft and create an ethical environment, especially with a history of employee theft or national security concerns, will strongly sway decisions in favor of adoption.

While Alder [13] and Panina and Aiello [14] proposed variation in the perception of employees at the national and organizational level, there are no studies that empirically examine the effect of subcultures on the acceptance of EPM and organizational

Table 1 Effect of culture on the acceptance of EPM

Scope of culture	Dimension of culture	Expectation
National	Power distance	Employees will tend to accept EPM in cultures that sustain a high-power distance
	Individualism/collectivism	EPM is likely to be accepted in an individualistic culture to support individual performance evaluation
	Risk Aversion	EPM is likely to be accepted in cultures that want to minimize risk
Organizational	Bureaucratic	Employees tend to accept the EPM as an objective measure of performance
	Supportive	Employees do not accept the EPM and attribute its use to mistrust when the organizational culture focuses on empowerment, teamwork, creativity, and information sharing

dynamics that converge the different views into one view aligned with the design of the EPM system. Previous research pointed to the difference in perception between employees and managers, where managers may see the EPM system as beneficial in providing feedback and enabling adaptation to organizational goals, promoting transparency and accountability [47]. On the other hand, employees may perceive it as a punitive monitoring system, violating employee privacy [6]. In this study, we attempt to avoid the "residual approach" to culture pointed out by Tayeb [48] by attributing differences between perceptions of an IT to culture without explaining why these differences exist. We also avoid the assumption that cultural beliefs are monolithic and enduring [25]. Instead, we adopt the view that individuals belong to several subcultures whose values and important shift over time and circumstances. While some of the beliefs may represent misfits with the technology, we also examine the organizational dynamics that mutually adapt cultural values and its design features, so convergence of subcultures toward a dominant culture is achieved and EPM is accepted.

In this study, we adopt a differentiation perspective toward organizational cultures, where culture is seen as a "series of overlapping, nested organizational subcultures" that may coexist in harmony or conflict [16, p. 83]. Janz and Wetherbe [49] defined "super-culture" as the firm's macro-culture, consisting of many national or ethnic "micro-cultures." Such a perspective will shed light on the power struggles between the subgroups and the emergence of a dominant culture [21]. The dynamics behind the convergence to a dominant culture are best understood through the "virtual onion" metaphor of culture introduced by Straub et al. [16]. The metaphor perceives members of an organization as having layers of the culture whose relevance and importance shift over time depending on time and circumstances. "At the individual level of analysis, a single person may be a member of several overlapping, nested subcultures, some of which may hold opposing views" [26, p. 301]. Thus, culture is "contested, temporal, and emergent" [50, p. 13].

3 Theoretical Approach

In this study, we embrace the Social Construction of Technology (SCOT) to shed light on the importance of revealing the different viewpoints that different social groups have of electronic monitoring and understanding how organizations can converge these views into one that accepts the technology. The approach highlights the role of "Relevant Social Groups" in rendering different interpretations of the technology and how the power asymmetry between the groups affects the negotiation of the meaning of the technology resulting in a system design that resolves the conflicting interpretations and reaches stability without posing a threat to the powerful social groups [50]. The approach enables us to examine how social factors shape the design of technologies [51]. Power relationships negotiate the meaning of the technology through exchanging of resources allowing the emergence of a shared meaning [24].

EPM systems denote a certain distribution of power among managers and employees, yielding a loss of employee power, which would typically trigger resistance. SCOT is useful in showing how culture, power, and politics interact and form a solid and interlinked perspective of EPM systems. The approach allows us to examine how RSGs differ in their ideas about the EPM and how powerful RSGs have a decisive influence when constructing a dominant "official" perspective for the whole organization serving their interests. We also present the changes in the view of RSGs to analyze the discrepancies between the old and new perceptions. Thus, our theoretical stance is based on the following constructs:

(1) The meaning of monitoring and performance management to RSGs.
(2) The organizational power of the different RSGs.
(3) The cultural background of the different RSGs.

4 Research Methodology

The construction of meaning, power relationships, and cultural values surrounding information technology are best achieved through qualitative research due to the complex interaction between the different dimensions of IT and the dimensions of culture. The data is collected primarily through the qualitative approach using semi-structured interviews in two organizations: in Qatar and the US between January and August of 2018. We sliced through the organizational structures and interviewed respondents across all levels: entry, mid, and senior levels and whose jobs entail self or subordinate evaluation. To identify all the relevant social groups, we interviewed employees—natives and expats—at different organizational levels and with diverse cultural backgrounds. After all relevant groups were included, we ended up with the following informants: senior managers who oversee the entire work operations; HR managers who oversee policies that govern monitoring; directors of functional departments who evaluate the performance of employees; and staff employees who perform daily business processes. Most of the interviews were conducted in Qatar were in English. Interviews with Arabic speakers were translated by a professional service company and cross-checked by the researchers.

In order to collect the necessary data for this research, interviews that lasted 45 and 60 minutes were conducted. We recorded and transcribed the interviews. To analyze the data, we followed Krippendorf's [52] approach of "content analysis," assigning words from the interviews to an emerging concept using QSR NVivo qualitative analysis software. First, two of the researchers coded a sample of the interviews separately. The researchers frequently discussed and revised the code to reach a consensus. A research assistant was then trained to code using the codes developed by the researchers. The research assistant was also encouraged to create new codes when existing codes were inadequate to address emerging concepts from the interviews. Given the difference in the cultural background of the research team, codes were frequently revised unearthing conditions outside the literature that helped explain differences among the relevant social groups and identified factors that shape the perceptions. In the final phase of the analysis, we used template analysis [53] through which codes were categorized as design features of the EPM; perceptions regarding electronic monitoring; type of culture affecting these perceptions; differences among members of the same organization; and shift in perceptions.

Our goal was to understand how the final design of the system reflected the power asymmetry among the groups and how the organization reached closure. The template analysis helped us generate themes that are summarized in Table 2.

5 Findings

At a national level, Qatar is ranked high on power distance, uncertainty avoidance, and individualism [54]. The orientation within the country is neither masculine nor feminine. The society is known for preserving the cultural identity in terms of appearance and respect to religious values. The case selected is of a petrochemical company (referred to in the paper as QCHEM) in the Oil and Gas sector: an industry known for its highly competitive environment and fast adoption of monitoring systems to reduce cost and lower the risks of regulatory non-compliance. The company operates in several countries and the Qatar branch hires professionals from twenty-seven different nationalities. The company strives to provide the necessary training and career development to have an incident-free and clean workplace. Monitoring systems are installed in factories and offices to run a very lean organizational structure that is safe and productive.

In the US, we studied USOIL, an oilfield service provider for reservoir characterization, drilling, production, and processing. The company takes pride in its culture of science and innovation leveraging, its collective experience, and the diverse background and knowledge of its employees. There are several monitoring systems in place within office buildings, on equipment and mobile devices for monitoring employees' safety, behavioral awareness, and performance evaluation.

It became evident from the interviews that participants do not attach the same meaning to monitoring and performance management within or across the two cases. At QCHEM, employees saw monitoring as a given fact of living in Qatar.

> Monitoring is everywhere. It is inside and outside of work. You cannot do anything about it. It is a given fact.

While some employees saw the scope of monitoring as pervasive and inessential as it inadvertently shifts focus away from the output and may lead to micro-managing and focusing on non-productive dimensions of performance indicators like time spent at work rather than output generated.

> It feels like someone is breathing on your neck, trust must be there between the person and the supervisor... I could be sitting on my desk working or daydreaming without doing any work. As long as the output is right, if you do it in 1 hour or 8 hours, it shouldn't matter. I'm supposed to be judged by my own output. ... one could finish the work in 2 days, and another can finish it in 10 days, would you say that the person who did it in 2 days is bad, while the other person is good? No, the 10 days employee is bad because he is slow.

Table 2 Effect of Culture on the Acceptance of EPM

Theme	Codes
EPM is pervasive in both organizations however the perceived goals of the systems are different	***Pervasiveness:*** both in Qatar and the US employees believed the system is tracking all employee activities. Employees in Qatar are aware of the analytics that run behind the scenes, whereas in the US. Employees believe the data are just used for security
	Objectivity: EPM in Qatar is perceived by management as an objective decision-maker, a cross-validator against human bias, and a safeguard of organizational interests. In the US, both employees and managers found fault with the EPM objectivity and highlighted problems with its design
	Centrality: EPM in Qatar plays a central role in the distribution of work and recognition of talent. The employees give much importance to the system. In the US, the system has a limited role in providing feedback to employees
The interplay between national and corporate culture	***Dominance:*** In the US, the corporate culture dominates over the national culture leading some non-US employees to shift their perception concerning the EPM. In Qatar, the dominance of national over corporate culture is experienced due to the power difference between locals and expats. The corporate culture dominates for expats. Several expats shifted their initial perception of the EPM in allegiance with the national culture of Qataris
	Convergence: In both countries, we see convergence toward the national culture of the dominant group
Differences within	***Degree of variability:*** There is a higher degree of variability within Qatar's national and corporate cultures than within the US. In Qatar, not all locals share the culture of entitlement. Highly educated locals feel they have to speed up to the world pace and push themselves to take on bigger challenges. Given the higher number of multinational employees within Qatar, not all share the same perception of the EPM system
Organizational Mechanisms	**Competition vs. Collaboration:** (compete with expats and collaborate with locals)
	Empowerment vs. Enablement: (nationals need empowerment and expats in Qatar need enablement—should have necessary resources to achieve necessary objectives)

Others saw it as the fastest way to align employees' different perspectives regarding work.

> My privacy is at home only, I come here to execute a certain job, whatever are the policies of the company, I'd be happy to follow. Monitoring actually benefit me more than it'll harm me...instead of focusing or rejecting technologies, it is better to invest the energy to develop yourself.
>
> We live and abide by those [rules enforced by the EPM system]. It is the most expedite way of controlling a multinational eclectic workforce, tell us what you are monitoring we'll understand that rule and there is very little latitude for swaying out of it.

HR managers (Qatari nationals) saw monitoring as an important control mechanism that helps create an ethical environment.

> There's a proverb that says unguarded money teaches people to steal, so there is control and audits, that leaves no scope for fraud. But if someone isn't being monitored, it becomes a problem. Mistakes could happen from the lowest level up to the highest level.
>
> The problem is that some employees spend most of their time chatting, and they are not generating any benefit to the company, the advantage of this technology is that it does not depend on the evaluation of the person or the manager, it is objective and unbiased.

While managers from western countries disagree with the level of monitoring, they felt that in Qatar it was justified given that nationals only make less than twenty percent of the population and there is a significant knowledge asymmetry between nationals and foreigners.

> Here, it is a different scenario. The rules of this company were very much written with best practices in mind. I recognize a lot of the things I do here are things that I cannot do in the UK, however we are not in the UK, we are in Qatar, therefore one has to be aware of the local culture, and sometimes what appears a black and white policy isn't necessarily sympathetic to the culture of the country, so one has to strike a balance between the two issues. I can't run an HR business in Qatar the same way I would run it in the UK, there are far more players in that particular arena perhaps the trade unions and employer relations.

The same pattern was recognized in the data collected from USOIL, the difference in the meaning of the EPM and its acceptance was apparent. American employees are known to value privacy and advocate personal boundaries, they perceive monitoring as a lack of trust.

> If they monitor every single thing that he/she does, that means there are trust issues, so if you do not trust him or her first do you think you can gain any positive feeling, not really, Employees want that ability to think on their own, take their time to work, and chances are when you leave them to do that, they are probably going to deliver than when you monitor them. The best performing people are those that you have empowered to say I know you can deliver it.

While US employees value privacy, they are aware of the monitoring system and appear to be desensitized:

> Basically, they track everything, if you have your work email on your device, they track everything you do on your device, receiving and sending emails, text messages, pictures, phone calls, computer is the same thing, all the files.

> I don't have any issues with them to track whatever they want. Personally, I would love it if they did. There's nothing that would put me to shame.
>
> Employees, with time, got to understand and know why the tool is important.

However, they believe that monitoring is an ineffective control and creates a sub-optimal environment:

> You have people who are engaged, have a very high technical proficiency and because of that inclination sometimes do not come the regular 9–5 and are not always on their desk, and that's not truly a reflection of their capabilities so much as it is a proper balance or prioritization of how they should do their work.

HR managers acknowledge that USOIL uses a monitoring system but reject tagging the process as monitoring and instead refer to it as performance management.

> Generally, we do not monitor people, we manage their performance. Obviously, almost every major organization today has some form of performance management system. however, it is not meant to monitor, it is meant to give feedback to the employee regarding their performance.

Yet, they saw a need to monitor behavior as well as output:

> Performance has to do not just with the technical proficiency but also how you engage within the team itself so there are other components that we have to manage when evaluating an employee.

While much of the electronic monitoring literature tends to portray the negative effects on employees in terms of stress [6], lack of trust [6], and drop in citizenship behavior [17], little attention has been paid to the effect of national and corporate culture on employee acceptance of electronic monitoring. Several conceptual frameworks highlighting the effect of culture on employee perception of electronic monitoring [13, 14] have been proposed. However empirical testing of the role of culture in shaping employee perceptions is lagging. With the current focus on big data and analytics, organizations can benefit from EPM data if they design the proper mechanisms to tackle employee resistance to monitoring. Simply renouncing EPM and HR analytics is not in the best interest of the organization or the employee. Designing an effective monitoring program will provide the necessary feedback to employees to readjust and meet performance expectations. As employees perceive the benefits of monitoring, they will develop their own self-monitoring strategies to regulate behavior and meet job expectations. High self-monitors will tend to adapt quickly to organizational and national culture.

6 Conclusion

In this study, we examine how different groups have different perceptions of EPM technology. We conduct a comparative study within two organizations in the US and Qatar to examine how they can successfully reshape employee perceptions with different national and corporate cultures. The study highlights the role of technology in raising self-awareness and self-monitoring and increasing the chances of employees to meet performance objectives. Multinational organizations need to realize the fit between the design of electronic performance monitoring systems and the cultural setting. Additionally, results suggest that organizations who aspire to use analytics on employee monitoring data need to do more than providing justifications and create an environment where the analytics prove to be beneficial to the employee and the organization.

References

1. Alge, B.J. (2001). Effects of computer surveillance on perceptions of privacy and procedural justice. *Journal of Applied Psychology 86*(4), 797. doi:https://doi.org/10.1037//0021-9010.86.4.797.
2. Pierce, L., Snow, D. C., & McAfee, A. (2015). Cleaning house: The impact of information technology monitoring on employee theft and productivity. *Management Science, 61*(10), 2299–2319. https://doi.org/10.1287/mnsc.2014.2103
3. Wells, D. L., Moorman, R. H., & Werner, J. M. (2007). The impact of the perceived purpose of electronic performance monitoring on an array of attitudinal variables. *Human Resource Development Quarterly, 18*(1), 121–138. https://doi.org/10.1002/hrdq.1194
4. Alder, G. S., et al. (2008). Employee reactions to internet monitoring: The moderating role of ethical orientation. *Journal of Business Ethics 80*(3), 481–498. doi: https://doi.org/10.1007A10551-007-9432-2.
5. Wen, H. J., Schwieger, D., and Gershuny, P. (2007). Internet usage monitoring in the workplace: Its legal challenges and implementation strategies. *Information Systems Management 24*(2), 185–196. doi: https://doi.org/10.1080/10580530701221072.
6. Holt, M., Lang, B., & Sutton, S. G. (2017). Potential employees' ethical perceptions of active monitoring: The dark side of data analytics. *Journal of Information Systems, 31*(2), 107–124. https://doi.org/10.2308/isys-51580
7. Decramer, A., Smolders, C., & Vanderstraeten, A. (2013). Employee performance management culture and system features in higher education: Relationship with employee performance management satisfaction. *The International Journal of Human Resource Management, 24*(2), 352–371. https://doi.org/10.1080/09585192.2012.680602
8. Hamilton, R.H., Sodeman, W.A. (2020). The questions we ask: Opportunities and challenges for using big data analytics to strategically manage human capital resources. *Business Horizons, 63* (1), 85–95. doi: https://doi.org/10.1016/j.bushor.2019.10.001.
9. Tomczak, D. L., Lanzo, L. A., & Aguinis, H. (2018). Evidence-based recommendations for employee performance monitoring. *Business Horizons, 61*(2), 251–259. https://doi.org/10.1016/j.bushor.2017.11.006
10. Bhave, D. P., Teo, L. H., & Dalal, R. S. (2020). Privacy at work: A review and a research agenda for a contested terrain. *Journal of Management., 46*(1), 127–164. https://doi.org/10.1177/0149206319878254
11. Angrave, D., et al. (2016). HR and analytics: Why HR is set to fail the big data challenge. *Human Resource Management Journal 26*(1), 1–11. doi: https://doi.org/10.1111/1748-8583.12090.
12. Aiello, J. R., & Kolb, K. J. (1995). Electronic performance monitoring and social context: Impact on productivity and stress. *Journal of Applied Psychology, 80*(3), 339.
13. Alder, G. S. (2001). Employee reactions to electronic performance monitoring: A consequence of organizational culture. *The Journal of High Technology Management Research 12* (2), 323–342. doi: https://doi.org/10.1016/S1047-8310(01)00042-6.

14. Panina, D., & Aiello, J. R. (2005). Acceptance of electronic monitoring and its consequences in different cultural contexts: A conceptual model. *Journal of International Management, 11*(2), 269–292. https://doi.org/10.1016/j.intman.2005.03.009
15. Veiga, J. F., Floyd, S., & Dechant, K. (2001). Towards modelling the effects of national culture on IT implementation and acceptance. *Journal of Information technology, 16*(3), 145–158. https://doi.org/10.1080/02683960110063654
16. Straub, D., et al. (2002). Toward a theory-based measurement of culture. *Journal of Global Information Management (JGIM) 10* (1), 13–23.
17. Niehoff, B. P., & Moorman, R. H. (1993). Justice as a mediator of the relationship between methods of monitoring and organizational citizenship behavior. *Academy of Management journal, 36*(3), 527–556.
18. Moorman, R. H., & Wells, D. L. (2003). Can electronic performance monitoring be fair? Exploring relationships among monitoring characteristics, perceived fairness, and job performance. *Journal of Leadership & Organizational Studies, 10*(2), 2–16.
19. Bijker, W. E., Hughes, T. P., Pinch, T. (1987). The social construction of technological systems. Cambridge, MA: The MIT Press.
20. Orlikowski, W. J., & Gash, D. C. (1994). Technological frames: Making sense of information technology in organizations. *ACM Transactions on Information Systems, 12*(2), 174–207. https://doi.org/10.1145/196734.196745
21. Jackson, M. H., Poole, M. S., & Kuhn, T. (2002) The social construction of technology in studies of the workplace. In: *Handbook of new media: Social shaping and consequences of ICTs* (pp. 236–253).
22. Pinch, T. J., & Bijker, W. E. (1984). The social construction of facts and artefacts: Or how the sociology of science and the sociology of technology might benefit each other. *Social studies of science, 14*(3), 399–441.
23. Weber, Y., Pliskin, N. (1996). The effects of information systems integration and organizational culture on a firm's effectiveness. *Information & Management 30* (2), 81–90. doi: https://doi.org/10.1016/0378-7206(95)00046-1. ISSN 0378-7206
24. Bartis, E., & Mitev, N. (2008). A multiple narrative approach to information systems failure: A successful system that failed. *European Journal of Information Systems, 17*(2), 112–124. https://doi.org/10.1057/ejis.2008.3
25. Gallivan, M., Srite, M. (2005). Information technology and culture: Identifying fragmentary and holistic perspectives of culture. *Information and Organization* 15, 295–338. https://doi:https://doi.org/10.1016/j.infoandorg.2005.02.005.
26. Alder, G., & Tompkins, P. (1997). Electronic performance monitoring. *Management Communication Quarterly, 10*, 259–288. https://doi.org/10.1177/0893318997010003001
27. Al-, M., & Sherif, K. (2018). Employee perceptions of fairness toward IoT monitoring. *VINE Journal of Information and Knowledge Management Systems, 48*(4), 504–516. https://doi.org/10.1108/VJIKMS-01-2018-0007
28. Stanton, J. M. (2000). Reactions to employee performance monitoring: Framework, review, and research directions. *Human Performance, 13*(1), 85–113.
29. Zweig, D., & Webster, J. (2002). Where is the line between benign and invasive? An examination of psychological barriers to the acceptance of awareness monitoring systems. *Journal of Organizational Behavior: The International Journal of Industrial, Occupational and Organizational Psychology and Behavior, 23*(5), 605–633. https://doi.org/10.1002/job.157
30. Amick, B. C., & Smith, M. J. (1992). Stress, computer-based work monitoring and measurement systems: A conceptual overview. *Applied Ergonomics, 23*, 6–16. https://doi.org/10.1016/0003-6870(92)90005-g
31. Cooke, R., & Rousseau, D. (1988). Behavioral norms and expectations: A quantitative approach to the assessment of organizational culture. *Group and Organizational Studies, 13*, 245–273. https://doi.org/10.1177/105960118801300302
32. Wilkins, A. L., & Ouchi, W. G. (1983). Efficient cultures: exploring the relationship between culture and organizational performance. *Administrative Science Quarterly, 28*(9), 468–481. https://doi.org/10.2307/2392253

33. Schein, E. H. (1986). What you need to know about organizational culture. *Training & Development Journal 40*(1), 30–33. https://psycnet.apa.org/record/1986-28753-001.
34. Karahanna, E., Evaristo, R., & Srite, M. (2005). Levels of culture and individual behaviour: An integrative perspective. *Journal of Global Information Management., 13*(2), 1–20. https://doi.org/10.4018/jgim.2005040101
35. Fong, S., & Shaffer, M. (2003). The dimensionality and determinants of pay satisfaction: A cross-cultural investigation of a group incentive plan. *The International Journal of Human Resource Management, 14*, 559–580. https://doi.org/10.1080/0958519032000057592
36. Fischer, R., & Smith, P. (2004). Values and organizational justice: Performance- and seniority based allocation criteria in the United Kingdom and Germany. *Journal of Cross-Cultural Psychology, 35*, 669–688. https://doi.org/10.1177/0022022104270110
37. Kim, T., & Leung, K. (2007). Forming and reacting to overall fairness: A cross-cultural comparison. *Organizational Behavior and Human Decision Processes, 104*, 83–95. https://doi.org/10.1016/j.obhdp.2007.01.004
38. Long, C., Bendersky, C., & Morrill, C. (2011). Fairness monitoring: Linking managerial controls and fairness judgments in organizations. *Academy of Management Journal, 54*, 1045–1106. https://doi.org/10.5465/amj.2011.0008
39. Jeng, V., & Ross, W. H. (2005). The managerial decision to implement electronic surveillance at work: A research framework. *International Journal of Organizational Analysis, 13*(3), 244–268. https://doi.org/10.1108/eb029006
40. Yang, I. (2015). Cross-cultural perceptions of clan control in Korean multinational companies: A conceptual investigation of employees' fairness monitoring based on cultural values. *International Journal of Human Resource Management, 26*(8), 1076–1097. https://doi.org/10.1080/09585192.2014.922600
41. Zakaria, N., Stanton, J. M., & Sarkar-Barney, S. T. (2003). Designing and implementing culturally-sensitive IT applications: The interaction of culture values and privacy issues in the Middle East. *Information Technology & People, 16*(1), 49–75. https://doi.org/10.1108/09593840310463023
42. Sherif, K., Jewesimi, O., & El-, M. (2020). Empowering employees: The other side of electronic performance monitoring. *Journal of Information, Communication and Ethics in Society, 19*(2), 207–221. https://doi.org/10.1108/JICES-04-2020-0038
43. Rocha Flores, W., Antonsen, E., & Ekstedt, M. (2014). Information security knowledge sharing in organizations: Investigating the effect of behavioral information security governance and national culture. *Computers & Security*, 4390–110. doi: https://doi.org/10.1016/j.cose.2014.03.004.
44. Matejka, J. K., & Liebowitz, S. J. (1989). A commitment to Ex-sell. *Manage, 40*(4), 2–6.
45. Nutt, P. C. (1996). Views of implementation approaches by top managers in health service technologies: the case of electronic surveillance. *Journal of Business Ethics 10*, 519–526. https://pubmed.ncbi.nlm.nih.gov/10157962/.
46. Grant, R., & Higgins, C. (1989). Monitoring service workers via computer: The effect on employees, productivity, and service. *National Productivity Review, 8*, 101–112. https://doi.org/10.1002/npr.4040080203
47. Vaught, B. C., Taylor, R. E., & Vaught, S. F. (2000). The attitudes of managers regarding the electronic monitoring of employee behavior: Procedural and ethical considerations. *American Business Review, 18*, 107–114. https://doi.org/10.1177/2158244015580168
48. Tayeb, M. (1994). Organizations and national culture: Methodology considered. *Organization Studies., 15*(3), 429–445. https://doi.org/10.1177/017084069401500306
49. Janz, B. D., & Wetherbe, J. C. (1998). Information technology, culture, and learning at Federal Express. *Journal of Global Information Technology Management, 1*(1), 17–26. https://doi.org/10.1080/1097198X.1998.10856222
50. Klein, H. K., & Kleinman, D. L. (2002). The social construction of technology: Structural considerations. *Science, Technology, & Human Values., 27*(1), 28–52. https://doi.org/10.1177/016224390202700102

51. Wilson, M., & Howcroft, D. (2005). Power, politics and persuasion in IS evaluation: a focus on relevant social groups. *The Journal of Strategic Information Systems 14*(1), 17–43. doi: https://doi.org/10.1016/j.jsis.2004.11.007. ISSN 0963-8687.
52. Krippendorff, K. (1980). *Content analysis an introduction to its Methodology*. Sage.
53. King, N. (2004). Using templates in the thematic analysis of text. In C. Cassell & G. Symon (Eds.), *Essential guide to qualitative methods in organizational research*. London: Sage.
54. Dailan, & Aldulaimi, S. (2011). Examining national culture of Qatar. *Australian Journal of Basic and Applied Sciences 5*(10), 727–735. ISSN 1991-8178

Digital Transformation in the Pharmaceutical Sector: An Italian Overview

Sabrina Bonomi, Laura Pentassuglia, and Cecilia Rossignoli

Abstract Today industry is experiencing its fourth industrial revolution. "Industry 4.0" is based on Cloud systems and the Internet of Things, among others. This study examines the technological innovations applied to the pharmaceutical industry to analyze which organizational changes they bring with them. These new technologies aimed to improve the inter and intra-organizational processes lowering the time and the costs associated. The case of an Italian consortium shows that pharmaceutical industries favor the requalification of current personnel and adapt the organizational structure to accommodate better and exploit these new technologies. These changes affect the whole ecosystem, encouraging aggregation and partnerships between different players and outsourcing those processes that would require too much time and high costs to be internally developed.

Keywords Digital transformation · Pharmaceutical sector · Industry 4.0

1 Introduction

Technology has a strong influence in all industrial sectors. For example, production changed over time in many of them, during four distinct revolutions, from a manual/artisanal to almost entirely automated execution, making extensive use of robotics.

S. Bonomi (✉) · L. Pentassuglia
eCampus University, Via Isimbardi 10, 22060 Novedrate, CO, Italy
e-mail: sabrina.bonomi@uniecampus.it

L. Pentassuglia
e-mail: laura.pentassuglia@uniecampus.it

S. Bonomi
University of Ostrava, Ostrava, Czech Republic

C. Rossignoli
University of Verona, Caserma Santa Marta, Via Cantarane 24, 37129 Verona, Italy
e-mail: cecilia.rossignoli@univr.it

C. Dal Zotto et al. (eds.), *Smart Technologies for Organizations*, Lecture Notes in Information Systems and Organisation 60, https://doi.org/10.1007/978-3-031-24775-0_17

The custom of computers and the internet perhaps permeated every sector and level of the supply chains, facilitating and making use faster. In recent decades, we have witnessed extraordinary transformations that have brought about profound changes, even in our daily lives. Think, for example, how cell phones have become pocket computers, going well beyond their original function as a means of communication; we can use them for home automation, lighting control services, individual room temperature control, security from unwanted intrusions, and consumption.

Mainly thanks to the innovation generated by the pharmaceutical industry, we live longer and better. When we think about technological advancement in the pharmaceutical sector, typically, we associate it with discoveries in the scientific field and how these offer us to deepen the knowledge of essential diseases and improve treatment. However, scientific research is only one of the critical aspects that allowed this supply chain to reach new goals and bring new products in our homes and hospitals, superior in efficacy and with significant effects compared to the previous ones.

This work investigates how technological advancement transforms the pharmaceutical sector and, consequently, health care, explicitly identifying the leading technologies that are changing this industry at the organizational level, that are under investigated. However, as we are moving forward from the industry 4.0 to the society 5.0, it is paramount to fully understand the current state of the pharmaceutical sector and how we can implement the required technological and organizational changes. The ultimate goal is to pave the way toward the personalized medicine, which requires a higher value and quality of research but also of activities along the pharmaceutical supply chain.

Firstly, we will analyze some of the key technologies that permeate this sector; secondarily, we present technologies revolutionizing the pharmaceutical industry, such as engineering, robotics, and artificial intelligence. Then, a specific case study and the associated methodology are presented, showing an applied technology and how it transformed their organization. Finally, we discuss how new technologies can change the pharmaceutical industry and prospects for the entire sector.

2 Digital Transformation and the Pharmaceutical Industry: A Literature Review

The pharmaceutical industry is now living in the Industry 4.0 era, characterized by the introduction of state-of-the-art technologies in the informatics and engineering sectors. These innovations allow more immediate and reliable monitoring of the production processes, instantaneous communication between all the parties involved, the generation and analysis of large-scale data, a more efficient, clean, and flexible production, and last but not least, more efficient counterfeit technologies.

2.1 Internet of Things

Industry 4.0 is introducing new technologies, like the Internet of Things (IoT), robotics, and Artificial Intelligence (AI), bound to bring profound changes in the organization of every industrial sector, pharmaceutical included [1].

Technological innovations influence each department of the pharmaceutical industry. Total quality management (TQM) is bound to continuously improve the drugs produced, starting with the raw materials. They are often sensitive to temperature, humidity, and atmospheric pressure [2]; therefore, changes in these parameters may alter compounds' stability and organoleptic properties, with possible significant economic and health consequences. The possibility of discharging such altered compounds before the production phase becomes then paramount. To address such a problem, companies are introducing a network of wireless sensors that allows constant monitoring twenty-four/seven and can alert workers on any parameter deviation in real-time [3].

2.2 Information and Communication Technology

Information and communication technology (ICT) nowadays relies heavily on intranet [4], Internet, and associated services, such as e-mail [5], videoconferences [6], and cloud systems [7]. They allow efficient and prompt communication and coordination between all the business functions, conditions required for the active development, approval, production, and commercialization of drugs [8–12].

2.3 Research and Development

Research and development (R&D) is one of the functions that heavily rely on new and innovative technologies; it encompasses activities from the development of new drugs up to clinical trials and post-commercialization research. Some excellent example is the use of lasers for microscope imaging, the so-called confocal microscopy [13, 14], and the introduction of LED and photodiodes to detect even one single molecule of DNA [15, 16]. During clinical trials, particularly during phase III, massive data production requires a precise and intensive analysis to understand if and how a new drug works [17]. The big data analysis consists of a set of the data volume of petabytes, based on volume, speed, variety, veracity, and value [18]. This technology is progressively introduced together with machine learning, neuronal networks, and computational linguistics, an integral part of the analysis [19].

2.4 Drug Production

The constant need to increase the production quality for the new generation drugs, such as humanized monoclonal antibodies, is met with diverse technologies according to the volume required. For example, pharmaceutical companies rely on bottle-based protocols during a pilot phase, which involves only a few liters. In contrast, stainless still fermenters allow up to two thousand liters per lot [20]. If the former requires shaking incubators to maintain the temperature and the critical nutrients at the needed levels for optimal growth, the latter relies on the state of the heart sensor and injectors to ensure the highest quality possible of the final product [20]. Due to some disadvantages inherent to the system, like the requirement of in-depth sterilization to avoid bacterial contamination and cross-contamination from previous cell cultures and the shear stress that can lead to cell death, more and more industries are relying on disposable bioreactors. These last ones consist of unique plastic bags, called CellBag™, filled with the appropriate culture medium and sealed. Dedicated sensors monitor critical parameters, namely temperature, oxygen concentration, and pH of the solution, adjusting the first two to meet the physiological requirements. Once the culture reaches the optimal growth concentration, it empties the bag and purifies the drug [20]. This last requires the application of several technologies to ensure the highest possible purity of the final product.

2.5 Packaging

The packaging is often viewed by most as just a container of the drug when, in reality, it plays a critical role in assuring that the proper medication at the right concentration reaches our homes. The package is the primary source of information regarding the name, the active ingredient, the dosage, the producer, the lot number, and the expiring date, but it is also much more. Today it is designed to avoid forgery and any alteration of the drug, both of which pose a heavy burden on the social health, especially of every third-world country[1] (www.who.it). Several technologies are being applied, and more are being developed to fight back against such a growing health threat. For example, seals, labels, and films are the most common features against tempering. The unique design of the package and its hidden features, holographic and multilayer names, UV visible fibers, security wires, bar, 2D codes, and casual or semi-casual systems are among them used to recognize a genuine product against a forged one. There is also a whole set of innovations, called forensic makers, currently under research to improve further the security of the drugs that reach our homes. These markers can be a uniquely designed chemical substance, a biological compound, a unique DNA sequence hidden among several random ones, the unique print of

[1] A forged drug is a compound whose identity has been deliberately and fraudulent, whereas alteration is defined as lower quality or properties (www.fda.gov and www.who.int).

a natural radioisotope, or micro-tags that only through a microscope is possible to codify and read [21, 22].

2.6 *Blockchain Technology as an Anti-Forgery Measure*

The packaging design is but one of the possible ways of tracing forged or altered drugs. Recent studies are employing the blockchain technology first developed for bitcoins. Three are the main characteristics of a blockchain: decentralization of the structure, verifiability of each user, and immutability of each block, which can be only rejected or accepted but not modified [23]. Once the blockchain is created, it is possible to trace the post-production history, from the moment it leaves the pharmaceutical company up to the moment that it reaches the last distributor, would that be a pharmacy or a hospital [24]. In this setting, it is also possible to create a system that alerts companies and authorities when a package is recognized as illicit [25].

3 Unique Technologies Applied in the Pharmaceutical Sector

Engineering and artificial intelligence (AI) are becoming more and more prominent in the pharmaceutical sector, addressing different needs both in the medical and research field.

3.1 *Engineering*

Among the most notable engineering applications, we can find medical devices and research equipment of all kinds. A medical device can be defined as any instrument, machine, implant, and similarly used to help to improve the health of patients (www.who.int). Cardiac devices are some of the most notable, like pacemakers, to correct bradycardias and arrhythmias, essential alterations in the initiation and propagation of the electrical impulse in the heart, which lead to anomalies in the cardiac frequency [26]. Today models exploit Bluetooth technology to ensure continuous control of the cardiac functions and status of the device in real-time (www.medtronic.com). It cannot say the same for heart failure, which is possible to cure only through the autologous heart transplant, a compatible human donation, which inevitably implies the death of another person. For understandable reasons, the number of available hearts every year is far below the actual need, and these data lead to different research solutions [27, 28]. The only one with FDA and EMA approval is the total artificial heart Syncardia®, which enables patients to survive until the transplantation. The ultimate goal is to build a permanent artificial heart to avoid the need for a compatible donor (www.syncardia.com).

According to the form, the most common application concerns the design and construction of prostheses of different kinds and materials. The most common ones are bone implants, like hip replacement, a joint surgery in older people. These implants must address several mechanical and biological issues to ensure biocompatibility and high resistance [29, 30]. Most of the current implants are engineered, starting from the assumption that there are no bone malformations, but this is not always true. To address such unique medical conditions, China is studying to introduce 3D printers to create implants to implant under such exceptional circumstances [31, 32].

3.2 *Artificial Intelligence*

The application of AI and robotics are ever more prominent at all levels of the pharmaceutical sector. They can be helpful to develop new drugs (www.digitalauthority.me), exploit orphan drugs (Novartis), improve the diagnostic power (Merk), monitor patients (Tencent) remotely, efficiently select the most suitable patients to recruit during the clinical research (www.digitalauthority.me). The neuronal networks and their evolved version, deep learning, are based on the human brain's mechanism of action to learn and process new information [33] and aim to reproduce such ability artificially. They comprise input, the information given to the systems, and output, which elaborates the data. In between there, we can find one (neuronal networks) or more (deep learning) hidden layers, whose elements are called neurons and, like the human ones, they are interconnected with each other [34, 35]. Improving the diagnostic power for cardiac diseases is critical to better tailor the cure to the patient's status. In this regard, AI allows us to efficiently harmonize data generated through imagines, such as biopsies, x-rays, genetic tests, and electrocardiograms, with those derived from physical examinations or lab reports. The former can be fed directly to the neuronal networks, whereas the latter requires another form of AI, called Natural Language, to convert them into a compatible format [36].

One last exciting innovation permeating laboratories is introducing robotics to ensure the automatization of several research steps, the so-called High Thoroughput (HTP) research [37]. HTP aims to reduce costs and time and, at the same time, to increase the level of output, mainly the number and quality of potential drugs. Some examples are the Syrrx from Takeda, which can purify and crystalize one hundred proteins in only five hours [38]. More recently, AstraZeneca, one of the biggest pharma companies globally, developed the CoLab High, a Throughput Screening System capable of handling forty million chemicals per year to treat forty to fifty diseases. It has introduced a modular and robotized research station controlled by using a mobile phone or a tablet. The modularization allows more than one researcher to use it simultaneously, becoming an extremely efficient tool for the drug discovery phase.

At the same time, this company is massively introducing Artificial Intelligence technologies to efficiently analyze clinical trials as data become available (the Big Data analysis). This fact increases their ability to discover new targets, study diseases, and develop personalized medicine (http://www.astrazeneca.com). Other critical applications are related to improving diagnosis capability (Merk), remote monitoring of patients (Tencent), treatment of rare diseases (Healx), and creating new drugs or study orphaned drugs[2] (Novartis).

4 Case Study: The Delta Consortium

4.1 Methodology

This "how" question requires a qualitative analysis [39, 40] through an explorative case study; data were collected through the access of the chosen company's website and fifteen semi-structured interviews, based on a predetermined set of questions.

The case study methodology allows us to collect information from the people and institutions that live such changes through interviews recorded and transcribed. According to prior studies [41], our dataset retained a certain degree of flexibility, along with our research questions, which were often updated according to ongoing feedback and unexpected events. In this instance, we chose a consortium with a central role in the Italian pharmaceutical district and a tight interaction with the various players to collect data spanning the whole ecosystem of the sector [42]. The semi-structured interviews lasted 50 min on average and were conducted with Delta's CEO and some managers of associate companies.

In this case, researchers used computer-assisted qualitative data analysis through the Atlas.ti software to code and summarize the gathered data. The data and results were presented to the main actors in the organization and its directors. The case study analysis commenced in March 2019 and finished in December 2019; we undertook it without preconceptions and without the need to prove anything in advance; only the disinterested aim of contributing to the pharmaceutical sector and the academic literature moved us.

We first explored the case study on the information available on the internet to understand its role in the pharmaceutical sector, its organization, and the services offered. We interviewed the CEO and associated members with several questions aiming to deepen, broaden, and complete our knowledge on the matter. These interviewed people allowed us to study the consortium itself and have an insight into the other organizations of the ecosystem. The interview has three parts: the first, introduction, describes the company's culture, structure, and organizational project; the

[2] Drugs that do not yet have a known target.

second in-depth explores knowledge of the intra-organizational aspects, focusing on ICT technologies and regulatory changes; finally, the third set of questions allows the interviewed to express their opinions based on their experience.

4.2 History and Organizational Structure

According to research conducted by the FarmaIndustria (www.farmaindustria.it) and based on data obtained from ISTAT and Eurostat, the Italian pharmaceutical sector has seen a steep growth in the past decades, becoming the first country in the European Union, head-to-head with Germany, for production and export. Employment, in fact, grew by 4.5% between 2015 and 2017. It was the highest among all sectors, with a more predominant role of younger people (age below 35) and women. The Italian pharmaceutical industries also committed significant investments in state-of-the-art technologies to protect the environment (47.2% between 2011 and 2015 compared to 32% of all manufacturing industries). This growth coupled with the introduction of the Industry 4.0 may require the introduction of significant changes in the organization of pharmaceutical industries.

The Delta consortium was born in 1991 as a non-profit organization to facilitate and improve the efficiency of communication between the healthcare sector actors, from the pharmaceutical industries to hospitals and pharmacies, passing through distributors and dealers [43]. Since its founding, it has focused on providing "digital solutions to exchange information regarding the order cycle" (CEO). The consortium can now count 751 participating organizations throughout the sector, 151 members: 108 are industries, whereas 43 are intermediate distributors.

The consortium relies on a slim organizational structure comprising only the CEO, an emeritus CEO, and three internal collaborators. All of them have experience in digital innovations for the B2B. There are also a Board of Directors, a Board of Auditors, and a technical committee. The members of the Board of Directors reflect the characteristics of the consortium. Every three years, six representatives from the industries and six from the distributors are selected to ensure equal representation of each category.

On top of its administrative governance, the board can endorse or veto investments and develop new projects. The board has the primary role of warranting the excellent administration of the consortium, ensuring the interests of all types of organizations that shape it, and facilitating the communication between industry and distributors.

The Board of Auditors plays the crucial role of ensuring that the consortium complies with the current laws and the company statute.

The consortium also relies on the technical board, made up of managers from the whole healthcare sector, to continually improve the efficiency of the consortium and address the needs of its members and associates. The technical board relies on face-to-face and webinar meetings to ensure that everyone interested can participate in the discussion, independently of their location. The consortium also relies on external professional people, according to the specific project in development.

4.3 Role and Technologies

Delta tasked itself to be the communication intermediary between each sector element, providing an internet-based platform through which companies can send orders using a standard format. The producers can redirect all the documents to their system and evade orders immediately. The platform plays the role of exchanging nodes for all the documents regarding the order cycle, decoupling sender and receiver, and facilitating access for those relying on basic structures, such as smaller companies.

In 2018, it decided to introduce new and innovative technologies more incisively to become a more specialized B2B organization. Therefore, highly skilled people with a background outside one of health care were introduced to critical positions. Such a change brought the consortium to step ahead from an organization dedicated only to exchanging documents to one with a central place in the healthcare sector through new initiatives and projects borrowed for the experience and competence derived from other industries. This role is to carry on projects from which the whole ecosystem can benefit and address its members' needs. One of the primary needs is to continue abiding by new rules, laws, and technologies that can potentially change the organizational processes. Once the critical points are established, and projects need to be developed, the consortium will address them directly or through an external party. The membership fees cover the costs.

At present, the consortium relies on Internet and Cloud technologies renewed in 2019 and selected based on a tender. It had to offer a product able to maintain the present functionality. Still, it had to be, at the same time, innovative, flexible, and to be easily adapted according to the emerging needs of the members and to the new regulations.

The platform relies on four key elements: 1. Delta's account creates a virtual space for the documents exchange (each user can create more than one account according to their needs; all documents can be codified in a universal or licensed format). 2. The communication channels, which can be either structured or manual; 3. Essential services, like direct access to the institutional nodes, electronic billing, the digital storage of all documents, and digital files. 4. Other valued services to develop projects for the whole community.

The main advantage introduced by the consortium is a significant scale economy. The result depends on various factors: the non-profit formula does not require the generation of profits; Internet and Cloud technologies necessitate low costs; a large number of participants from the whole sector creates a critical mass of economic advantages for every participant, small companies, in particular, ensuring innovations in compliance with all new rules and laws.

The high number and variety of members allow for addressing new projects and initiatives proactively and spontaneously. On top of it, a technical board's presence ensures that no single company has the burden of developing a new plan, but it falls on the whole ecosystem voluntarily. This organization ensures a representative of all members, so every need of each organization in the consortium, actively participating or not, is met and appropriately addressed.

Three main positive aspects come from the structure and working of the consortium. First of all, the organization can more quickly and efficiently outsource some essential functions reducing its fixed expenses while maintaining a certain level of flexibility in the choice of the third party, still maintaining its own identity. Also, the consortium's solutions and projects follow the logic of the personalized answer, which, if on one side, will solve specific needs; on the other, it is hard to introduce them on an international level.

The platform promotes more efficient and timely communication between departments inside the same organization because the documentation is made promptly available to those who need to be informed to keep the internal requirements updated. Digitalization influenced the members and, consequently, even the organization processes not directly linked to the consortium resources have to be rethought and adapted.

The consortium renewed its platform to be flexible and adaptable, according to the needs of the participants, with the idea of introducing new technologies, which will allow a more accessible, safe, and timely traceability of the orders. It will also take continuing measurements of the most critical environmental parameters during the transport and delivery of goods from the producer to the final retailer.

Several products are sensitive to changes in temperature and pressure, sun exposure, and vibrations, to name a few. The consortium is planning to introduce new technologies based on the Internet of Things, which allows constant monitoring of the goods at any stage of processing to ensure the delivery of the products in the best possible conditions. If the system detected significant variations, it would be possible to correct the situation, preserving the quality of the products or discharging only the affected lots.

The new technology of great interest is the aggregation code, which can enormously simplify tracing the goods if shipped as multiple packages. Unique code on the box will allow people to know the exact content before it is opened. This technology can simplify downstream activities. It is known immediately the quantity that will be stored, decommissioned, or sold, that will be fed back to the market, lost, or used for production.

A new trend is Blockchain technology, and, as of today, the consortium is evaluating how it will improve the notarization and availability of crucial information to the public. This solution would allow tight control on the selling history of the drug sold or administrated.

One of the main negative aspects that the consortium is working on ratifying is the limited involvement of the public healthcare organizations: they have not expressed an active interest in being involved. Therefore, a key element is not present when new projects are carried on. The public healthcare sector has not yet addressed such changes, but it will face similar or even more profound changes when implemented.

5 Discussion

Introducing a flexible cloud-based platform required specific skills suitable for the implementation and activation of the novelty introduced. The contest is Industry 4.0, which is still in the making. Nevertheless, understanding how it influences the healthcare sector will allow us to anticipate the needs of the next step, Industry 5.0, or "Super Smart Society" [44]. It also will create new activities and jobs, whereas others will disappear, like the one at low added value. The preserved businesses will also be affected, requiring retraining and introducing new activities for the worker involved. The changes will include several, if not all, departments, such as customer care, back office, receipt and fulfillment of the orders, payment of the invoices, purchasing office, inbound logistics, warehouse, and naturally the ICT, no matter the type of organization.

The study emphasizes the need for expertise in three distinct categories: technical, managerial, and social, even though only the last two were considered fundamental by all the people interviewed [45, 46]. "*All technological innovations are introduced through collaboration with other companies in the sector because that is where the need arises.*" Moreover, another "*when requirements and functionalities are shared, they can be researched in a joint and coordinated way; collaboration with other companies ranges from selecting suppliers to incorporating skills.*"

A second critical aspect is the retraining of the employees by organizing refresher courses inside the company itself. The training should focus on six issues, technology, industry, software, industrial operations, and transversal and job skills, focusing on the first one [47, 48]. A manager said: "*Technology frees up much time for many people who must be accompanied by process of requalification and training to carry out activities with high added value. It is also possible to introduce activities that they could not do before because there was no real-time. They are often the most valuable assets.*"

One of the main driving business forces is the constant need to reduce the costs, especially in the pharmaceutical industry, where only one drug on ten thousand will obtain approval for commercialization by the FDA or the EMA. Patented drugs are also warranted intellectual protection only for a limited time, after which it will be possible for other companies to produce the generic version. These two aspects together call for constant improvement in the production and research of new compounds. Otherwise, the costs would be too high to sustain the company in the long run. To increase the number of approved drugs or reduce the amount that gets drop off, a first approach is to create a network of partnerships with universities and cloud-based organizations for the data analysis [49]. "*Here, we found a different approach: the creation of consortium. The non-profit formula and the Internet-based technologies minimize the costs not only for the consortium itself but also for the members and the participants. Moreover, because the participants cover every step of the distribution, single companies do not have to create specific connections with each player from the producers to the distributors.*"

Still, they can access the platform to connect with hundreds of organizations. A manager told us, "*We are looking at the issues and technologies related to the traceability of the chain, and the detection of relevant parameters in the transport and delivery processes of goods, from the industry to the final dispenser. To the hospital or the pharmacy, we trace the movements and changes of ownership of sensitive products and relevant parameters.*"

The digitalization of the order cycle is required not only by introducing the latest innovations but mainly by the new laws at the local and European levels, which aim to move all the documentation to a digital level to improve the transparency of each transaction. Studies have identified some key ingredients that paved the way for the successful integration of the new technologies, and digitalization is one [50, 51].

"*Automation and digitalization are the two main subjects developed and implemented, for us,*" told us one of the biggest pharmaceutical companies.

At the organizational level, we observed different types of adaptations.

> *The consortium decided to introduce specialized personnel to better exploit those Internet technologies currently available. Consequently, they chose collaborators with a similar background and kept a flat but highly specialized structure.*

However, some companies opted for a profound reorganization of their departments. One company decided to change the organization of the departments based on the advancement of the research (Discovery, preclinical, clinical, post-commercialization). Now the departments are divided by type of drug and encompass all stages. This approach facilitates communication and coordination along the whole line of the development of a drug.

The introduction of Blockchain technologies to trace the sector's products requires that each player collaborates in building up the information flux required and stay updated. The new platform introduced by the Delta consortium was created, keeping in mind to expand it and to introduce this new technology. "*The aim is to detect forged or altered drugs through the information made available by the blockchain,*" said the CEO. Other companies are also working toward this implementation. Sigma, for example, a company in the ophthalmology sector, is promoting specialized partnerships to support such processes with the final goal of extending and sharing with every other player of the industry those new technologies that can improve the quality of the product offered to the customers [52]. "*We gave priority to the implementation of the Blockchain technology through the location and involvement of suitable partners,*" said again. Sharing resources is a leitmotif in the healthcare sector. "*The sharing of resources gives the peace of mind of enjoying a solution that is coherent and has all the requirements; it gives the possibility to see new project initiatives born in a propositional and proactive way. Proposals, starting from a significant critical mass and not from the single company, are discussed within the consortium, with plural participation of several actors.*"

Outsourcing is another increasingly chosen practice by companies, which rely on the expertise of the so-called Contract Research Organization (CRO), companies specialized in specific tasks. CROs can offer services at a lower price than the investments required by the company to develop and implement a one-time project.

CRO expertise spans preclinical to clinical research, such as animal experimentation, DNA and RNA sequencing or mutation, drug discovery and development, clinical data analysis, and others. In the case of the Delta consortium, they relied on software that heavily trusts machine learning technologies.

All the interviewed people agree that the pharmaceutical industry's main challenge is developing personalized medicine, which can potentially change the Healthcare sector's production paradigms. "*It is not simply the acknowledgment of an obligation, but it can be an opportunity to review the processes, capture collateral elements that rise in the decision-making agenda.*" The modern industry needs to transition from 4.0, which is currently undergoing, to society 5.0 [44], based on implementing the introduced technologies, like the Internet of Things, extensive data analysis, and Blockchain to implement this new approach [53]. "*Industry 4.0 represents an opportunity because the request of the legislator obligates us to question some practices, procedures, and models; there are areas for improvement.*" Furthermore, another: "*The pharmaceutical industry cannot ignore a pushed adoption of the most innovative, more frontier technological solutions, precisely to ensure maximum speed and, at the same time, the level of precision and guarantee of compliance required with these standards; only by relying on technological solutions can it be achieved.*"

They think that Society 5.0 will heavily rely on robotics and Artificial Intelligence and the tight interaction between them and human workers. Such cooperation requires machines to learn and adapt according to the circumstances and the inputs provided by the human counterpart [54]. It will inevitably bring further changes in the organization of companies and, according to Saeid Narharvandi, even to the creation of new positions, like, for example, that of the Chief Robotics Officer, with the task of coordinating the interaction between humans and robots [54].

6 Conclusions

In this work, we tried to analyze the role of new technologies in the pharmaceutical industry and their consequent changes.

The literature review showed that the innovations brought by industry 4.0 are still bringing essential changes at all levels, involving the entire ecosystem of enterprises. Two aspects that forcefully enter this scenario are efficiency and cost reduction obtained from automation and digitization of processes.

Another fascinating aspect is the safety of drugs, which has led to tracking technologies to guarantee their origin, anti-counterfeiting, and quality.

It is necessary to consider also the time. The ability to operate automatically and speed up different processes through digitalization allows people to dedicate themselves to tasks with high added value. For example, as one of the interviewed people said, "*a worker costs more than most of the equipment.*"

As the case study confirmed, the pharmaceutical industry takes full advantage of the technologies of the 4.0 era. It can offer more and higher quality products, as well as improving their traceability. It also can meet the growing sensitivity of the population, and the knowledge of employed people. There is the opportunity to develop medicines for rare diseases, which would have too much development cost with classical methods but are not sufficiently profitable to be marketed.

This explorative study, even if in its infancy, allows us to highlight some of the critical aspects of this sector, and we did not analyze others equally important, such as clinical research, production, and interaction with end-user. This aspect should be more deeply investigated. Clinical research requires continuous communication and interaction with several other organizations, like hospitals and public administrations. Its results can also lead to the necessity of introducing changes in the production of new drugs; it can also spur new research linked to unexpected side effects (for example, Viagra was initially developed as a cardiovascular drug). Another limit is that the production and stock depend on the needs of end-users and their ability to access a drug; there can also be significant differences from country to country according to their national health system and the pro-capita GDP. Papers relating to the changes induced by industry 4.0 at the level of the healthcare supply chain, for example, are limited.

However, in the future, it would be interesting to deepen the study of how technology can improve products, services, and, at the same time, decrease the workload; or the change of inter-organizational relationships, especially when it involves public administrations.

Industry 4.0 is paving the way for the next industrial revolution that affects society itself. It could be interesting to wonder if and what other changes the supply chain will be encountered with the advent of industry 5.0 and also, for example, how it can meet the need for personalized medicine and rare diseases, how it will change job requirements, and how it can improve the interaction with end-users.

References

1. Xu, L. Da, Xu, E.L., Li, L. (2018). Industry 4.0: State of the art and future trends. *Int. J. Prod. Res. 56*, 2941–2962. doi:https://doi.org/10.1080/00207543.2018.1444806.
2. Kumar, N., & Jha, A. (2017). Temperature excursion management: A novel approach of the quality system in the pharmaceutical industry. *Saudi Pharmaceutical Jounal*. https://doi.org/10.1016/j.jsps.2016.07.001
3. Potdar, M., Sharif, A., Potdar, V., Chang, E. (2009). Applications of wireless sensor networks in the pharmaceutical industry. *Proc. – Int. Conf. Adv. Inf. Netw. Appl. AINA*, 642–647. doi:https://doi.org/10.1109/WAINA.2009.193.
4. Lai, V. S. (2001). Intra organizational communication with intranets. *Communications of the ACM*. https://doi.org/10.1145/379300.379324
5. Gimenez, J. C. (2006). Embedded business e-mails: Meeting new demands in international business communication. *English for Specific Purposes*. https://doi.org/10.1016/j.esp.2005.04.005
6. Sprey, J. A. (2015). *Videoconferencing as a Communication Tool.*. https://doi.org/10.1002/9781119134633.ch61

7. Yen, C.T., Liu, Y.C., Lin, C.C., Kao, C.C., Wang, W. Bin, Hsu, Y.R. (2014). Advanced manufacturing solution to industry 4.0 trend through sensing network and Cloud Computing technologies. In *IEEE International Conference on Automation Science and Engineering* (2014). doi:https://doi.org/10.1109/CoASE.2014.6899471.
8. Singh, G. (2017). Preclinical Drug Development. In *Pharmaceutical Medicine and Translational Clinical Research* (2017). doi:https://doi.org/10.1016/B978-0-12-802103-3.00004-3.
9. Kennedy, T. (1997). Managing the drug discovery/development interface. *Drug Discovery Today*. https://doi.org/10.1016/S1359-6446(97)01099-4
10. Jatto, E., & Okhamafe, A. (2002). An overview of pharmaceutical validation and process controls in drug development. *Tropical Journal of Pharmaceutical Research*. https://doi.org/10.4314/tjpr.v1i2.14592
11. Atkinson, J. D., & Jones, R. (2009). Intellectual property and its role in the pharmaceutical industry. *Future Medical Chemistry, 1*, 1547–1550.
12. Becker, M. C., & Lillemark, M. (2006). Marketing/R&D integration in the pharmaceutical industry. *Research Policy*. https://doi.org/10.1016/j.respol.2005.09.005
13. Paddock, S. W. (1999). Confocal laser scanning microscopy. *Biotechniques*. https://doi.org/10.2144/99275ov01
14. Olivier, T., Moine, B. (2013). Confocal laser scanning microscopy. In: *Optics in Instruments*. doi:https://doi.org/10.1002/9781118574386.ch1.
15. Peham, J.R., Vellekoop, M.J., Nohammer, C., Wiesinger-Mayr, H. (2011). PCR product detector with LED-photodiode fluorescence sensing in a nanoliter flow-cell for the high-throughput detection of double-stranded DNA. In: *Procedia Engineering*. doi:https://doi.org/10.1016/j.proeng.2011.12.230.
16. Ahrberg, C. D., Ilic, B. R., Manz, A., & Neužil, P. (2016). Handheld real-time PCR device. *Lab on a Chip*. https://doi.org/10.1039/c5lc01415h
17. Lee, C.H., Yoon, H.J. (2017). Medical big data: Promise and challenges. *Kidney Research Clinical Practice 36*, 3–11. doi:https://doi.org/10.23876/j.krcp.2017.36.1.3.
18. Demchenko. (2012). In: *Proceeding of the 4th IEEE Conference on Cloud Computing Technologies and Science*.
19. Docherty, A. B., & Lone, N. I. (2015). Exploiting big data for critical care research. *Current Opinion in Critical Care, 21*, 467–472. https://doi.org/10.1097/MCC.0000000000000228
20. Carvalho, L.S., Silva, O.B. da, Almeida, G.C. da, Oliveira, J.D. de, Parachin, N.S., Carmo, T.S. (2017). Production processes for monoclonal antibodies. In *Fermentation Processes* (2017). doi:https://doi.org/10.5772/64263.
21. Zadbuke, N., Shahi, S., Gulecha, B., Padalkar, A., & Thube, M. (2013). Recent trends and future of pharmaceutical packaging technology. *Journal of Pharmacy and Bioallied Science, 5*, 98–110. https://doi.org/10.4103/0975-7406.111820
22. Bansal, D., Malla, S., Gudala, K., & Tiwari, P. (2013). Anti-counterfeit technologies: A pharmaceutical industry perspective. *Scientia Pharmaceutica*. https://doi.org/10.3797/scipharm.1202-03
23. Hackius, N., Petersen, M., Kersten, W. (2017). Blockchain in logistics and supply chain: Trick or treat? Epubli.
24. Schöner, M., Kourouklis, D., Sandner, P., Gonzalez, E., Förster, J. (2017). Blockchain technology in the pharmaceutical industry. FSBC Working Paper, 1–9.
25. Tseng, J. H., Liao, Y. C., Chong, B., & Liao, S. W. (2018). Governance on the drug supply chain via gcoin Blockchain. *International Journal of Environmental Research and Public Health*. https://doi.org/10.3390/ijerph15061055
26. Lilly, L.S. (2013). Pathophysiology of heart disease: A collaborative project of medical students and faculty (5th ed.).
27. Alnajar, A., Frazier, O.H. (2019). The state of artificial heart therapy. *Texas Heart Institute Journal 46*, 77–79. doi:https://doi.org/10.14503/THIJ-18-6756.
28. Weber, B. N., Kobashigawa, J. A., & Givertz, M. M. (2017). Evolving areas in heart transplantation. *JACC Heart Failure, 5*, 869–878. https://doi.org/10.1016/j.jchf.2017.10.009

29. Alt, V., Rupp, M., Lemberger, K., Bechert, T., Konradt, T., Steinrücke, P., Schnettler, R., Söder, S., & Ascherl, R. (2019). Safety assessment of microsilver-loaded poly(methyl methacrylate) (PMMA) cement spacers in patients with prosthetic hip infections. *Bone Joint Research, 8*, 387–396. https://doi.org/10.1302/2046-3758.88.bjr-2018-0270.r1
30. Aherwar, A., K Singh, A., Patnaik, A. (2015). Current and future biocompatibility aspects of biomaterials for hip prosthesis. *AIMS Bioengineering 3*, 23–43 (2015). doi:https://doi.org/10.3934/bioeng.2016.1.23.
31. Wang, S., Wang, L., Liu, Y., Ren, Y., Jiang, L., Li, Y., Zhou, H., Chen, J., Jia, W., & Li, H. (2017). 3D printing technology used in severe hip deformity. *Experimental and Therapeutic Medicine, 14*, 2595–2599. https://doi.org/10.3892/etm.2017.4799
32. Xia, R., Zhai, Z., Chang, Y., & Li, H. (2019). Clinical applications of 3-dimensional printing technology in hip joint. *Orthopaedic Surgery, 11*, 533–544. https://doi.org/10.1111/os.12468
33. Bear, M.F., Connors, B.W., Paradiso, M.A. (2016). Neuroscienze, Esplorando il cervello. *Edra Masson.*
34. Waldrop, M. M. (2019). News feature: What are the limits of deep learning? *Proceedings of the National Academy of Sciences USA, 116*, 1074–1077. https://doi.org/10.1073/pnas.1821594116
35. Mitchell Waldrop, M. (2019). What are the limits of deep learning? *Proceedings of the National Academy of Sciences USA, 116*, 1074–1077. https://doi.org/10.1073/pnas.1821594116
36. Jiang, F., Jiang, Y., Zhi, H., Dong, Y., Li, H., Ma, S., Wang, Y., Dong, Q., Shen, H., & Wang, Y. (2017). Artificial intelligence in healthcare: Past, present and future. *Stroke Vascular Neurology, 2*, 230–243. https://doi.org/10.1136/svn-2017-000101
37. Russell, S., Norvig, P. (2011) *Artificial intelligence: A moder approach.* Pearson Global Edition.
38. Chapman, T. (2003). Lab automation and robotics: Automation on the move. *Nature.* https://doi.org/10.1038/421661a
39. Bryman, A., Bell, E. (2011). *Business research methods.* Oxford University Press (2011).
40. Gupta, M., Shaheen, M., & Reddy, K. P. (2019). Qualitative techniques for workplace data analysis. *IG Global.* https://doi.org/10.4018/978-1-5225-5366-3
41. Leonard-Barton, D. (1998). Synergistic design for case studies: longitudinal single-site and replicated multiple-site. In *Proceedings of the National Science Foundation Conference on Longitudinal Research Methods in Organizations, Austin, USA.*
42. Yin, R. K. (2013). Validity and generalization in future case study evaluations. *Evaluation.* https://doi.org/10.1177/1356389013497081
43. Payaro, A. (2006). La catena del farmaco. *Largo Consumo 12.*
44. Kobelev, B. (2017) On the way from Industry 4.0 to Industry 5.0: From digital manufacturing to digital society. International Science Journal.
45. Grzybowska, K., Łupicka, A. (2017) Key competencies for Industry 4.0. In: *Economics and Management Innovations (ICEMI).* doi:https://doi.org/10.26480/icemi.01.2017.250.253.
46. Łupicka, A., Grzybowska, K. (2018) Key managerial competencies for Industry 4.0 – Practitioners', Researchers' and Students' Opinions. *Logistics and Transport.*
47. Fitsilis, P., Tsoutsa, P., Gerogiannis, V. (2018) Industry 4.0 : Required personnel competencies. *International Science Journal "Industry 4.0".*
48. Rüssmann, M., Lorenz, M., Gerbert, P., Waldner, M., Justus, J., Engel, P., Harnisch, M. (2015) *Industry 4.0: The Future of Productivity and Growth in Manufacturing Industries.* doi:https://doi.org/10.1007/s12599-014-0334-4.
49. Khanna, I. (2012). Drug discovery in the pharmaceutical industry: Productivity challenges and trends. *Drug Discovery Today.* https://doi.org/10.1016/j.drudis.2012.05.007
50. Sony, M., Naik, S. (2019). *Key ingredients for evaluating Industry 4.0 readiness for organizations: A literature review.* doi:https://doi.org/10.1108/BIJ-09-2018-0284.
51. Agostini, L., Filippini, R. (2019). Organizational and managerial challenges in the path toward Industry 4.0. European Journal of Innovation Management. doi:https://doi.org/10.1108/EJIM-02-2018-0030.

52. Chiacchio, F., D'Urso, D., Compagno, L., Chiarenza, M., Velardita, L. (2019) Towards a blockchain based traceability process: a case study from pharma industry. In: *IFIP Advances in Information and Communication Technology*. doi:https://doi.org/10.1007/978-3-030-30000-5_56.
53. Branke, J., Farid, S.S., Shah, N. (2016) Industry 4.0: a vision for personalized medicine supply chains? *Cell and Gene Therapy Insights*. doi:https://doi.org/10.18609/cgti.2016.027.
54. Nahavandi, S. (2019) Industry 5.0-a human-centric solution. *Sustainability* (2019). doi:https://doi.org/10.3390/su11164371.